Cocrystals in Pharmaceutical Sciences

Scrivener Publishing
100 Cummings Center, Suite 541J
Beverly, MA 01915-6106

Publishers at Scrivener
Martin Scrivener (martin@scrivenerpublishing.com)
Phillip Carmical (pcarmical@scrivenerpublishing.com)

Cocrystals in Pharmaceutical Sciences

Design to Applications

Edited by

Inderbir Singh

Chitkara College of Pharmacy, Chitkara University, Punjab, India

Subrata Mallick

School of Pharmaceutical Sciences, Siksha 'O' Anusandhan (Deemed to be University), Odisha, India

and

Ritu Rathi

Chitkara College of Pharmacy, Chitkara University, Punjab, India

Scrivener Publishing

WILEY

This edition first published 2025 by John Wiley & Sons, Inc., 111 River Street, Hoboken, NJ 07030, USA and Scrivener Publishing LLC, 100 Cummings Center, Suite 541J, Beverly, MA 01915, USA

For more information about Scrivener publications please visit www.scrivenerpublishing.com.

Wiley Global Headquarters
111 River Street, Hoboken, NJ 07030, USA

For details of our global editorial offices, customer services, and more information about Wiley products visit us at www.wiley.com.

Library of Congress Cataloging-in-Publication Data

ISBN 978-1-394-30247-5

Front cover image courtesy of Inderbir Singh
Cover design by Russell Richardson

Set in size of 11pt and Minion Pro by Manila Typesetting Company, Makati, Philippines

Printed in the USA

10 9 8 7 6 5 4 3 2 1

Contents

Preface

Pharmaceutical cocrystals represent an expanding field that has the potential to enhance drug safety, efficacy, and patient compliance. Cocrystals are multicomponent crystalline materials composed of two or more compounds, typically a drug and a coformer, held together by non-covalent interactions such as hydrogen bonding or van der Waals forces. Cocrystals offer numerous advantages, including improved solubility, bioavailability, stability, taste masking, drug release, hygroscopicity, and mechanical properties. However, the utilization of cocrystals requires careful consideration of factors such as scalability, regulatory compliance, and industrial feasibility. Addressing these challenges can help bridge the gap between laboratory research and commercialization.

This book provides a comprehensive overview of pharmaceutical cocrystals, addressing fundamental principles, practical considerations, and the regulatory landscape. The content is organized into distinct chapters covering the historical background, design, structural characteristics, and preparation techniques of cocrystals. Additionally, a designated chapter discusses computational methods for screening and characterizing pharmaceutical cocrystals. The book also explores the concept of herbal cocrystals and the commercialization of pharmaceutical cocrystals.

Various applications of pharmaceutical cocrystals—such as enhancing solubility, stability, mechanical properties, and therapeutic efficacy of active pharmaceutical ingredients—are thoroughly discussed using the latest updated literature. Moreover, the regulatory perspectives surrounding cocrystals are also examined in detail.

We would like to extend our heartfelt thanks to Martin Scrivener and Scrivener Publishing for their belief in this project and their kind cooperation throughout the process. We also gladly acknowledge all the sources utilized in the curation of this book. Furthermore, we extend our gratitude to the authors, colleagues, and others who contributed to and supported the compilation of this book. The content has been deliberately chosen to cater to a diverse audience, including academics, pharmaceutical experts, and students interested in drug formulation and delivery. We hope this book will inspire further study and innovation in the field of pharmaceutical cocrystals.

Prof. (Dr.) Inderbir Singh
Prof. (Dr.) Subrata Mallick
Ms. Ritu Rathi

1

Pharmaceutical Cocrystals: Introduction, History and Applications

Oluwatoyin A. Odeku* **and Olufunke D. Akin-Ajani**

Department of Pharmaceutics and Industrial Pharmacy, University of Ibadan, Ibadan, Nigeria

Abstract

Cocrystals represent an expanding field of pharmaceutical sciences that offers an innovative approach to improving the physicochemical properties of active pharmaceutical ingredients (APIs). Cocrystals are a unique type of solid-state crystalline substance created by the interaction of two or more molecular species, usually an API and a coformer, joined together by noncovalent interactions, without forming chemical bonds. Unlike traditional drug formulations, cocrystals use purposeful molecular design to manufacture precise interactions between components, resulting in distinct crystal shapes and characteristics. Cocrystals offer the potential to increase drug solubility, bioavailability, and stability, mask taste, boost flavor, and control the release of drugs from formulations. In addition, cocrystals can facilitate the co-formulation of multiple drugs with complementary pharmacological activities, and they serve as versatile carriers or excipients in drug delivery systems such as nanoparticles, liposomes, or microspheres, enabling drug targeting. This chapter gives a background of cocrystals and their application in various aspects of drug development and delivery, as well as some marketed cocrystal formulations.

Keywords: Cocrystals, bioavailability, solubility, stability, marketed cocrystal formulations

**Corresponding author*: pejuodeku@yahoo.com; o.odeku@ui.edu.ng

Inderbir Singh, Subrata Mallick and Ritu Rathi (eds.) Cocrystals in Pharmaceutical Sciences: Design to Applications, (1–28) © 2025 Scrivener Publishing LLC

1.1 Introduction

Cocrystal is a supramolecular chemistry idea that is gaining popularity among pharmaceutical and chemical researchers, as well as drug regulatory bodies. The primary reason for this is its capacity to affect the physicochemical characteristics of active medicinal components [1]. During pharmaceutical product development, formulators must maximize the physicochemical qualities of active medicinal components.

The European Medicines Agency (EMA) has defined cocrystals as "homogenous (single phase) crystalline structures made up of two or more components in a definite stoichiometric ratio where the arrangement in the crystal lattice is not based on ionic bonds (as with salts) and the components of a cocrystal may nevertheless be neutral as well as ionized" [2]. Cocrystals have been defined by the United States Food and Drug Administration (USFDA) as "Crystalline materials composed of two or more different molecules, typically API and cocrystal formers (coformers), in the same crystal lattice in a defined stoichiometric ratio" [3]. They were characterized as a discrete category of innovative, crystalline compounds that may affect the physicochemical properties of APIs, signifying the start of a new era in the domains of crystal engineering and manufacture [4]. Cocrystals consist of an active pharmaceutical ingredient and one or more distinct coformers that are solid at ambient temperature [5]. They are preferred over amorphous API forms or solid dispersions because they offer the solubility benefits of high-energy solids and a crystalline structure with strong thermal stability [6].

A coformer is described as "a component that interacts non-ionically with the active medicinal substance in the crystal lattice, which is typically non-solvent and is not volatile" [7]. The coformer selection has a significant impact on the final cocrystal properties. When synthesized as a cocrystal, the coformers can alter the API's stability and solubility by causing modifications to the structure of its crystals [8]. Unlike traditional drug formulations, cocrystals use purposeful molecular design to manufacture precise interactions between components, resulting in distinct crystal shapes and characteristics. Cocrystals can be formed by interactions, including van der Waals forces, p-stacking, and hydrogen bonding. This excludes API solvates and hydrates from the category of cocrystals. However, cocrystals may comprise one or more water molecules or solvents in the crystal lattice [9]. Cocrystals frequently rely on hydrogen-bonded assemblies formed by neutral drug molecules and additional components [4]. For nonionizable substances, cocrystals increase medicinal qualities by modifying their mechanical behavior, solubility, water absorption, dissolution rate, bioavailability, and stability [10–12].

The major distinction between solvates and cocrystals rests in the physical condition of their constituents [13]. Crystals are known as solvates if one of the constituents is liquid at ambient temperature and cocrystals if both constituents are solid at ambient temperature. Solvates are prevalent because they develop as an accidental outcome of crystallization from solution and can boost the medicine dissolution rate, as demonstrated by spironolactone solvate [14]. Cocrystals have a more stable crystalline form without having to make or break covalent bonds, and both weakly ionizable and nonionizable drug molecules can be combined to form cocrystals. They have several potential commercial pharmaceutical and food applications [15]. Cocrystals have the potential to improve medication solubility, stability, bioavailability, and other essential properties important for medicinal efficacy [16].

1.2 History

Cocrystals have a history dating back to the early nineteenth century, yet their notion and understanding have developed greatly over time (Figure 1.1). The development of cocrystals may be traced back to 1844 when Friedrich Wöhler mixed quinone and hydroquinone solutions to create quinhydrone, a green solid with a 1:1 stoichiometry between the reactants

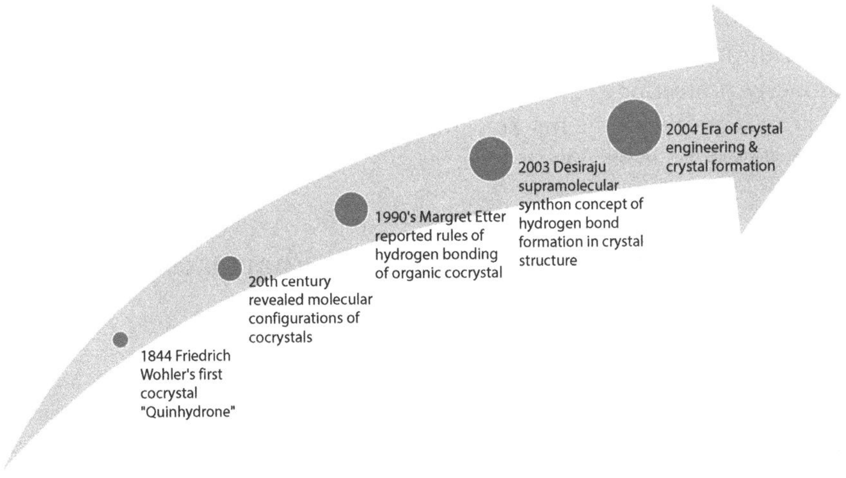

Figure 1.1 The evolution of cocrystals over the years.

[17, 18]. For several decades, there remained debate regarding the actual nature of the chemical composition of Wöhler's product, and it took more than a century to verify the content and structure of this crystalline substance using single-crystal X-ray diffraction [19]. The introduction of X-ray crystallography in the early 20th century transformed the study of crystal structures, revealing insights into the molecular configurations inside cocrystals.

A close evaluation of the literature from the latter part of the 19th and early 20th centuries reveals the existence of practically hundreds of cocrystals (even though they would continually appear under various names). The concept of cocrystals evolved in the later part of the 20th century when researchers defined them as "solid-state structures made up of two or more components held together by noncovalent interactions." The area of crystal engineering sprang to prominence, with an emphasis on designing and manipulating crystal formations to attain certain qualities and functionality. Researchers investigated the use of cocrystals in medicines, noting their capacity to increase medication solubility, stability, and other characteristics. Stahly has published an excellent comprehensive summary of cocrystals reported before the year 2000 [18].

The 21st century saw an increase in research on cocrystals, notably in the pharmaceutical business. The emphasis on cocrystals as an easily recognized research owes a great deal to Etter's foundational work in the late 1980s and early 1990s [20, 21]. In 1984, Desiraju and Steiner presented a systematic definition of hydrogen-bonded molecular complexes, which helped to formalize cocrystal notions in a book on "crystal engineering" [22].

Cocrystals attracted attention for their capacity to handle issues such as medication solubility, bioavailability, polymorphism, and intellectual property protection [15]. Several successful examples of cocrystal-based pharmaceuticals hit the market, emphasizing the practical applicability of cocrystals in pharmaceutical sciences.

1.3 Applications of Cocrystals in Pharmaceutical Sciences

Cocrystals have a wide range of pharmaceutical uses, providing answers to difficulties in standard medication formulations. Some important pharmaceutical uses of cocrystals include the following.

1.3.1 Enhanced Solubility

Many drugs have been revealed to have low solubility, which can reduce their bioavailability and therapeutic effectiveness. Approximately 70% of new drugs belong to the class II Biopharmaceutical Classification System (BCS; low solubility/high permeability) or class IV BCS (poor solubility/low permeability) [23, 24]. Since the gastrointestinal (GI) system has varied pH in different parts, drugs taken orally exhibit varying solubility in GI fluids at different pH, resulting in unpredictable and variable absorption and the inability to evaluate the drug's efficacy and safety appropriately. Hence, restricted drug solubility is a key challenge in developing oral drug delivery systems [25].

Cocrystallization with proper coformers can improve API solubility by changing the crystal lattice structure, resulting in faster dissolution and greater absorption in the body [26]. Synthesizing salts and cocrystals boosted the solubility of ketoconazole, an antifungal medication, by 53 and 100 times, respectively, compared to ketoconazole [27]. Thus, cocrystals resulted in better drug solubility than the salt form.

The formulation of pterostilbene cocrystals with piperazine resulted in a sixfold increase in solubility, while pterostilbene-glutaric acid cocrystals led to the fast precipitation of the drug due to glutaric acid's high solubility [28]. Cocrystals of the anticancer drug 6-mercaptopurine with nicotinamide exhibited two times greater solubility than pure drug [29]. Apixaban cocrystals exhibited improved solubility of about twofold, and the cocrystals showed a faster dissolution rate when compared to pure drugs [30]. Studies have shown that combining curcumin with resorcinol and pyrogallol enhanced its solubility significantly [31]. Curcumin-resorcinol and curcumin-pyrogallol cocrystals were shown to dissolve 5 and 12 times faster than pure curcumin, respectively.

1.3.2 Improved Bioavailability

Currently, several newly marketed drugs experience the challenges of poor water solubility as well as poor bioavailability [32, 33]. The limited oral bioavailability of medications makes developing new drugs a considerable issue. Crystal engineering is critical in developing and manufacturing pharmaceutical cocrystals with increased oral bioavailability and aqueous solubility, leading to therapeutic efficacy [34].

In vivo studies of the bioavailability of baicalein in rats showed that the formation of cocrystals of baicalein with nicotinamide resulted in increased oral bioavailability of baicalein [35]. The peak plasma concentration

(Cmax) and area under the curve (AUC) for the baicalein-nicotinamide cocrystals were 2.49 and 2.80 times higher, respectively, than the pure drug [36]. The hydrotropic solubilization of nicotinamide improves the bioavailability of cocrystals as a result of the concurrent improvement in solubility and dissolution of baicalein. In a pharmacokinetic investigation carried out on beagle dogs, apixaban-oxalic acid cocrystals demonstrated better oral bioavailability than pure drugs [30]. Meloxicam-aspirin cocrystals exhibited a 12-fold increase in onset of action compared to the pure drugs in rats, as well as higher oral bioavailability [37]. Furthermore, cocrystals of 6-mercaptopurine, categorized as a BCS class II medicine, with isonicotinamide, exhibited faster dissolution and greater oral bioavailability (168.7%) than the pure drug in rats [29].

1.3.3 Stability Enhancement

Cocrystallization can increase an API's physical stability, eliminating polymorphism concerns such as unwanted phase shifts during storage and processing. Cocrystals can improve the chemical stability of APIs, shielding them from deterioration caused by environmental variables such as moisture, heat, and light.

Stability studies are usually conducted throughout the development of cocrystals, including solution, chemical, and thermal stability, relative humidity (RH) stress, and photostability. Indomethacin-saccharin cocrystals demonstrated negligible water sorption in RH investigations, and theophylline cocrystals also showed RH stability at varied RH with distinct coformers (maleic, glutaric, malonic, and oxalic acid) for diverse periods. The study confirmed an increase in the physical quality and stability of the drugs, particularly by avoiding hydrate formation [38].

The chemical stability of the cocrystal is usually done using accelerated stability conditions to investigate whether there are any changes or chemical deteriorations in the formulation. There were very few publications on cocrystals' chemical stability. Cocrystals of glutaric with API exhibited no degradation and high physical and chemical stability under thermal stress and RH for 2 months [39]. Indomethacin-saccharin cocrystals also demonstrated negligible moisture sorption during stability studies, and no separation or conversion arose under trial circumstances [40].

Solution stability is a crucial parameter for cocrystal growth and it provides a better knowledge of cocrystal behavior in the release medium [41]. A study of the stability of carbamazepine cocrystals in water showed that cocrystals made with poorly water-soluble coformers stayed as such in the solution for 20 to 48 hours, whereas cocrystals with highly water-soluble coformers transformed to dihydrates [42]. Caffeine/oxalic acid cocrystals demonstrated superior stability at all RH up to 98% for 7 days with no noticeable change occurring in the physical structure compared to when the materials were suspended in aqueous medium for two days at ambient temperature [43]. Carbamazepine and saccharine cocrystals suspended in equal parts in water for 24 hours showed good stability as indicated by the presence of only the cocrystals in the solution, and no other form was discovered [44]. Nitrofurantoin cocrystals prepared with various coformers showed greater photostability than pure drug and physical mixtures. Cocrystallization can minimize the photodegradation of light-sensitive drugs since all cocrystals demonstrated little deterioration (<3%) after 168 hours [45].

1.3.4 Taste Masking and Palatability

Taste-masking methods are utilized to mask or overcome the bitter or unpleasant taste of active medicinal constituents/drugs, resulting in patient acceptance and compliance, particularly for pediatric and geriatric patients [46]. Sweetener-based coformers have been used to mask the taste of bitter drugs. The coformer interacts with medicinal molecules to produce single-phase crystalline assemblies via intermolecular interactions [47]. An innovative cocrystal of paracetamol and trimethylglycine with a sweeter taste enhanced compressibility, and dissolution rate than pure paracetamol has been reported [48]. Likewise, the sweetness of theophylline was boosted by saccharin by producing a cocrystal, thereby enhancing stability and assuring continuous drug release [49]. However, increased sweetness does not always indicate excellent masking of bitter taste; rather, the flavor is a mix of bitter and sweet, which could be of inferior quality. Ogata *et al.* discovered that certain acids that are not sweeteners can have beneficial effects on the taste masking of bitter drugs like propiverine [50]. Hence, cocrystals can mask the taste of bitter drugs due to the inability of human taste receptors to register the taste of cocrystals formed from the molecular aggregation of drug and coformer [51]. As a result, some coformers, despite their lack of sweetness, may successfully conceal the bitter taste of medications due to their propensity to form tiny transitory clusters with drug particles when dissolved in the mouth.

The strongly unpalatable and bitter taste of nevirapine, an antiretroviral drug, used in newborns and children, particularly in sub-Saharan Africa, has been masked with the cocrystals using maleic acid, benzoic acid, and glutaric acid as coformers [51]. The data showed that the taste of the pure coformer and solution aggregation play essential roles in taste masking.

1.3.5 Controlled Release

Cocrystals can be designed to alter drug release rates, enabling controlled or sustained release formulations that have improved therapeutic outcomes. Cocrystallization has been shown as a promising strategy for producing controlled-release formulations of olaparib, a PARP1/2 inhibitor with blockbuster anticancer activity, that have a shorter half-life and limited dose-limiting toxicity than marketed immediate-release formulations [52]. Sustained-release cocrystals of olaparib with kaempferol and the cocrystal solvate of olaparib with quercetin showed 21.4% and 18.6% reduction in the dissolution rate, respectively, in comparison to the original drug. Furthermore, the cocrystals exhibited increased pharmacological activity and inhibited ovarian cancer cells significantly more effectively than the pure drug.

Human insulin has been cocrystallized at predetermined proportions with octanoyl-Nε-LysB29-human insulin (C8-HI), a lipophilic modified insulin derivative [53]. The cocrystal mimics the neutral protamine Hagedorn (NPH) cocrystals generated by human insulin, generally used as a long-acting insulin element in diabetes treatment. The release rates of the cocrystal can be controlled, *in vitro* and *in vivo*, by altering the quantities of the two insulin contents. The cocrystal ratio-dependent release rate provides a successful method to regulate insulin pharmacodynamics, suggesting that a crystalline protein matrix is capable of supporting a chemical alteration that changes the rate of dissolution of the crystals in an intensely beneficial manner while remaining adequate to maintain the medicinal quality and the therapeutic effects of the parent drug.

1.3.6 Increased Tabletability

Tabletability denotes a material's capacity to be converted to a compact. Tabletability, compatibility, and crystal packing are critical characteristics in preformulation. Cocrystallization with appropriate coformers can alter these characteristics to improve the tabletability of poorly compressible pharmaceutical powders. The tabletability of resveratrol, a poorly compressible drug, has been increased by cocrystallization [54]. Resveratrol cocrystals with coformers 4-amino benzamide and isoniazid exhibited improved tabletability, leading to the formation of tablets with acceptable tensile strength even at low compression pressure.

Cocrystallization can be used to alter the mechanical characteristics of medicinal compounds. Vanillin isomers cocrystals with the coformer have demonstrated greater tabletability than single materials [55]. Paracetamol is poorly compressible, and wet granulation which is a tedious process requiring the use of binders is used to overcome this problem. Studies have shown that paracetamol cocrystals prepared using caffeine as a coformer using techniques such as solvent evaporation, liquid-assisted grinding, dry grinding, and the addition of anti-solvent improved the compression and the mechanical properties of paracetamol tablets [56].

1.3.7 Combination Therapy

Cocrystals allow for the formulation of multiple drugs in a unit dosage form, facilitating combination therapy and synergistic outcomes. Cocrystals consisting of two or more active ingredients, generally known as drug-drug, multi-drug, or multi-API cocrystals, are made feasible by noncovalent connections between the active drugs, which enables the drugs to retain their activity [57]. The two or more therapeutically useful elements are bound in a stoichiometric equilibrium in a single crystal lattice structure where they can interact primarily through nonionic interactions and, on some occasions, through a hybrid interaction with or without the inclusion of solvate molecules [58].

Furthermore, drug-drug cocrystals provide possible answers to restrictions like solubility, stability, and chemical instabilities between the APIs, which are frequently encountered in standard combination therapy [57]. Table 1.1 presents some multi-drug cocrystals developed to date, as well as their preparation techniques and uses.

Table 1.1 Drug-drug and nutraceuticals cocrystals developed to date [58].

Drug combination	Method of preparation	Indication	Refs
Theophylline–phenobarbital (2:1)	Distillation	Sedative-hypnotic and Antiasthmatic	[58]
Sulfadimidine–aspirin (1:1)	Solvent evaporation	NSAID and Antibacterial	[74]
Sulfadimidine–4-ASA (1:1)		Antibacterial	[74]
Theophylline–5-FU (2:1)		Antiasthmatic and anticancer	[75]

(Continued)

Table 1.1 Drug-drug and nutraceuticals cocrystals developed to date [58]. (*Continued*)

Drug combination	Method of preparation	Indication	Refs
Trimethoprim–sulfadimidine (1:1, 1:2)		Antibacterial	[76, 77]
Trimethoprim–sulfamethoxy-pyridazine (1:1)	Heating at boiling point followed by instant cooling	Antibacterial	[78]
Tetroxoprim–sulfametrole (1:1)	Cogrinding and solvent evaporation	Antibacterial	[79, 80]
Piracetam–gentisic acid (1:1)	Co-grinding and solvent evaporation	Nootropic agent and NSAID	[81]
Amoxicillin trihydrate–potassium clavulanate (3:7, 5:5, 7:3)	Melting at 50°C for 30 min	Antibacterial and β-lactamase inhibitor	[82]
Lamivudine–zidovudine (1:1)	Solvent evaporation	Antiviral	[83]
Theophylline–gentisic acid (1:1)	Thermal-assisted solvent evaporation	Antiasthmatic and NSAID	[84]
Ethenzamide–gentisic acid (1:1)	Solvent evaporation	Both drugs are NSAIDs, the latter also has anti-ageing properties	[49]
Sulfamethazine–theophylline (2:1)		Antibacterial and antiasthmatic	[85]
Meloxicam–aspirin (1:1)	Solution crystallization, solvent and slurry drop grinding methods	NSAIDs	[37]

(*Continued*)

Table 1.1 Drug-drug and nutraceuticals cocrystals developed to date [58]. (*Continued*)

Drug combination	Method of preparation	Indication	Refs
Isoniazid–4-ASA (1:1)	Solvent drop grinding	Antitubercular drugs	[86]
Pyrazinamide–4-ASA (1:1)			
Carbamazepine–salicylic acid (1:1)	Unexpected in the presence of moisture	Antiepileptic and anti-inflammatory	[87]
Pyrazinamide–diflunisal (1:1)	Ball mill grinding	Antitubercular and NSAID	[88]
Curcumin–pyrogallol (1:1)	Liquid-assisted manual grinding	Anticancer	[83]
Aceclofenac–paracetamol (1:1)	Various methods	NSAIDs	[89]
Isoniazid–2-chloro-4-nitro benzoic acid (1:1)	Solvent evaporation	Antitubercular and antiviral compounds	[90]
Piracetam–lithium chloride (1:1)	Solvent evaporation and grinding	Nootropic agent and mood-stabilizing agent	[91]
Furosemide–pentoxifylline (1:1)	Solvent evaporation	Loop diuretic; the latter is used for the treatment of intermittent claudication	[92]
Pyrimethamine–carbamazepine (1:1)		Antimalarial and antiepileptic	[93]
Pyrimethamine–theophylline (1:1)	Solvent evaporation	Antimalarial and antiasthmatic	

(*Continued*)

Table 1.1 Drug-drug and nutraceuticals cocrystals developed to date [58]. (*Continued*)

Drug combination	Method of preparation	Indication	Refs
Ciprofloxacin–norfloxacin (1:1)		Antibacterial	[94]
Paracetamol–indomethacin and mefenamic acid (1:1)		NSAIDs	[95]
Sildenafil–aspirin (1:1)		Antihypertensive and NSAID	[96]
Diclofenac and Diflunisal–Theophylline (1:1)	Solvent evaporation and Solvent drop grinding	Antiasthmatic and NSAID	[97]
Dapsone–sulfanilamide flavone, luteolin, caffeine and benzothiazolone (1:1)	Solution crystallization	Antileprotic, antibacterial and antioxidants	[98]
Carbamazepine–ibuprofen	Solution crystallization	Antiepileptic and NSAID	[99]

Abbreviations: 4-ASA, 4-aminosalicyic acid; NSAID, Non-steroidal anti-inflammatory agent.

1.4 Challenges in the Development of Cocrystals

The fundamental problem in developing therapeutic cocrystals is the choice of appropriate coformers. Given the vast number of potential coformers, it is necessary to design an approach to screening that can predict the most probable coformers [59]. The prospective coformers should subsequently be evaluated experimentally to see whether cocrystals occur [60]. Owing to the expanded study on the subject of cocrystals over the past few decades, enough evidence has been amassed to estimate the likely coformers.

Some of the effective methods for screening coformers include hydrogen bond propensity, Hansen solubility parameter, Cambridge Structural Database, pKa rule, and supramolecular synthon approach. However, there is a need to develop more efficient screening techniques for the fruitful implementation of the use of cocrystals in drug development in the pharmaceutical sector [61].

A variety of cocrystallization procedures can be employed to screen cocrystals. Solvent-based cocrystallization procedures present obstacles, such as the selection of the proper solvent, adjustments in the solubility of active drug and coformer in the selected solvent (compatible and incompatible), concentration effects, and the selection of appropriate cooling and heating profiles [62]. The solid-state grinding approach of screening cocrystals outperforms solvent-based methods. However, occasionally phase transitions in medicinal cocrystals occur in cocrystals prepared by solid-state grinding [63].

The chance of cocrystal separation owing to the contact with other excipients, the substitution of coformers by excipients, alteration of the cocrystal stoichiometry, and its transformation to a poorly soluble form of the parent drug during dissolution needs to be addressed during the cocrystal preparation [64, 65].

Sometimes, the selected coformers may not generate cocrystals with appropriate processability, and physical, chemical, and pharmacokinetic qualities [41]. The obstacles encountered in the production and evaluation of cocrystals include the formation of hybrids, salts, or solvates, polymorphism of cocrystals, and alteration to the less soluble form of the parent drug [66–68]. In addition, pH, surface active agent concentration, solubility and dissolution rates, and inherent instability of cocrystals in the solution phase can hinder the successful formulation [69, 70].

Industrial cocrystallization processes must be scalable, flexible, and environmentally friendly, and the cocrystal quality should not be jeopardized by commercial-scale manufacture. However, there are no recognized strategies for scaling up the production of medicinal cocrystals. Methods such as hot-melt extrusion, supercritical fluid technologies spray congealing, and spray drying are now being used for commercial scale-up of cocrystal production. An additional obstacle in the large-scale production of cocrystals is the reliable regulation of stoichiometry [71] and the lack of correlation features *in vitro* and *in vivo*, which can drastically shorten the development period [72, 73].

Table 1.2 Some commercial pharmaceutical cocrystals.

Commercial name	Approval	API	Coformer	Improved property	Indication	Manufacturer	References
Suglat®	Japan 2014	Ipragliflozin	l-Prolin	Stability against hydrate formation	Diabetes	Kotobuki Pharmaceuticals, Nishina, Shizuoka, Japan, and Astellas Pharma, Tokyo, Japan	[100]
Cafcit®	USFDA 1999	Caffeine	Citric acid	Improved dissolution behavior and lower hygroscopicity	Infantile apnoea	Hikma Pharmaceuticals Plc, London, UK	[100–102]
Entresto®	USFDA 2015	Valsartan	Sacubitril	Improved pharmacokinetics and bioavailability	Heart failure	Novartis, Basel, Switzerland	[103, 104]
Lamivudine/ zidovudine Teva®	EMA 2011	Lamivudine	Zidovudine	Improved solubility, stability, and bioavailability	HIV infection	Teva Pharma B.V., Tel Aviv-Yafo, Israel	[76, 102, 105]
Steglatro®	USFDA 2017	Ertugliflozin	Z-Pyroglutamic acid	Improved stability	Diabetes	Pfizer, New York, USA	[106, 107]
Depakote®	USFDA 1983	Valproic acid	Valproate sodium	Improved stability	Epilepsy	Abbott Laboratories, Illinois, USA	[100, 108]

(*Continued*)

Table 1.2 Some commercial pharmaceutical cocrystals. (*Continued*)

Commercial name	Approval	API	Coformer	Improved property	Indication	Manufacturer	References
Lexapro®	USFDA 2002	Escitalopram	Oxalate	Improved stability of API	Anxiety and depression	Allergan Inc., Dublin, Ireland	[109]
ESIX-10®	USFDA 2009	Escitalopram	Oxalate	Improved stability of API	Anxiety and depression	Sag Health Science Pvt Ltd., New Delhi, India	[59, 83]
Beta-chlor®	USFDA 1963	Chloral hydrate	Betaine	Improved thermal stability	Sedation	Mead Johnson, Illinois, USA	[100, 110]
Abilify®	USFDA 2002	Aripiprazole	Fumaric acid	Improved the dissolution rate and solubility	Schizophrenia	Otsuka Pharmaceuticals, Tokyo, Japan	[111, 112]
Odomzo®	USFDA 2015	Sonidegib	Phosphoric acid	Improve its solubility and bioavailability	Basal cell carcinoma	Sun Pharma Global, Mumbai, India	[83, 113, 114]
Mayzent®	USFDA 2019	Siponimod	Fumaric acid	Improved bioavailability and aqueous solubility	Multiple sclerosis	Novartis, Basel, Switzerland	[115, 116]
Seglentis®	USFDA 2021	Celecoxib	Tramadol	Improved stability	Acute pain	Kowa Pharmaceuticals, Alabama, USA	[116, 117]

(*Continued*)

Table 1.2 Some commercial pharmaceutical cocrystals. (*Continued*)

Commercial name	Approval	API	Coformer	Improved property	Indication	Manufacturer	References
Dimenhydrinate	USFDA 1982 (ANDA)	Diphenhydramine	8-Chlorotheophylline	Enhanced solubility and pharmacokinetics	Motion sickness	Watson Laboratories Inc., New Jersey, USA	[37, 118]
Mobic®	USFDA 2000	Meloxicam	Aspirin	Enhanced solubility, improved bioavailability, stabilized the API, and provided synergistic therapeutic effects	Osteoarthritis and rheumatoid arthritis.	Boehringer Ingelheim Pharmaceuticals	[37]
Zafatek®	Japan 2015	Trelagliptin	Succinic acid	Improved solubility	Diabetes	Takeda Pharmaceutical Company Limited, Tokyo, Japan	
Ibrutinib	Tentative approval	Ibrutinib	Fumaric acid	Improved stability	Anticancer-chronic lymphocytic leukemia	Teva Pharmaceutical Industries Ltd., Tel Aviv-Yafo, Israel	[59]

1.5 Marketed Products

The market acceptance of cocrystallized pharmaceuticals demonstrates the effective implementation of cocrystallization in the pharmaceutical business. Some of the marked products are presented in Table 1.2.

1.6 Future of Pharmaceutical Cocrystals

The research conducted over the previous decade has helped to shape several elements of cocrystallization, including cocrystal screening, characterization, manufacturing processes, and development. The versatility of cocrystals offers opportunities beyond traditional pharmaceuticals. Exploring their use in nutraceuticals, cosmetics, and agrochemicals can open new markets and applications. Tailor-made cocrystals can find applications in customized medicines for individual patients based on their unique genetic and metabolic profiles, improving personalized treatment plans. Cocrystals could be used in precision medicine to create more precise dosing regimens and drug combinations, thereby optimizing therapeutic outcomes and minimizing side effects. It could also be used as a multifunctional carrier that can deliver multiple drugs simultaneously, improving treatment efficacy for complex diseases. They can also be integrated into advanced drug delivery systems like nanoparticles, liposomes, and microspheres for targeted and controlled release.

In addition, the recent advances in green and sustainable chemistry can be utilized to develop sustainable green and eco-friendly synthesis methods for cocrystals, thereby reducing the environmental impact of pharmaceutical manufacturing. These innovative processes minimize waste and energy consumption, contributing to a more sustainable pharmaceutical industry.

While the potential of cocrystals is immense, regulatory hurdles and market acceptance remain significant challenges. A continuous dialog with regulatory bodies and transparent clinical data can help mitigate these challenges. The regulatory strategies from agencies (USFDA and EMA) on medicinal cocrystals demonstrate their recognized significance in the creation of better candidates with superior properties [59]. There is a need to establish comprehensive regulatory guidelines for the approval and commercialization of cocrystals, thereby facilitating their entry into the market.

The pharmaceutical manufacturing and investigations of cocrystals should also focus on the standardization of the synthetic procedures, identification, and improvement of the properties that impact the quality of cocrystals including the reproducibility, yield and purity, and integration of cocrystal screening and research into the drug development process. Technological advancements in crystallography, computational chemistry, and material science will continue to drive the evolution of cocrystals, enabling more sophisticated and effective drug formulations. Cocrystals have the potential to make a global impact by improving drug accessibility and efficacy in developing regions. Addressing affordability and production scalability will be crucial for this vision.

The future of cocrystals in pharmaceutical sciences is promising, with the potential to revolutionize drug development and delivery. By embracing interdisciplinary collaboration, leveraging advanced technologies, and addressing regulatory and market challenges, the pharmaceutical industry can harness the full potential of cocrystals to improve patient outcomes and create a more sustainable and innovative future in drug development.

1.7 Conclusion

Cocrystals exhibit promising outcomes in altering and enhancing the physicochemical properties of API. Several cocrystal formation techniques are available, ranging from lab-scale synthesis to continuous operations at the industrial level. This article provides a brief overview of several approaches for cocrystal formation, coformer selection, and cocrystal characterization, along with appropriate examples. Cocrystals are gaining popularity in the pharmaceutical industry owing to their enhanced physicochemical qualities, as well as the availability of patent rights for cocrystals, which is critical for huge pharma businesses to earn money. Pharmaceutical cocrystals are projected to become a more common method of pharmaceutical research once the way for their creation is established and their advantages have been demonstrated.

Acknowledgment

The authors are very thankful to the Department of Pharmaceutics and Industrial Pharmacy, University of Ibadan for providing infrastructural and library facilities.

References

1. Rathi, R., Kaur, S., Singh, I., A Review on Co-crystals of Herbal Bioactives for Solubility Enhancement: Preparation Methods and Characterization Techniques. *Cryst. Growth Des.*, 22, 3, 2023–42, 2022.
2. E. M. Agency, European Medicines Agency, Amsterdam, The Netherlands, 2015.
3. USFDA, 2015, https://www.fda.gov/files/drugs/published/Regulatory-Classification-of-Pharmaceutical-Co-Crystals.pdf.
4. Almarsson, O. and Zaworotko, M.J., Crystal engineering of the composition of pharmaceutical phases. Do pharmaceutical co-crystals represent a new path to improved medicines? *Chem. Commun. (Camb.)*, 7, 1889–1896, 2004.
5. Yadav, A.V., Shete As Fau - Dabke, A.P., Dabke Ap Fau - Kulkarni, P.V., Kulkarni Pv Fau - Sakhare, S.S., Sakhare, S.S., Co-crystals: a novel approach to modify physicochemical properties of active pharmaceutical ingredients. *Indian J. Pharm. Sci.*, 71, 359–370, 2009.
6. Issa, N., Barnett, S.A., Mohamed, S., Braun, D.E., Copley, R.C., Tocher, D.A., Price, S.L., Screening for cocrystals of succinic acid and 4-aminobenzoic acid. *CrystEngComm*, 14, 2454–2464, 2012.
7. Bhutani, P., Joshi, G., Raja, N., Bachhav, N., Rajanna, P.K., Bhutani, H., Paul, A.T., Kumar, R., US FDA approved drugs from 2015–June 2020: a perspective. *J. Med. Chem.*, 64, 2339–2381, 2021.
8. Singh, M., Barua, H., Jyothi, V., Dhondale, M., Nambiar, A.G., Agrawal, A., Kumar, P., Shastri, N.R., Kumar, D., Cocrystals by Design: A Rational Coformer Selection Approach for Tackling the API Problems. *Pharmaceutics*, 15, 1161, 2023.
9. Peterson, M.L., Hickey Mb Fau - Zaworotko, M.J., Zaworotko Mj Fau - Almarsson, O., Almarsson, O., Expanding the scope of crystal form evaluation in pharmaceutical science. *J. Pharm. Pharm. Sci.*, 9, 317–326, 2006.
10. Rodríguez-Hornedo N, Nehm SJ, Jayasankar A, Cocrystals: design, properties and formation mechanisms, in: *Encyclopedia of Pharmaceutical Technology, 3rd ed*, pp 615–635, Taylor & Francis, London, 2007.
11. Taylor, L.S. and Zografi, G., Spectroscopic characterization of interactions between PVP and indomethacin in amorphous molecular dispersions. *Pharm. Res.*, 14, 1691–1698, 1997.
12. Tong, P. and Zografi, G., A study of amorphous molecular dispersions of indomethacin and its sodium salt. *J. Pharm. Sci.*, 90, 1991–2004, 2001.
13. Morissette, S.L., Fau, A.O., Peterson, M.L., Remenar, J.F., Read, M.J., Lemmo, A.V., Ellis, S., Cima, M.J., Gardner, C.R., High-throughput crystallization: polymorphs, salts, co-crystals and solvates of pharmaceutical solids. *Adv. Drug Deliv. Rev.*, 56, 275–300, 2004.

14. Hickey, M.B., P.M.L., S.L.A., M.S.L., V.A., Guzmán, H., Remenar, J.F., Zhang, Z., Tawa, M.D., Haley, S., Zaworotko, M.J., Almarsson, O., Performance comparison of a co-crystal of carbamazepine with marketed product. *Eur. J. Pharm. Biopharm.*, 67, 112–119, 2007.
15. Steed, J.W., The role of co-crystals in pharmaceutical design. *Trends Pharmacol. Sci.*, 34, 185–193, 2013.
16. Gadade, D.D. and Pekamwar, S.S., Pharmaceutical Cocrystals: Regulatory and Strategic Aspects, Design and Development. *Adv. Pharm. Bull.*, 4, 479–494, 2016.
17. Wohler, F., Untersuchungen über das Chinon. *Ann. Chem. Pharm.*, 51, 145–163, 1844.
18. Stahly, G.P., A Survey of Cocrystals Reported Prior to 2000. *Cryst. Growth Des.*, 9, 4212–4229, 2009.
19. Matsuda, H., Osaki, K., Nitta, I., Crystal Structure of Quinhydrone, C12H10O4. *Bull. Chem. Soc. Jpn.*, 31, 611–620, 1958.
20. Etter, M.C., Urbanczyk-Lipkowska, Z., Zia-Ebrahimi, M., Panunto, T.W., Hydrogen bond-directed cocrystallization and molecular recognition properties of diarylureas. *J. Am. Chem. Soc.*, 112, 8415–8426, 1990.
21. Aakeröy, C.B. and Sinha, A.S., *Co-crystals: Preparation, Characterization and Applications*, C.B. Aakeröy and A.S. Sinha (Eds.), pp. 1–32, The Royal Society of Chemistry, London, UK. 2018.
22. Desiraju, G.R. and Parshall, G.W., Crystal engineering: the design of organic solids. *Mater. Sci. Monogr.*, 54, XIV–312, 1989.
23. Babu, N.J. and Nangia, A., Solubility advantage of amorphous drugs and pharmaceutical cocrystals. *Cryst. Growth Des.*, 11, 2662–2679, 2011.
24. Fong, S.Y.K., Ibisogly, A., Bauer-Brandl, A., Solubility enhancement of BCS Class II drug by solid phospholipid dispersions: Spray drying versus freeze-drying. *Int. J. Pharm.*, 496, 382–391, 2015.
25. Hisada, N., Takano, R., Takata, N., Shiraki, K., Ueto, T., Tanida, S., Kataoka, M., Yamashita, S., Characterizing the dissolution profiles of supersaturable salts, cocrystals, and solvates to enhance *in vivo* oral absorption. *Eur. J. Pharm. Biopharm.*, 103, 192–199, 2016.
26. Kumar, S. and Nanda, A., Pharmaceutical Cocrystals: An Overview. *Indian J. Pharm. Sci.*, 79, 858–871, 2017.
27. Martin, F.A., Pop, M.M., Borodi, G., Filip, X., Kacso, I., Ketoconazole salt and co-crystals with enhanced aqueous solubility. *Cryst. Growth Des.*, 13, 4295–4304, 2013.
28. Bethune, S.J., Schultheiss, N., Henck, J.-O., Improving the poor aqueous solubility of nutraceutical compound pterostilbene through cocrystal formation. *Cryst. Growth Des.*, 11, 2817–2823, 2011.
29. Wang, J.-R., Yu, X., Zhou, C., Lin, Y., Chen, C., Pan, G., Mei, X., Improving the dissolution and bioavailability of 6-mercaptopurine via co-crystallization with isonicotinamide. *Bioorg. Med. Chem. Lett.*, 25, 1036–1039, 2015.

30. Chen, Y., Li, L., Yao, J., Ma, Y.-Y., Chen, J.-M., Lu, T.-B., Improving the solubility and bioavailability of apixaban via apixaban–oxalic acid cocrystal. *Cryst. Growth Des.*, 16, 2923–2930, 2016.
31. Sanphui, P., Goud, N.R., Khandavilli, U.B.R., Nangia, A., Fast dissolving curcumin cocrystals. *Cryst. Growth Des.*, 11, 4135–4145, 2011.
32. Guo, M., Sun, X., Chen, J., Cai, T., Pharmaceutical cocrystals: A review of preparations, physicochemical properties and applications. *Acta Pharm. Sin. B*, 11, 2537–2564, 2021.
33. Kalepu, S. and Nekkanti, V., Insoluble drug delivery strategies: Review of recent advances and business prospects. *Acta Pharm. Sin. B*, 5, 442–453, 2015.
34. Rathi, R., Kushwaha, R., Goyal, A., Singh, I., Oxaliplatin-flavone pharmaceutical co-crystal-CN111205332A: patent spotlight. *Pharm. Pat. Anal.*, 11, 5, 147–154, 2022.
35. Chettri, A., Subba, A., Singh, G.P., Bag, P.P., Pharmaceutical co-crystals: A green way to enhance drug stability and solubility for improved therapeutic efficacy. *J. Pharm. Pharmacol.*, 76, 1–12, 2023.
36. Huang, Y., Zhang, B., Gao, Y., Zhang, J., Shi, L., Baicalein–nicotinamide cocrystal with enhanced solubility, dissolution, and oral bioavailability. *J. Pharm. Sci.*, 103, 2330–2337, 2014.
37. Cheney, M.L., Weyna, D.R., Shan, N., Hanna, M., Wojtas, L., Zaworotko, M.J., Coformer selection in pharmaceutical cocrystal development: A case study of a meloxicam aspirin cocrystal that exhibits enhanced solubility and pharmacokinetics. *J. Pharm. Sci.*, 100, 2172–2181, 2011.
38. Trask, A.V., Motherwell, W.S., Jones, W., Physical stability enhancement of theophylline via cocrystallization. *Int. J. Pharm.*, 320, 114–123, 2006.
39. McNamara, D.P., Childs, S.L., Giordano, J., Iarriccio, A., Cassidy, J., Shet, M.S., Mannion, R., O'Donnell, E., Park, A., Use of a glutaric acid cocrystal to improve oral bioavailability of a low solubility API. *Pharm. Res.*, 23, 1888–1897, 2006.
40. Basavoju, S., Boström, D., Velaga, S.P., Indomethacin–saccharin cocrystal: design, synthesis and preliminary pharmaceutical characterization. *Pharm. Res.*, 25, 530–541, 2008.
41. Schultheiss, N. and Newman, A., Pharmaceutical cocrystals and their physicochemical properties. *Cryst. Growth Des.*, 9, 2950–2967, 2009.
42. Childs, S.L., Rodríguez-Hornedo, N., Reddy, L.S., Jayasankar, A., Maheshwari, C., McCausland, L., Shipplett, R., Stahly, B.C., Screening strategies based on solubility and solution composition generate pharmaceutically acceptable cocrystals of carbamazepine. *CrystEngComm*, 10, 856–864, 2008.
43. Trask, A.V., Motherwell, W.S., Jones, W., Pharmaceutical cocrystallization: engineering a remedy for caffeine hydration. *Cryst. Growth Des.*, 5, 1013–1021, 2005.

44. Oswald, I.D., Allan, D.R., McGregor, P.A., Motherwell, W.S., Parsons, S., Pulham, C.R., The formation of paracetamol (acetaminophen) adducts with hydrogen-bond acceptors. *Acta Crystallogr., Sect. B: Struct. Sci*, 58, 1057–1066, 2002.
45. Vangala, V.R., Chow, P.S., Tan, R.B., Co-crystals and co-crystal hydrates of the antibiotic nitrofurantoin: structural studies and physicochemical properties. *Cryst. Growth Des.*, 12, 5925–5938, 2012.
46. Mohamed, M.A., Nasr, M., Elkhatib, W.F., Eltayeb, W.N., Elshamy, A.A., El-Sayyad, G.S., Nanobiotic formulations as promising advances for combating MRSA resistance: susceptibilities and post-antibiotic effects of clindamycin, doxycycline, and linezolid. *RSC Adv.*, 11, 39696–39706, 2021.
47. Vishweshwar, P., McMahon, J.A., Bis, J.A., Zaworotko, M.J., Pharmaceutical co-crystals. *J. Pharm. Sci.*, 95, 499–516, 2006.
48. Maeno, Y., Fukami, T., Kawahata, M., Yamaguchi, K., Tagami, T., Ozeki, T., Suzuki, T., Tomono, K., Novel pharmaceutical cocrystal consisting of paracetamol and trimethylglycine, a new promising cocrystal former. *Int. J. Pharm.*, 473, 179–186, 2014.
49. Aitipamula, S., Wong, A.B., Kanaujia, P., Evaluating suspension formulations of theophylline cocrystals with artificial sweeteners. *J. Pharm. Sci.*, 107, 604–611, 2018.
50. Ogata, T., Tanaka, D., Ozeki, T., Enhancing the solubility and masking the bitter taste of propiverine using crystalline complex formation. *Drug Dev. Ind. Pharm.*, 40, 1084–1091, 2014.
51. Shen, Y., Aucamp, M., Abdelhakim, H.E., Li, X., Ghazali, Y., Edkins, K., Exploring co-crystallisation as a technique for taste-masking of nevirapine. *RSC Pharm.*, 1, 132–140, 2024.
52. Duan, C., Chen, Y., Zhang, Y., Liang, F., Liu, W., Xiao, X., Xu, C., Zhuang, T., Hao, C., Chen, Y., Two Cocrystals of Olaparib with Flavonoids toward Sustained Release: Structure, Dissolution Behavior, and Anticancer Activity Analysis. *Cryst. Growth Des.*, 22, 4885–4894, 2022.
53. Brader, M.L., Sukumar, M., Pekar, A.H., McClellan, D.S., Chance, R.E., Flora, D.B., Cox, A.L., Irwin, L., Myers, S.R., Hybrid insulin cocrystals for controlled release delivery. *Nat. Biotechnol.*, 20, 800–804, 2002.
54. Zhou, Z., Li, W., Sun, W.-J., Lu, T., Tong, H.H., Sun, C.C., Zheng, Y., Resveratrol cocrystals with enhanced solubility and tabletability. *Int. J. Pharm.*, 509, 391–399, 2016.
55. Rathi, R. and Singh, I., Multicomponent crystal compromising dasatinib and selected co-crystals formers: a patent evaluation of EP2861589B1. *Pharm. Pat. Anal.*, 11, 1, 15–21, 2021.
56. Latif, S., Abbas, N., Hussain, A., Arshad, M.S., Bukhari, N.I., Afzal, H., Riffat, S., Ahmad, Z., Development of paracetamol-caffeine co-crystals to improve compressional, formulation and *in vivo* performance. *Drug Dev. Ind. Pharm.*, 44, 1099–1108, 2018.

57. Srivastava, D., Fatima, Z., Kaur, C.D., Multicomponent Pharmaceutical Cocrystals: A Novel Approach for Combination Therapy. *Mini-Rev. Med. Chem.*, 18, 1160–1167, 2018.
58. Thipparaboina, R., Kumar, D., Chavan, R.B., Shastri, N.R., Multidrug co-crystals: towards the development of effective therapeutic hybrids. *Drug Discov. Today*, 21, 481–490, 2016.
59. Rathi, R., Guggal, Y., Twinkle, Sandhu, V., Singh, I., Exploring the Patent Landscape and Regulatory Prospective on Pharmaceutical Cocrystals. *J. Pharm. Technol. Res. Manag.*, 11, 49–61, 2023.
60. Jones, W., Motherwell, W.S., Trask, A.V., Pharmaceutical cocrystals: An emerging approach to physical property enhancement. *MRS Bull.*, 31, 875–879, 2006.
61. Kumar, S. and Nanda, A., Approaches to design of pharmaceutical cocrystals: A review. *Mol. Cryst. Liq. Cryst.*, 667, 54–77, 2018.
62. Chow, P.S., Lau, G., Ng, W.K., Vangala, V.R., Stability of pharmaceutical cocrystal during milling: a case study of 1: 1 caffeine–glutaric acid. *Cryst. Growth Des.*, 17, 4064–4071, 2017.
63. Kaur, N., Duggirala, N.K., Thakral, S., Suryanarayanan, R., Role of lattice disorder in water-mediated dissociation of pharmaceutical cocrystal systems. *Mol. Pharmaceutics*, 16, 3167–3177, 2019.
64. Qiu, S. and Li, M., Effects of coformers on phase transformation and release profiles of carbamazepine cocrystals in hydroxypropyl methylcellulose based matrix tablets. *Int. J. Pharm.*, 479, 118–128, 2015.
65. Duggirala, N.K., Vyas, A., Krzyzaniak, J.F., Arora, K.K., Suryanarayanan, R., Mechanistic insight into caffeine–oxalic cocrystal dissociation in formulations: Role of excipients. *Mol. Pharmaceutics*, 14, 3879–3887, 2017.
66. Porter Iii, W.W., Elie, S.C., Matzger, A.J., Polymorphism in carbamazepine cocrystals. *Cryst. Growth Des.*, 8, 14–16, 2008.
67. Rathi, R., Guggal, Y., Twinkle, Sandhu, V., Singh, I., Exploring the Patent Landscape and Regulatory Prospective on Pharmaceutical Cocrystals. *J. Pharm. Technol. Res. Manag.*, 11, 49–60, 2023.
68. Guerain, M., Guinet, Y., Correia, N., Paccou, L., Danède, F., Hédoux, A., Polymorphism and stability of ibuprofen/nicotinamide cocrystal: The effect of the crystalline synthesis method. *Int. J. Pharm.*, 584, 119454, 2020.
69. Lange, L., Schleinitz, M., Sadowski, G., Predicting the effect of pH on stability and solubility of polymorphs, hydrates, and cocrystals. *Cryst. Growth Des.*, 16, 4136–4147, 2016.
70. Ren, S., Liu, M., Hong, C., Li, G., Sun, J., Wang, J., Zhang, L., Xie, Y., The effects of pH, surfactant, ion concentration, coformer, and molecular arrangement on the solubility behavior of myricetin cocrystals. *Acta Pharm. Sin. B*, 9, 59–73, 2019.
71. Rathi, R., Sanshita, Singh, I., Hot Melt Extrusion Technique for Developing Pharmaceutical Co-crystals: A Review. *Drug Deliv. Lett.*, 14, 2, 97–108, 2024.

72. González-García, I., Mangas-Sanjuán, V., Merino-Sanjuán, M., Bermejo, M., *In vitro–in vivo* correlations: general concepts, methodologies and regulatory applications. *Drug Dev. Ind. Pharm.*, 41, 1935–1947, 2015.
73. Yousef, M.A. and Vangala, V.R., Pharmaceutical cocrystals: molecules, crystals, formulations, medicines. *Cryst. Growth Des.*, 19, 7420–7438, 2019.
74. Desiraju, G.R., Supramolecular synthons in crystal engineering—a new organic synthesis. *Angew. Chem. Int. Ed. Engl.*, 34, 2311–2327, 1995.
75. Khan, M., Enkelmann, V., Brunklaus, G., O– H··· N heterosynthon: a robust supramolecular unit for crystal engineering. *Cryst. Growth Des.*, 9, 2354–2362, 2009.
76. Bhatt, P.M., Azim, Y., Thakur, T.S., Desiraju, G.R., Co-crystals of the anti-HIV drugs lamivudine and zidovudine. *Cryst. Growth Des.*, 9, 951–957, 2009.
77. Delori, A., Galek, P.T., Pidcock, E., Patni, M., Jones, W., Knowledge-based hydrogen bond prediction and the synthesis of salts and cocrystals of the anti-malarial drug pyrimethamine with various drug and GRAS molecules. *CrystEngComm*, 15, 2916–2928, 2013.
78. Fábián, L., Cambridge structural database analysis of molecular complementarity in cocrystals. *Cryst. Growth Des.*, 9, 1436–1443, 2009.
79. Issa, N., Karamertzanis, P.G., Welch, G.W., Price, S.L., Can the formation of pharmaceutical cocrystals be computationally predicted? I. Comparison of lattice energies. *Cryst. Growth Des.*, 9, 442–453, 2009.
80. Bhogala, B.R., Basavoju, S., Nangia, A., Tape and layer structures in cocrystals of some di- and tricarboxylic acids with 4,4′-bipyridines and isonicotinamide. From binary to ternary cocrystals. *CrystEngComm*, 7, 551–562, 2005.
81. Chadha, R., Saini, A., Arora, P., Jain, D.S., Dasgupta, A., Guru Row, T.N., Multicomponent solids of lamotrigine with some selected coformers and their characterization by thermoanalytical, spectroscopic and X-ray diffraction methods. *CrystEngComm*, 13, 6271–6284, 2011.
82. Suresh, K., Minkov, V.S., Namila, K.K., Derevyannikova, E., Losev, E., Nangia, A., Boldyreva, E.V., Novel Synthons in Sulfamethizole Cocrystals: Structure-Property Relations and Solubility. *Cryst. Growth Des.*, 15, 3498–3510, 2015.
83. Sanphui, P., Bolla, G., Nangia, A., Chernyshev, V., Acemetacin cocrystals and salts: Structure solution from powder X-ray data and form selection of the piperazine salt. *IUCrJ*, 1, 136–150, 2014.
84. Sarma, B., Nath, N.K., Bhogala, B.R., Nangia, A., Synthon competition and cooperation in molecular salts of hydroxybenzoic acids and aminopyridines. *Cryst. Growth Des.*, 9, 1546–1557, 2009.
85. Lu, J. and Rohani, S., Synthesis and preliminary characterization of sulfamethazine-theophylline co-crystal. *J. Pharm. Sci.*, 99, 4042–4047, 2010.
86. Childs, S.L., Stahly, G.P., Park, A., The salt-cocrystal continuum: The influence of crystal structure on ionization state. *Mol. Pharmaceutics*, 4, 323–338, 2007.

87. Cruz-Cabeza, A.J., Acid-base crystalline complexes and the pKa rule. *CrystEngComm*, 14, 6362–6365, 2012.
88. Mohammad, M.A., Alhalaweh, A., Velaga, S.P., Hansen solubility parameter as a tool to predict cocrystal formation. *Int. J. Pharm.*, 407, 63–71, 2011.
89. Chandel, N., Gupta, V., Pandey, A., Saxena, S., Sciences, L., Co-crystalization of aceclofenac and paracetamol and their characterization. *Int. J. Pharm.*, 2, 1020–1028, 2011.
90. Lu, E., Rodríguez-Hornedo, N., Suryanarayanan, R., A rapid thermal method for cocrystal screening. *CrystEngComm*, 10, 665–668, 2008.
91. Braga, D., Grepioni, F., Maini, L., Capucci, D., Nanna, S., Wouters, J., Aerts, L., Quéré, L., Combining piracetam and lithium salts: Ionic co-crystals and co-drugs? *Chem. Commun.*, 48, 8219–8221, 2012.
92. Friščić, T., Childs, S.L., Rizvi, S.A., Jones, W., The role of solvent in mechanochemical and sonochemical cocrystal formation: a solubility-based approach for predicting cocrystallisation outcome. *CrystEngComm*, 11, 418–426, 2009.
93. Hong, C., Xie, Y., Yao, Y., Li, G., Yuan, X., Shen, H., A novel strategy for pharmaceutical cocrystal generation without knowledge of stoichiometric ratio: myricetin cocrystals and a ternary phase diagram. *Pharm. Res.*, 32, 47–60, 2015.
94. Chiarella, R.A., Davey, R.J., Peterson, M.L., Making co-crystals - The utility of ternary phase diagrams. *Cryst. Growth Des.*, 7, 1223–1226, 2007.
95. Nehm, S.J., Rodríguez-Spong, B., Rodríguez-Hornedo, N., Phase solubility diagrams of cocrystals are explained by solubility product and solution complexation. *Cryst. Growth Des.*, 6, 592–600, 2006.
96. Žegarac, M., Lekšić, E., Šket, P., Plavec, J., Bogdanović, M.D., Bučar, D.-K., Dumić, M., Meštrović, E., A sildenafil cocrystal based on acetylsalicylic acid exhibits an enhanced intrinsic dissolution rate. *CrystEngComm*, 16, 32–35, 2014.
97. Surov, A.O., Voronin, A.P., Manin, A.N., Manin, N.G., Kuzmina, L.G., Churakov, A.V., Perlovich, G.L., Pharmaceutical cocrystals of diflunisal and diclofenac with theophylline. *Mol. Pharmaceutics*, 11, 3707–3715, 2014.
98. Jiang, L., Huang, Y., Zhang, Q., He, H., Xu, Y., Mei, X., Preparation and solid-state characterization of dapsone drug-drug co-crystals. *Cryst. Growth Des.*, 14, 4562–4573, 2014.
99. Abramov, Y.A., Current computational approaches to support pharmaceutical solid form selection. *Org. Process Res. Dev.*, 17, 472–485, 2013.
100. Kavanagh, O.N., Croker, D.M., Walker, G.M., Zaworotko, M.J., Pharmaceutical cocrystals: from serendipity to design to application. *Drug Discov. Today*, 24, 796–804, 2019.
101. Gromila, R., Mooibroek, T., Frontera, A., *Hot Topics in Crystal Engineering*, Elsevier Amsterdam, The Netherlands, 2021.
102. Rissanen, K., *Hot Topics in Crystal Engineering*, Elsevier, Amsterdam, 2021.

103. Emami, S., Siahi-Shadbad, M., Adibkia, K.A.-O., Barzegar-Jalali, M., Recent advances in improving oral drug bioavailability by cocrystals. *Bioimpacts*, 8, 305–320, 2018.
104. Feng, L., Karpinski, P.H., Sutton, P., Liu, Y., Hook, D.F., Hu, B., Blacklock, T.J., Fanwick, P.E., Prashad, M., Godtfredsen, S., LCZ696: a dual-acting sodium supramolecular complex. *Tetrahedron Lett.*, 53, 275–276, 2012.
105. Katlama, C., Ingrand, D., Loveday, C., Clumeck, N., Mallolas, J., Staszewski, S., Johnson, M., Hill, A.M., Pearce, G., McDade, H., Safety and efficacy of lamivudine-zidovudine combination therapy in antiretroviral-naive patients: a randomized controlled comparison with zidovudine monotherapy. *JAMA*, 276, 118–125, 1996.
106. Duggirala, N.K., LaCasse, S.M., Zaworotko, M.J., Krzyzaniak, J.F., Arora, K.K., Pharmaceutical cocrystals: formulation approaches to develop robust drug products. *Cryst. Growth Des.*, 20, 617–626, 2019.
107. Vasoya, J.M., Shah, A.V., Serajuddin, A., Investigation of possible solubility and dissolution advantages of cocrystals, I: Aqueous solubility and dissolution rates of ketoconazole and its cocrystals as functions of pH. *ADMET DMPK*, 7, 106–130, 2019.
108. Brittain, H.G., Pharmaceutical cocrystals: The coming wave of new drug substances. *J. Pharm. Sci.*, 102, 311–317, 2013.
109. Harrison, W.T., Yathirajan, H., Bindya, S., Anilkumar, H., Escitalopram oxalate: co-existence of oxalate dianions and oxalic acid molecules in the same crystal. *Acta Crystallogr., Sect. C: Cryst. Struct. Commun.*, 63, o129–o131, 2007.
110. O'Nolan, D., Perry, M.L., Zaworotko, M.J., Chloral hydrate polymorphs and cocrystal revisited: solving two pharmaceutical cold cases. *Cryst. Growth Des.*, 16, 2211–2217, 2016.
111. Videla, S., Gascón, N., Plata-Salamán, C., Commentary on the "Co-Crystal of Tramadol-Celecoxib in Patients with Moderate to Severe Acute Post-Surgical Oral Pain: A Dose-Finding, Randomised, Double-Blind, Placebo- and Active-Controlled, Multicentre Phase II Trial". Implications for Cardiovascular Safety. *J. Cardiol. Cardiovasc. Sci.*, 3, 1–3, 2019.
112. Banerjee, M., Nimkar, K., Naik, S., Patravale, V., Unlocking the potential of drug-drug cocrystals – A comprehensive review. *J. Control. Release*, 348, 456–469, 2022.
113. San Martín, O., Llombart, B., Hernández, G.C., Menéndez, A.F., Botella-Estrada, R., Ceballos, E.H., Puig, S., Sonidegib in the treatment of locally advanced basal cell carcinoma. *Actas Dermo-Sifiliogr.*, 112, 295–301, 2021.
114. Alvani, A. and Shayanfar, A., Solution Stability of Pharmaceutical Cocrystals. *Cryst. Growth Des.*, 22, 6323–6337, 2022.
115. Díaz-Villamarín, X., Piñar-Morales, R., Barrero-Hernández, F.J., Antúnez-Rodríguez, A., Cabeza-Barrera, J., Morón-Romero, R., Pharmacogenetics of siponimod: A systematic review. *Biomed. Pharmacother.*, 153, 113536, 2022.

116. O'Sullivan, A., Long, B., Verma, V., Ryan, K.M., Padrela, L., Solid-state and particle size control of pharmaceutical cocrystals using atomization-based techniques. *Int. J. Pharm.*, 621, 121798, 2022.
117. USFDA, https://www.accessdata.fda.gov/drugsatfda_docs/label/2021/213426s000lbl.pdf.
118. Putra, O.D., Yoshida, T., Umeda, D., Higashi, K., Uekusa, H., Yonemochi, E., Crystal structure determination of dimenhydrinate after more than 60 years: solving salt–cocrystal ambiguity via solid-state characterizations and solubility study. *Cryst. Growth Des.*, 16, 5223–5229, 2016.

2

Design and Structural Characteristics of Pharmaceutical Cocrystals

Harpreet Singh[1*], Bhuvnesh Kumar Singh[2], Anil Kumar[2], Arvind Kumar[1], Arun Kumar Mishra[3], Ritu Rathi[4] and Inderbir Singh[4]

[1]School of Pharmaceutical Sciences (Faculty of Pharmacy), IFTM University, Moradabad, Uttar Pradesh, India
[2]Moradabad Educational Trust Group of Institutions, Faculty of Pharmacy, Moradabad, Uttar Pradesh, India
[3]SOS School of Pharmacy (Faculty of Pharmacy), IFTM University, Moradabad, Uttar Pradesh, India
[4]Chitkara College of Pharmacy, Chitkara University, Punjab, India

Abstract
This chapter deals with an in-depth focus on pharmaceutical cocrystals, basically describing their design, synthesis, and structural features. Cocrystals, which are the products of combining an API (active pharmaceutical ingredient) with a solid molecular/ionic compound, have widespread impacts on solubility and stability, leading to bioavailability essential for efficacy improvement. The chapter starts with an in-depth analysis of cocrystallization *via* heteromolecular interactions, including hydrogen bonds, halogen bonds, and charge transfer (CT) interactions. It highlights the essential role of noncovalent synthesis in designing functional molecules. Polymorphism and Form Selection of Cocrystals This chapter discusses complexities in polymorphs and the salt-cocrystal continuum that affect cocrystallization outcomes. Iondosekie *et al.* discussed key considerations in choosing a coformer and how it leads to intermolecular interactions from both points of view, highlighting that the designing strategies must always amalgamate these aspects for the efficient design of solid-state forms as well. Structural characterization of cocrystals, including X-ray diffraction and hydrogen bonding analysis, helps to determine the stability and functionalities using various physical methods. This chapter discusses the regulatory and patent aspects of cocrystals, which is indicative of their increasing role in medicinal chemistry. The techniques included in

**Corresponding author*: harpreetproctor@rediffmail.com

Inderbir Singh, Subrata Mallick and Ritu Rathi (eds.) Cocrystals in Pharmaceutical Sciences: Design to Applications, (29–54) © 2025 Scrivener Publishing LLC

this chapter are of great advantage to researchers working in pharmaceutical sciences, bringing explicit methodologies for designing and synthesizing cocrystals. The process of cocrystallization allows APIs to become more soluble and stable, which, in turn, leads to better therapeutic results. This results in better drugs for people, which have a potentially lower side effect as cocrystals outperform drug delivery and bioavailability. The insights presented in this chapter are anticipated to be instrumental in the development of innovative pharmaceutical products that meet the stringent standards of contemporary medical practice.

Keywords: Pharmaceutical cocrystals, active pharmaceutical ingredient, polymorphism, x-ray diffraction, hydrogen bonding analysis

2.1 Introduction

Pharmaceutical cocrystals are a specific kind of crystalline structure formed through the precise combination of an API (active pharmaceutical ingredient) with other solid molecular or ionic compounds. The compounds are formulated to maintain their solid state even when exposed at room temperature [1]. The creation of these cocrystals requires the use of a variety of intermolecular forces such as hydrogen bonding, halogen bonding, π-π stacking, and charge transfer interactions [2]. The processes involved in this process make it possible to synthesize functional molecules through noncovalent synthetic operations. Some of the reasons why this formula is advantageous include that it is easy to process, manageable structures, and affordable to implement. Organic cocrystals have been of interest in the field of molecular optoelectronics [3]. It is also relevant to appreciate the importance of these frameworks concerning the general aim of relating structure at the molecular level to the properties of the material [4]. They are not solely held by ionic bonds but held together through different types of molecular interactions [5, 6]. Cocrystals in pharmaceuticals are an interesting class of medicinal materials found to enhance physical properties and stability and even may offer new chances for the solid-state form under patent protection [7]. Crystal issues may arise due to the formation of solvates, hydrates, and salts within the structure. Strategies such as hydrogen bonding control and adjustments of solubility parameters are crucial in the formulation of cocrystals. In the field of design and construction of new cocrystals and cocrystals, the incorporation of CT interactions is extensively used as a driving force for the construction of supramolecular cocrystals. These interactions discussed in the study help in enhancing the stability and quality of the crystals, leading to closely packed morphology and size [8]. In organic cocrystals, CT interactions between a donor and an acceptor have led to the development of new forms of organic D-A system at a fast pace [9]. Pharmaceutical researchers are obsessed with enhancing the physicochemical characteristics of APIs so that they can deliver increased therapeutic potential. Some of the changes elicited by altering the solid states of an API are changes in its melting point, solubility, stability, permeability, bioavailability, and ability to form a tablet [10]. A key challenge to the formulation of oral drugs is the capability of APIs, which depend on their solubility and bioavailability. The enhancement of the solubility of APIs has been the subject of several research studies, and significant advances have been achieved by scientists in the development of techniques to improve solubility. Such strategies include prodrug formation, solid dispersion, particle size reduction, cyclodextrin complexation, self-emulsifying system, salt formation, surfactant application, polymorphism, nanoparticle formation, and multicomponent molecular cocrystal formation [11]. Both techniques show advantages and limitations, depending on the nature of APIs and polymers being used; the efficiency of each of these methods can be greatly affected by the API's characteristics [12]. In the US and Europe, there are some rules concerning the regulatory classification of API cocrystals. These guidelines give basic information about product registration [13]. In recent years, the focus has shifted toward new crystalline forms like salts, polymorphs, hydrates, solvates, and, especially, cocrystals. Pharmaceutical cocrystals are gaining recognition as an effective strategy for enhancing various physicochemical attributes, including melting point, dissolution rate, stability, solubility, permeability, bioavailability, and the ability to form tablets [14].

Cocrystals are formed when a drug and coformer or several other components combine in an exact molar ratio and crystallize together. The formation of cocrystals is facilitated by hydrogen bonding. These components are held together by different noncovalent interactions, such as hydrogen bonds, van der Waals forces, and π-π interactions. The era of understanding the structure, design, and bonding interactions will enable the enhancement of the properties of pharmaceuticals in applications. Molecular cocrystals, which are also referred to as molecular CC, are formed by associating a neutral drug and a neutral coformer in a definite proportion be it 1:1 or 2:1. In the design strategy, one has to choose coformers that are capable of forming similar hydrogen bonds or any other forms of bond with the drug, regarding their functional groups. The main types of bonding interactions involve hydrogen bond, which relates to carboxylic acids, amides, and alcohols. van der Waals forces present in the secondary model play some role in stability and packing, while more stable packing is achieved through aromatic interactions between π-rings [15–18].

Ionic cocrystals (also referred to as ionic CCs) are formed from definite stoichiometric proportions of ionic species; in most cases, it has the drug in its salt state and another species, which can be anionic or cationic in nature. During the design strategy, it is mandatory to select ionic coformers that are capable of forming an ionic or stable hydrogen bond with the ionic drug. Strong ion/ion compensations, hydrogen bonds, and additional secondary van der Waals forces are responsible for bond interaction that contributes to structural integrity. Such relations are vital in the overall stability of the system of such interactions. Cocrystals or solvates are drugs solvated with solvent molecules, building a crystal lattice in which the ratio of the solvent molecules to the drug molecules is fixed, for instance, 1:1 or 1:2. The design strategy means the choice of the solvents that can form certain interactions with the drug and stabilize the lattice [19–21].

Hydrated cocrystals, otherwise called hydrates, encompass a drug, a coformer, and water molecules. These cocrystals are resolved based on the number of water molecules in relation to the involvement of the drug-coformer pair. The design strategy leverages water to facilitate hydrogen bonds with the material, in particular with the polar functional groups. The bonding interactions in this system are multiple hydrogen bonds between water molecules and the coformer as well as the drug. Also in view are van der Waals forces contributing to stability, and sometimes there may be π-π interactions in the case of the components present. Hydrated cocrystals (or simply hydrates) consist of a drug, a coformer, and water molecules. The stoichiometry of these cocrystals is determined by the ratio of water molecules with respect to the drug-coformer combination. A design methodology based on polar functional groups takes advantage of the presence of water to assist in hydrogen bond formation using a supramolecular motif based on a virtually infinite network for hydrogen bonding involving water molecules, the drug, and the coformer. In turn, the system's stability also depends on van der Waals forces that could lead to unexpected π-π interactions. In addition, the molecular arrangements of the cocrystals make solid forms of patent protection worthy of protection. Additionally, cocrystals can increase the solubility of drugs, hence increasing the delivery of drugs and, therefore, leading to more therapeutic benefits [23–25]. This chapter provides a comprehensive analysis of the design and structural characteristics of pharmaceutical cocrystals, along with their diverse range of applications. Furthermore, it explores the difficulties and potential future developments of cocrystal research [22].

2.1.1 Polymorphism

Polymorphism refers to the ability of a substance, typically a solid, to exist in various crystalline shapes or formations [26]. Figure 2.1 [27] illustrates that various polymorphic forms of a drug can exhibit unique physical properties, including solubility, melting temperature, and stability. Different polymorphic forms can display varying levels of bioavailability and rates of dissolution [28]. Below are a few illustrations: Sulfathiazole exists in multiple polymorphic forms, each exhibiting a unique crystal structure. Carbamazepine exists in many polymorphs, which can affect its dissolution [29]. The molecular configurations inside the solid state are responsible for the fundamental physical properties of crystalline materials. Modifying these configurations or interconnections can significantly influence the characteristics of the solid. Scientists in solid-state chemistry employ many methodologies to alter the chemical and physical characteristics of APIs. The approaches encompass the generation of salts, polymorphs, hydrates, solvates, and cocrystals, as seen in Figure 2.1.

Salt formation is a prevalent method for altering API properties, with many medicines marketed as salts. This approach requires the API to have an ionizable site. Cocrystals, however, are multicomponent structures stabilized by noncovalent interactions and do not require ionizable groups, thus broadening the scope for modifying APIs with nonionizable functional groups. It is important to acknowledge that there is no universal approach to enhancing the characteristics of every API. Each API's molecular structure and desired properties must be individually evaluated. Guidelines for cocrystal design through supramolecular synthesis are crucial in this assessment process.

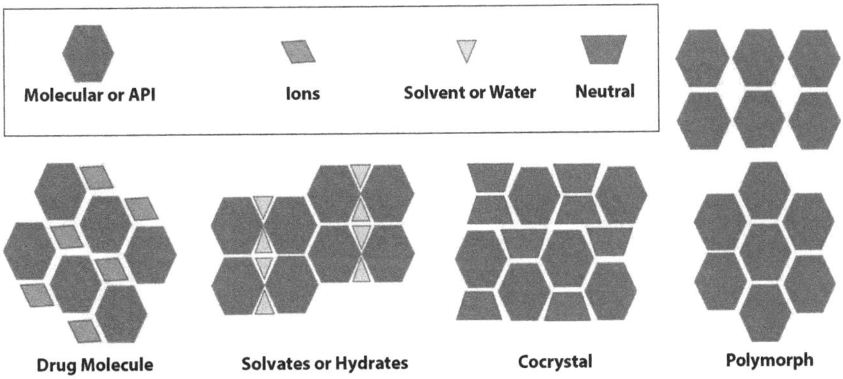

Figure 2.1 Solid state polymorphism.

2.1.2 Salt Cocrystal Continuum

Crystalline structures known as salt forms are created when an API combines with an ion, typically a counterion, resulting in the formation of a salt (see Figure 2.1). Salts are crucial for improving the solubility, stability, and bioavailability of APIs. Commonly used salt forms in pharmaceuticals include hydrochloride salts (e.g., diphenhydramine hydrochloride), sulfate salts, and citrate salts [30]. Cocrystals are crystalline substances consisting of an API and one or more distinct cocrystal formers. These cocrystals can alter the properties of APIs while preserving their biological activity. Examples include caffeine-theophylline and carbamazepine-saccharin cocrystals [31]. The salt-cocrystal continuum is a fascinating area of study, where the distinction between salts and cocrystals becomes less clear [32]. The spectrum ranges from pure salts to pure cocrystals, encompassing a range of hybrid structures in between [33]. Factors such as ionic character, hydrogen bonding, and crystal packing determine the position on this continuum. The diversity of solid forms significantly affects drug properties and offers exciting opportunities for advancements in the field of pharmaceuticals [34].

2.2 Design of Pharmaceutical Cocrystals

The different parameters to be considered in designing cocrystals are shown in Figure 2.2 [35].

The physicochemical properties of the API and coformer are the most important factors to consider when selecting a coformer for cocrystal development. The functional groups present in the API and coformer will

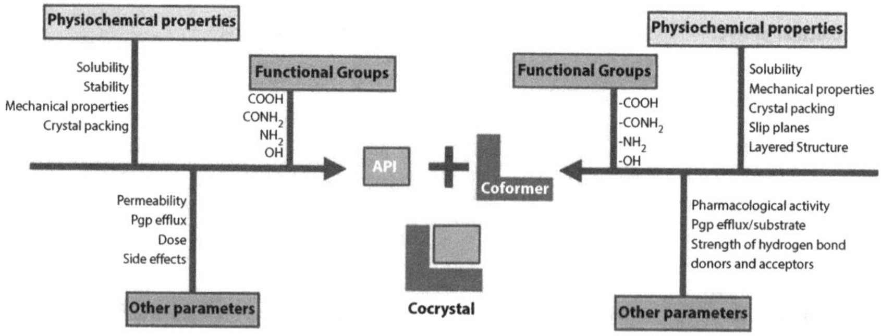

Figure 2.2 Parameters to be considered in coformer selection (adopted from Singh *et al.*, 2023; published under CC4 license) [35].

determine the type of interactions that can occur between them, which will, in turn, affect the crystal structure and properties of the cocrystal. Solubility, mechanical properties, crystal packing, and other parameters are also important to consider, as they can affect the stability, bioavailability, and efficacy of the cocrystal [35].

2.2.1 Molecular Recognition and Supramolecular Interactions

One of the key principles underlying the rational design of pharmaceutical cocrystals is molecular recognition and the use of supramolecular interactions [36]. Pharmaceutical cocrystals aim to manipulate properties like solubility, dissolution rate, and bioavailability through supramolecular synthon formation between the API and a cocrystal former at the molecular level [37]. A fundamental understanding of the noncovalent interactions driving molecular recognition events is thus crucial for the development of improved cocrystal systems. The most commonly exploited supramolecular synthons in pharmaceutical cocrystals include hydrogen bonding and π-stacking interactions. APIs often contain functional groups like carboxylic acids and amides that are amenable to hydrogen bond donation/acceptance. Systematic screening of potential coformers containing complementary hydrogen bonding functionalities allows the identification of suitable molecular recognition events between the API and coformer. For example, functionalities like amines/pyridines in coformers can act as hydrogen bond acceptors to the acidic hydrogens in an API [38]. π-stacking is another ubiquitous noncovalent interaction that can direct molecular assembly in crystalline solids. Planar aromatic groups in APIs and coformers contribute to stabilizing molecular packing through π-π stacking. Such interactions influence the preferred crystal packing and physical properties [39]. Understanding the preferred orientation, interplanar distance, and relative displacement of aromatic systems through computational methods aids in selecting coformers.

By considering both the hydrogen bonding properties and the tendency for π-stacking, computational tools in combination with high-throughput screening allow for the identification of coformer:API pairs that form robust and directional supramolecular synthons. Such synthons can template the assembly of cocrystal structures possessing optimal properties like improved aqueous solubility, leading to enhanced oral bioavailability and therapeutic efficacy. The principles of molecular recognition serve as a foundation for rational cocrystal design and the development of new solid forms with improved performance [40]. Carbamazepine (CBZ) is a BCS class II drug used as an anticonvulsant and antidepressant. However, its oral bioavailability is only 65% to 80% due to poor aqueous solubility. A screening of CBZ with dicarboxylic acids as coformers using solid-state NMR identified malonic acid as forming strong O-H···N and C-H···O hydrogen bonds based on chemical shift perturbations. X-ray crystallography of the CBZ-malonic acid cocrystal confirmed the presence of a reinforced synthon consisting of two N-H···O hydrogen bonds and one C-H···O interaction. Dissolution rate studies showed a 400% increase in IDR compared to CBZ due to improved wettability imparted by the synthon. Oral administration of the cocrystal to rats resulted in a 30% increase in bioavailability compared to CBZ alone [41]. These experimental findings demonstrate how identifying a robust synthon through molecular recognition principles can yield a cocrystal with enhanced solubility and oral absorption. Felodipine (FDP) is an antihypertensive drug limited by poor aqueous solubility. Cocrystals were designed and synthesized between FDP and nicotinamide (NAM) or isonicotinamide (INAM) using solvothermal methods. X-ray studies showed FDP formed π-stacked dimers that were further associated through N-H···O hydrogen bonds with NAM or INAM molecules. Dynamic vapor sorption experiments indicated a twofold increase in aqueous solubility for the FDP-NAM cocrystal compared to FDP alone. Permeability studies across a Caco-2 monolayer also demonstrated a 1.5-fold enhancement for the cocrystal [42]. These molecular recognition-directed cocrystals led to improved solubility and permeability profiles governed by self-assembled supramolecular synthons. Such properties indicate the propensity for enhanced oral absorption through cocrystal formation.

2.2.2 Hansen Solubility Parameters (HSPs)

HSPs help understand and predict solubility, dispersion, and formulation. They explain the thermodynamics of substance dissolution and are based on the idea that similar chemicals dissolve each other [36]. HSP characteristics are useful for studying solubility, including pharmaceutical cocrystals. When assessing solvent-solute compatibility, dispersion forces, polar interactions, and hydrogen bonding must be considered. These aspects are vital to assessment. This approach allows for more advanced solvents that dissolve the API and coformer, which is necessary for cocrystal formation. Researchers use HSPs to estimate solubility and miscibility, which are essential for synthesizing cocrystals with specified features [37, 38]. The HSP includes three main components: δD (dispersion), δP (polarity), and δH (hydrogen bonding). The δD value measures van der Waals force strength and describes interactions between nonpolar compounds. Chemistry relies on polarity to characterize dipole moments and polar interactions. The δH value expresses [39] hydrogen bonding capability. HSP offers a more whole explanation since it anticipates solubility events outside of "polar" and "hydrophobic." Unlike LogP [40, 41], HSP factors several solubility variables. HSP clarifies solvent-polymer interactions, guiding formulation for useful applications. It influences patent strategies and finds use in numerous contexts. HSP offers a whole paradigm for solubility prediction [42]. Forecasting cocrystal development calls for knowledge of HSPs. HSPs and miscibility depend on a knowledge of solvent-solute compatibility. The heat of solution parameters of medicine and coformer should match to show compatibility and mixing. Miscibility ensures that both components dissolve in the same solvent, which is essential for cocrystal formation [43]. HSP analysis helps researchers choose coformers by predicting medication compatibility. This forecast helps choose coformers before screening cocrystals. By selecting coformers with drug-compatible HSP, the selection procedure is simplified [44]. HSPs are essential for rational cocrystal design because they identify compatible coformers. Scientists can make informed selections about prospective cocrystal partners, improving screening efficiency and precision.

2.2.3 Coformers Selection and Combinatorial Approaches

The selection of an appropriate coformer is one of the most critical steps in the rational design and development of cocrystals. The coformer must be able to form supramolecular heterosynthons with the API through favorable noncovalent interactions like hydrogen bonding. It should also improve the physicochemical properties of the API like solubility, dissolution rate, and bioavailability without compromising stability and safety. A combinatorial approach involving the screening of multiple coformers can help identify the most suitable candidate for further development [35].

2.2.3.1 Theoretical Methods

Advances in crystal structure databases and computational modeling techniques have enabled the application of a structure-based selection of potential coformers. Tools like the Cambridge Structural Database (CSD) allow the identification of structural moieties that can serve as complementary hydrogen bond donors/acceptors for a given API. Density functional theory (DFT) calculations can provide insights into relative binding affinities and interaction energies of different coformer:API complexes. Such *in silico* screening helps prioritize candidates for experimental evaluation [35].

2.2.3.2 GRAS Coformer Libraries

Combinatorial screening typically utilizes libraries of FDA-approved Generally Recognized as Safe (GRAS) coformers. Common classes of coformers evaluated include carboxylic acids, amides, amines, alcohols, sugars, and carbohydrates. Commercial databases containing characterizations of >100 approved ingredients facilitate the selection of coformers with desired physicochemical properties like solubility and pKa. Multiple coformers are also evaluated in combination to discover multicomponent frameworks [43].

2.2.3.3 High-Throughput Experimentation

Automated platforms allow rapid evaluation of large coformer sets under identical conditions. Methods like slurry mixing, liquid-assisted grinding, and thermal treatments are adapted for parallel synthesis of hundreds of cocrystal samples in microtiter plates [44]. Analytical techniques like thermal analysis and PXRD are miniaturized for direct evaluation in multi-well formats. Data mining identifies formulation-property relationships to guide the selection of lead coformer candidates.

Combining theoretical predictions with high-throughput synthesis and screening provides a powerful approach for the rational design of cocrystals with optimized therapeutic potential. Identification of leading coformer candidates through such combinatorial strategies helps focus subsequent optimization efforts on the most promising systems.

2.2.3.4 Coformer Selection

Rational selection of a suitable coformer is one of the most critical steps in the design of pharmaceutical cocrystals. The coformer must be able to form favorable noncovalent interactions like hydrogen bonds with the API within the crystal lattice. It should improve key properties of the API, like solubility, dissolution, and bioavailability, without compromising stability. A systematic approach to coformer selection can facilitate the design of cocrystals with optimized therapeutic potential (Figure 2.3) [35].

2.2.3.4.1 Molecular Compatibility

Coformers are selected based on their ability to serve as complementary hydrogen bond donors/acceptors for the functional groups on the API. Tools like the CSD can identify structural motifs from existing crystal

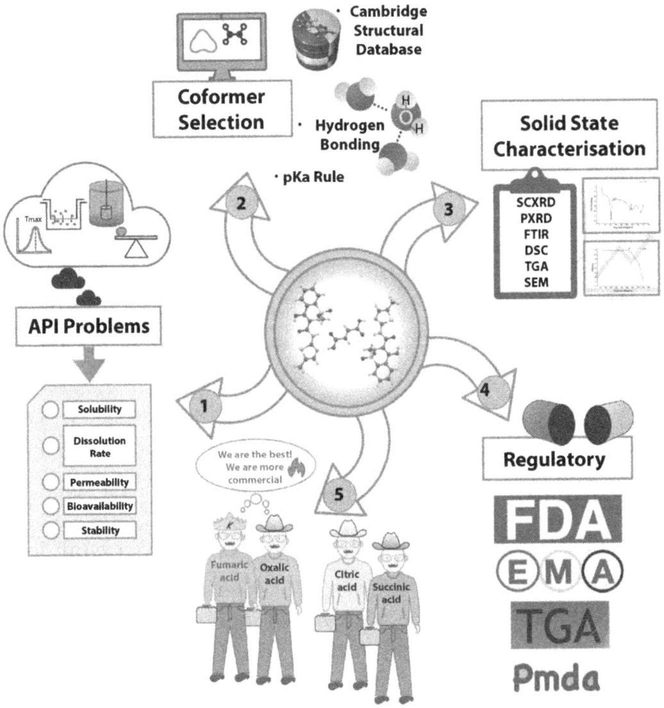

Figure 2.3 Schematic representation of the coformer selection process for the development of cocrystals to improve therapeutic efficiency. The process starts with the identification of the API problems that need to be solved, followed by coformer selection based on the CSD and literature data, and then solid-state characterization. The most promising coformers are then selected for regulatory approval. API: active pharmaceutical ingredient, CSD: cambridge structural database. Adopted from (Singh *et al.*, 2023, published under CC4 license) [35].

forms that are likely to interact favorably. Physical properties like pKa are also considered to guide the choice of acidic/basic/neutral coformers [35].

2.2.3.4.2 Intermolecular Interaction Energy

Computational tools enable the evaluation of interaction energies between possible coformer:API complexes. DFT calculations provide insights into relative binding affinities to guide prioritization for experimental screening. Strongly bound complexes are more likely to form stable cocrystals [35].

2.2.3.4.3 Physicochemical Properties

Selected coformers should improve the solubility, dissolution, or permeability of the API without compromising stability. Their own physicochemical properties, like solubility and melting point, are considered to influence the properties of the cocrystal formed.

2.2.3.4.4 Coformer Safety

Coformers approved for use in food, medicines, and neutraceuticals by regulatory agencies are preferred. This allows leveraging of existing non-clinical safety data to speed up development. Toxicity risks associated with non-GRAS coformers need careful evaluation [43]. Combining computational modeling with consideration of experimental factors like intermolecular interactions, physicochemical matching, and safety facilitates the rational selection of coformer candidates most likely to form pharmaceutically favorable cocrystals [35].

2.2.4 Combinatorial Approaches

A powerful design strategy for developing new cocrystal forms with enhanced properties is the use of combinatorial or high-throughput approaches. This allows the screening of a large number of potential coformer molecules systematically to explore the cocrystal landscape. By increasing the molecular and supramolecular diversity evaluated, rarer and higher energy forms may be isolated, which confers improved performance [45]. A key advantage of combinatorial methods is the ability to rapidly screen coformer libraries in parallel using automated crystallization techniques such as solvent-drop or thermal-gradient cocrystallization. This enables the evaluation of hundreds or even thousands of coformer-drug combinations to map out the structural landscape. Potent coformers can then be identified more efficiently for follow-up optimization.

Molecular diversity can be achieved by choosing coformer libraries containing a wide range of functional groups and architectures able to participate in multiple hydrogen bonding interactions with the drug. Supramolecular diversity results from the ability of a given drug to adopt multiple conformations interacting with coformers in different orientations. Together, this molecular and supramolecular variability allows for an expansive cocrystal landscape to be sampled [35]. The key principle underlying combinatorial crystal synthesis is competitive intermolecular interactions between suitable molecular components in solution. When multiple molecules possessing potential recognition sites like hydrogen bonding or π-π stacking functionality are combined in solution, they can engage in a dynamic equilibrium of reversible interactions. By tuning experimental parameters such as solvent composition, temperature, and stoichiometry, certain interacting subsets can be selectively amplified to direct formation of a stable crystalline phase [46].

This competitive selection mimics Darwinian evolution concepts, with the system itself choosing an interaction pattern or "fittest" supramolecular synthon based on relative interaction strengths in a given environment. It bypasses the need for rationally predetermining a step-wise buildup route based on interaction hierarchies. Instead, a library of molecular components and potential binding sites are provided, allowing the experimental conditions to guide supramolecular assembly A key requirement is that the competing interactions be marginally reversible near room temperature, permitting proof-correction toward low-energy conglomerates. This competitive dynamic helps avoid kinetic traps and optimizes the accommodation of multiple components. By systematically varying molecular precursors and crystallization protocols, a broad array of novel multicomponent crystals with diverse structures and properties can be accessed.

All the computational techniques described in Table 2.1 help narrow down large molecular libraries to high-priority combinations for physical screening in the laboratory [47–51].

Table 2.1 Various techniques required for the combinatorial approach to cocrystal synthesis.

S. no.	Technique	Description
1.	Compound libraries	Libraries of pharmaceutically relevant drug molecules and coformers with diverse chemical functionalities must be available.
2.	Liquid-assisted grinding	Technique used to intimately mix and grind drug and coformer mixtures in the presence of a small amount of solvent. Allows rapid screening.
3.	Solvent diversity	Ability to crystallize mixtures from a wide range of solvents and solvent mixtures increases the likelihood of crystallization and diversity of products
4.	Crystallization equipment	Tools like evaporation plates, sealed vials, etc. required to rapidly crystallize mixtures from various solvent conditions.

(Continued)

Table 2.1 Various techniques required for the combinatorial approach to cocrystal synthesis. (*Continued*)

S. no.	Technique	Description
5.	Automated systems	For high-throughput handling, mixing, crystallization, and data collection from a large number of experiments.
6.	Crystal screening	Techniques like microscopy for rapid evaluation of crystal formation in mixtures.
7.	Powder XRD	For identifying new crystalline phases and purity assessment of cocrystals.
8.	Solubility measurement	To screen for formulations with enhanced therapeutic properties like improved solubility.
9.	Thermal methods	TGA and DSC useful to understand the crystallinity and stability of cocrystals.
10.	Single-crystal XRD	Gold standard for determining molecular conformations and supramolecular synthons in cocrystals.
11.	Computational modeling	To predict favorable intermolecular interactions and guide cocrystal synthesis. Some common *in silico* and computational tools that can be used for the selection of candidate molecules for cocrystal synthesis *via* a combinatorial approach include the following:
i	Database searching	Screening crystal structure databases like CSD to find molecules with similar functional groups that have formed heterosynthons in previous crystal structures. This indicates recognition potential.
ii	Docking simulations	Molecular docking methods like Autodock can be used to predict favorable intermolecular interactions and binding poses between candidate drugs and coformer molecules. Provides insight into recognition patterns

(*Continued*)

Table 2.1 Various techniques required for the combinatorial approach to cocrystal synthesis. (*Continued*)

S. no.	Technique	Description
iii	Crystal structure prediction	Methods like CrystalPredictor and ELPASO can calculate lattice energies of putative cocrystal structures to identify thermodynamically stable cocrystal forms even before synthesis.
iv	H-bond propensity calculation	Tools like COSMOtherm allow the calculation of hydrogen bond donor/acceptor strengths of functional groups to guide the selection of complementary molecules.
v	Conformational searching	Methods like AutoDockTools and RaptorX are used to generate low-energy conformations of flexible molecules to capture important recognition conformation in cocrystals.
vi	Solubility prediction	Correlations with physicochemical properties allow *in silico* screening of cocrystal formulations with desired improvements in solubility/dissolution rate over the original drug form.
vii	Polymorph prediction	Predicting stable polymorphs of candidate drugs/coformers helps reduce experimental search space for cocrystal synthesis.
viii	Interaction energy calculation	Quantum chemical methods like DFT provide estimates of interaction strengths between molecular fragments for screening supramolecular synthons.

Fu *et al.* applied a combinatorial screening approach to discover new cocrystal forms of the poorly soluble antiepileptic drug carbamazepine (CBZ). The researchers selected four aromatic carboxylic acids as coformers: 2,3-dihydroxybenzoic acid (DBA), 1-naphthoic acid (NA), anthracene-9-carboxylic acid (AC), and systematically combined each with CBZ in 1:1, 2:1, and 1:2 molar ratios. Through liquid-assisted grinding of the mixtures with a small amount of solvent, they were able to rapidly crystallize and isolate four new CBZ cocrystals: CBZ-DBA (1:1), CBZ-DBA-CH_3OH (2:1), CBZ-NA-H_2O (1:1), and CBZ-AC (1:1). Single-crystal X-ray diffraction (SCXRD) revealed that their structures contained different supramolecular heterosynthons, including amide-acid, amide-amide, and amide-acid-H2O motifs. Solubility measurements found CBZ-NA-H_2O displayed a 1.9-fold increase in aqueous solubility (304 µg/ml) compared to CBZ alone (161 µg/ml). Additionally, neat grinding experiments demonstrated that CBZ-DBA (1:1) could be reversibly transformed to the more stable CBZ-DBA-CH_3OH (2:1) form upon the addition of CBZ or solvent [52].

This case study illustrates how systematically varying drug:coformer ratios in a combinatorial manner allows for rapid discovery and optimization of new cocrystal forms with improved properties through modulating their supramolecular assembly.

2.3 Structural Characteristics of Pharmaceutical Cocrystals

The effectiveness and usefulness of pharmaceutical cocrystals heavily rely on their unique characteristics.

2.3.1 Crystal Structure Analysis

In order to comprehend the formation process and characteristics of medicinal cocrystals, it is essential to investigate their crystal structures. Determining the molecular packing in crystalline structures and establishing the constitution of cocrystals are important activities that must be accomplished by employing SCXRD or powdered X-ray diffraction (PXRD). An understanding of the molecular arrangement in cocrystals is important, especially since it provides sufficient information on the positions of atoms in the given crystal structure and the nature of the bonds between these atoms. X-ray crystallography is an inalienable asset in this process. The arrangement of the molecules as they are packed in the crystal lattice influences such properties as density, hardness, and solubility. The measurements of the unit cell width of the crystal give insight into crystal symmetry and lattice nature thereby providing an understanding of the repetitive structural unit of the crystal [53–55].

2.3.2 A Perspective on Cocrystal Hydrogen Bonding

Hydrogen bonding is the major factor that contributes to the stability and structure of cocrystals. The phenomenon of hydrogen bonding between a donor and an acceptor affects the solubility and dissolution mechanisms of therapeutically active substances. The formation of cocrystals is due to hydrogen bonding and solubility, leading to the formation of stable structures, and efficiency. Related to this definition, hydrogen bonding relates to the attractive forces existing between a hydrogen atom and an electronegative atom, such as oxygen or nitrogen in the other molecule. The stability of cocrystals is defined by the contact of hydrogen atoms of the first compound with the oxygen or nitrogen atoms in the second compound. That is due to the fact that the strength and organization of chemical connections have an impact on the features of cocrystals. Improved bond strength improves the stiffness of the structure while a weaker bond improves the flexibility of the structure [10, 56, 57].

2.3.3 Characterization Techniques

The different instrumental analytical techniques used in the cocrystal analysis are shown in Figure 2.4.

A number of methods provide detailed information about the structural features of pharmaceutical cocrystals. The equipment used in this study is SCXRD (single-crystal X-ray diffraction), PXRD (powder X-ray diffraction), FTIR (Fourier transform infrared spectroscopy), DSC (differential

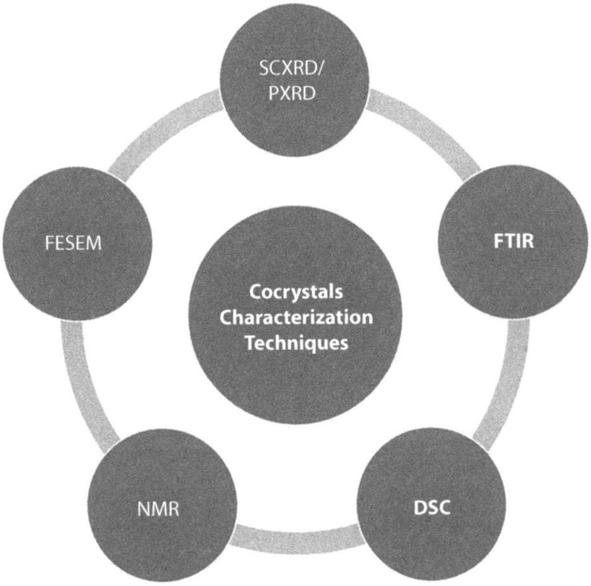

Figure 2.4 Methods for characterizing pharmaceutical cocrystals.

scanning calorimetry), solid-state NMR (nuclear magnetic resonance), and FESEM (field emission scanning electron microscopy). Due to the usage of many novel techniques, each of which gives a unique view of the molecular and crystal structures of the cocrystals, their organization can be studied. The qualitative analysis of FTIR is perfectly well sorted for functional group identification and considering intermolecular forces. The technique used to achieve this goal is a quantification of the absorption of light at the infrared spectrum, which provides valuable information on the vibrational modes, as well as a means for the detection of hydrogen bonding and other interactions [58, 59]. The principle of NMR spectroscopy is based on the magnetic properties of the atomic nuclei, giving informative insight into the spatial arrangement and motion of molecules. The ability to monitor alterations in the chemical environment is linked to the formation of cocrystals due to interactions, such as hydrogen bonding or π-π stacking.

Recognizing crystal atomic arrangement within three dimensions is challenging one and particularly well accomplished by SCXRD. X-rays may image the atomic arrangement of crystalline lattice revealing interactions such as van der Waals forces, π-π interactions, and hydrogen bonds [58]. Application of FTIR to determine hydrogen bonds for coffee and of succinic acid, NMR to map out the chemical shift for coffee and succinic acid, and SCXRD to confirm the π-π interactions between two aromatic rings and carboxyl groups for coffee and succinic acid. H bonds present in the carbamazepine and nicotinamide are confirmed by the NMR chemical shifts; FTIR reveals these bonds; SCXRD explains how these interactions maintain and improve the cocrystal. These enable identification of cocrystal structures as well as bonding allowing for the rational design of the new cocrystals with the right characteristics. These results are useful for drug delivery and cocrystal formation [60, 61].

2.4 Challenges and Future Directions in Pharmaceutical Cocrystal

Pharmaceutical cocrystals are an area of promising research that can enhance the solubility, stability, and biological availability of APIs. As a developing area of expertise, a significant problem in this realm is the proper scaling of cocrystal synthesis. Furthermore, approval of these products in different countries remains a challenge in this segment since the regulatory authorities of different countries do not approve certain products easily. Recent developments have been centered around continuous manufacturing techniques, which hold the potential for maintaining consistent quality and enabling real-time monitoring using process analytical technology (PAT). Nevertheless, there remain challenges to be addressed, including the need to improve the predictability of cocrystal formation and the ability to control their physical properties on a larger scale [62]. In anticipation of future developments, the industry is focused on overcoming these obstacles by enhancing design approaches, selecting appropriate coformers, and refining cocrystallization techniques. There is also a growing interest in exploring cocrystals for alternative dosage forms beyond oral administration. The successful commercialization of cocrystals, like the valsartan-sacubitril cocrystal (Entresto and Novartis), demonstrates the potential for cocrystals to become a more mainstream option in pharmaceutical development [63, 64].

The market success of pharmaceutical cocrystals depends on effectively managing their strategic and regulatory aspects. Strategically, the design and development process is crucial, requiring the careful selection of coformers that enhance the desired properties of the API. This task necessitates a strong understanding of molecular interactions and crystal engineering principles [64].

The formation of cocrystals means a decent and cost-effective strategy in drug design and development since the physicochemical properties of APIs can be improved without profound chemical modifications [25]. Both the US and European regulatory bodies have established criteria that streamline the registration of pharmaceutical products to ease the use of cocrystals to develop pharmaceutical products. The recommendations given here detail the conditions for the registration of pharmaceutical products containing cocrystals so as to meet safety, effectiveness, and quality requirements. The regulatory authorities appreciate the high potential of cocrystals in enhancing the properties of APIs; therefore, they encourage new developments and advances in formulation and therapy. Current research and development in the use of cocrystals as formulation tools in developing pharmaceuticals largely depends on regulatory support [65].

Ensuring quality and consistency while meeting the industrial needs for pharmaceutical cocrystals presents a great difficulty. Mechanochemical developments provide promise for scalable, solvent-free cocrystal synthesis, therefore providing the means of more affordable and environmentally friendly manufacture [66–70]. Apart from raising solubility, cocrystals are attracting more and more use in taste masking, mechanical property improvement, and intellectual property generation. Researchers are looking at many methods to create cocrystals with enhanced physicochemical qualities. Among other things, better bioavailability, flowability, solubility, and tabletability could emerge from these advancements [71]. Researchers are investigating cocrystals, especially for young and elderly patients as they can enhance the taste of specific APIs, therefore considerably boosting the palatability of medications [72].

2.5 Conclusion

In this chapter, we provide a comprehensive analysis of the developments and advantages that pharmaceutical cocrystals have brought to the field of drug development. One of the most promising approaches to enhancing the effectiveness of drugs is represented by these cocrystals. The efficacy of the medicine is much enhanced by the structural elements and design ideas. In the realm of medical research, the usage of pharmaceutical cocrystals presents an interesting opportunity to surpass the restrictions imposed by conventional drug forms and to transform drug generation. This creative solution opens the path for major developments as it can efficiently address solubility and stability problems. The potential of cocrystals is driving studies on them. Future use and manufacturing of medicinal cocrystals should be much improved by ongoing research and development. Combining pharmacology, materials science, and chemistry gives a fascinating field. Emphasizing the relevance of these molecules in the pharmaceutical industry, this work presents a thorough study of the many characteristics of medicinal cocrystals. Further improvements will be shaped by the use of advanced characterization techniques, environmentally friendly synthesis methods, customized medication, combination therapies, and legislative framework construction. Constant research, creativity, and industry-wide cooperation among pharmaceutical companies will help us to fully appreciate the advantages of cocrystals for patients all around. Whatever challenges have been overcome; the future seems bright. Researchers and doctors are quite interested in them as they have the ability to improve the quality of drugs.

Acknowledgment

We extend our heartfelt gratitude to IFTM University, Moradabad, and Moradabad Educational Trust Group of Institutions, Faculty of Pharmacy for providing the necessary facilities and resources to complete this chapter.

References

1. Rathi, R., Kaur, S., Singh, I., A Review on Co-crystals of Herbal Bioactives for Solubility Enhancement: Preparation Methods and Characterization Techniques. *Cryst. Growth. Des.*, 22, 3, 2023–42, 2022.
2. Das, A. and Ghosh, S., Supramolecular Assemblies by Charge-Transfer Interactions between Donor and Acceptor Chromophores. *Angew. Chem. Int. Ed.*, 53, 8, 2038–2054, 2014.
3. Sun, L., Zhu, W., Yang, F., Li, B., Ren, X., Zhang, X., Hu, W., Molecular cocrystals: design, charge-transfer and optoelectronic functionality. *Phys. Chem. Chem. Phy.*, 20, 9, 6009–6023, 2018.
4. Zhu, W., Zheng, R., Fu, X., Fu, H., Shi, Q., Zhen, Y., Dong, H., Hu, W., Revealing the charge-transfer interactions in self-assembled organic cocrystals: two-dimensional photonic applications. *Ang Chem. Int. Ed. Eng.*, 54, 23, 6785–6789, 2015.
5. Park, S.J. and Seo, M.K., Solid-Solid Interfaces. *Interface Science and Technology*, Elsevier., 18, 253–331, 2011.
6. Schön, C.F., van Bergerem, S., Mattes, C., Yadav, A, Grohe, M., Kobbelt, L., Wuttig, M., Classification of properties and their relation to chemical bonding: Essential steps toward the inverse design of functional materials. *Sci. Adv.*, 8, 47, eade0828, 2022.
7. Gruss., M., Aspects for Developing and Processing Solid Forms. Solid State Development and Processing of Pharmaceutical Molecules. *Salts Cocrystals Polymorphism*, 79, 79, 1–43, 2021.
8. Rathi, R., Kushwaha, R., Goyal, A., Singh, I., Oxaliplatin-flavone pharmaceutical co-crystal-CN111205332A: patent spotlight. *Pharm. Pat. Anal.*, 11, 5, 147–154, 2022.
9. Yang, F., Sun, L., Han, J., Li, B., Yu, X., Zhang, X., Ren, X., Hu, W., Low-Voltage Organic Single-Crystal Field-Effect Transistor with Steep Subthreshold Slope. *ACS App. Mater Inter.*, 10, 31, 25871–25877, 2018.
10. Karagianni, A., Malamatari, M., Kachrimanis, K., Pharmaceutical Cocrystals: New Solid Phase Modification Approaches for the Formulation of APIs. *Pharmaceutics*, 10, 1, 18, 2018.
11. Bhalani, D.V., Nutan, B., Kumar, A., Singh Chandel, A.K., Bioavailability Enhancement Techniques for Poorly Aqueous Soluble Drugs and Therapeutics. *Biomed.*, 10, 9, 2055, 2022.
12. Seo, K.S., Bajracharya, R., Lee, S.H., Han, H.K., Pharmaceutical Application of Tablet Film Coating. *Pharmceu.*, 12, 9, 1–20, 2020.
13. Ringle, D.S., Regulatory Implications Using API Cocrystals for Generic Medicinal Products within the EU and US, Rhenish Friedrich Wilhelm University of Bonn, Germany, 2018, Master thesis.
14. O'Sullivan, A., Long, B., Verma, V., Ryan, K.M., Padrela, L., Solid-state and particle size control of pharmaceutical cocrystals using atomization-based techniques, *Int. J. Pharm.*, 621, 121798, 2022.

15. Buddhadev, S.S. and Garala, K.C., Pharmaceutical Cocrystals—A Review, Proceedings, 62, 14, 2020.
16. Rathi, R., Guggal, Y., Twinkle, V., Singh, I., Exploring the Patent Landscape and Regulatory Prospective on Pharmaceutical Cocrystals. *J. Pharm. Technol. Res. Manag.*, 11, 49–60, 2023.
17. Kumar, S., and Nanda, A., Pharmaceutical Cocrystals: An Overview, *Indian J. Pharm. Sci.*, 79, 6, 858–871, 2017.
18. Wang, X., Du, S., Zhang, R., Jia, X., Yang, T., Zhang, X., Drug-drug cocrystals: Opportunities and challenges. *Asian J. Pharm. Sci.*, 16, 3, 307, 2021.
19. Karimi-Jafari, M., Padrela, L., Walker, G.M., Croker, D.M., Creating cocrystals: A review of pharmaceutical cocrystal preparation routes and applications, *Cryst. Growth Des.*, 18, 6370–87, 2018.
20. European Medicines Agency, Reflection paper on the use of cocrystals of active substances in medicinal products, 2015, https://www.ema.europa.eu/en/documents/scientific-guideline/reflection-paper-use-cocrystals-active-substances-medicinal-products_en.pdf.
21. Fda, Cder, Stewart, and Felicia, Regulatory Classification of Pharmaceutical Cocrystals Guidance for Industry, 2018, http://www.fda.gov/Drugs/GuidanceComplianceRegulatoryInformation/Guidances/default.html.
22. Akki, R., Ramya, M.G., Krishna, K.C., Pharmaceutical Cocrystals: An Overview on Synthesis and Regulatory Aspects. *J. Drug Del. Ther.*, 9, 4, 623–628, 2019.
23. Rathi, R. and Sanshita, I.S., Hot Melt Extrusion Technique for Developing Pharmaceutical Co-crystals: A Review. *Drug Deliv. Lett.*, 14, 2, 97–108, 2024.
24. Chettri, A., Subba, A., Singh, G.P., Bag, P.P., Pharmaceutical cocrystals: A green way to enhance drug stability and solubility for improved therapeutic efficacy. *J. Pharm. Pharm.*, 76, 1, 1–12, 2024.
25. Thayyil, A.R., Juturu, T., Nayak, S., Kamath, S., Pharmaceutical Co-Crystallization Regulatory Aspects Design Characterization and Applications. *Adv. Pharm. Bull.*, 10, 2, 203–212, 2020.
26. Lee, A.Y., Erdemir, D., Myerson, A.S., Crystal polymorphism in chemical process development. *Ann. Rev. Chem. Biomol. Eng.*, 2, 259–280, 2011.
27. Saifee, M., Inamdar, N., Dhamecha, D.L., Rathi, A., Drug polymorphism: a review. *Inter. J. Heal. Res.*, 2, 4, 291–306, 2009.
28. Censi, R. and Di Martino, P., Polymorph Impact on the Bioavailability and Stability of Poorly Soluble Drugs. *Mol.*, 20, 10, 18759–18776, 2015.
29. Deng, Y., Liu, S., Jiang, Y., Martins, I.C.B., Rades, T., Recent Advances in Co-Former Screening and Formation Prediction of Multicomponent Solid Forms of Low Molecular Weight Drugs. *Pharmaceu.*, 15, 9, 2174, 2023.
30. Handa, M., Almalki, W.H., Shukla, R., Afzal, O., Altamimi, A.S.A., Beg, S., Rahman, M., Active pharmaceutical ingredients (APIs) in ionic liquids: An effective approach for API physiochemical parameter optimization. *Drug Disc. Tod.*, 27, 9, 2415–2424, 2022.

31. Rathi, R. and Singh., I., Multicomponent crystal compromising dasatinib and selected co-crystal formers: a patent evaluation of EP2861589B1. *Pharm. Pat. Anal.*, 11, 1, 15–21, 2021.
32. Shemchuk, O., Grepioni, F., Leyssens, T., Braga, D., Chiral Resolution via Cocrystallization with Inorganic Salts. *Isr. J. Chem.*, 61, 9–10, 563–572, 2021.
33. Da Silva, C.C.P., De Oliveira, R., Tenorio, J.C., Honorato, S.B., Ayala, A.P., Ellena, J., The continuum in 5-fluorocytosine. toward salt formation. *Cryst. Growth Des.*, 13, 10, 4315–4322, 2013.
34. Steed, J.W., The role of cocrystals in pharmaceutical design. *Trends Pharmacol. Sci.*, 34, 3, 185–193, 2013.
35. Singh, M., *et al.*, Cocrystals by Design: A Rational Coformer Selection Approach for Tackling the API Problems. *Pharmaceutics*, 15, 1161, 2023.
36. Schultheiss, N. and Newman, A., Pharmaceutical Cocrystals and Their Physicochemical Properties. *Cryst. Growth Des.*, *9*, 6, 2950–2967, 2009.
37. Yadav, A.V., Shete, A.S., Dabke, A.P., Kulkarni, P.V., Sakhare, S.S., Co-Crystals: A Novel Approach to Modify Physicochemical Properties of Active Pharmaceutical Ingredients. *Indian J. Pharm. Sci.*, *71*, 4, 359–370, 2009.
38. Nugrahani, I. and Jessica, M.A., Amino Acids as the Potential Co-Former for Co-Crystal Development: A Review. *Molecules*, *26*, 11, 3279, 2021.
39. Haiduc, I., Tiekink, E., Schpector, J., The Importance of Pi-Interactions in Crystal Engineering. *Front. Cryst. Eng.*, 301–322, 2012.
40. Sugden, I.J., Braun, D.E., Bowskill, D.H., Adjiman, C.S., Pantelides., C.C., Efficient Screening of Coformers for Active Pharmaceutical Ingredient Cocrystallization. *Cryst. Growth Des.*, *22*, 7, 4513–4527, 2022.
41. Wasim, M., Mannan, A., Bin Asad, M.H.H., Amirzada, M., II, Shafique, M., Hussain, I., Fabrication of Carbamazepine Cocrystals: Characterization, in: *Vitro and Comparative In Vivo Evaluation. Biomed Res. Int*, p. 6685806, 2021.
42. Lou, B. and Velaga, S.P., Polymorph Control of Felodipine Form II in an Attempted Cocrystallization. *Cryst. Growth Des.*, *9*, 3, 1254–1257, 2009.
43. Juneja, P., Kaur, B., Odeku, O.A., Singh, I., Development of Corn Starch-Neusilin UFL2 Conjugate as Tablet Superdisintegrant: Formulation and Evaluation of Fast Disintegrating Tablets. *J. Drug Deliv.*, 2014, 1, 827035, 2014.
44. Douroumis, D., Ross, S., Nokhodchi, A., Advanced Methodologies for Cocrystal Synthesis. *Adv. Drug Deliv. Rev.*, *117*, 178–195, 2017.
45. Ngilirabanga, J.B. and Samsodien, H., Pharmaceutical Co-Crystal: An Alternative Strategy for Enhanced Physicochemical Properties and Drug Synergy. *Nano Sel.*, *2*, 3, 512–526, 2021.
46. Ludwig, A., Discovery of New Materials Using Combinatorial Synthesis and High-Throughput Characterization of Thin-Film Materials Libraries Combined with Computational Methods. *npj Comput. Mater.*, *5*, 1, 70, 2019.
47. Rosbottom, I., Ma, C.Y., Turner, T.D., O'Connell, R.A., Loughrey, J., Sadiq, G., Davey, R.J., Roberts, R.J., Influence of Solvent Composition on the

Crystal Morphology and Structure of P-Aminobenzoic Acid Crystallized from Mixed Ethanol and Nitromethane Solutions. *Cryst. Growth Des.*, *17*, 8, 4151–4161, 2017.

48. Fábián, L., Cambridge Structural Database Analysis of Molecular Complementarity in Cocrystals. *Cryst. Growth Des.*, *9*, 3, 1436–1443, 2009.
49. Meng, X.-Y., Zhang, H.-X., Mezei, M., Cui, M., Molecular Docking: A Powerful Approach for Structure-Based Drug Discovery. *Curr. Comput.-Aided Drug Des.*, *7*, 2, 146–157, 2011.
50. Hyttinen, N. and Prisle, N.L., Improving Solubility and Activity Estimates of Multifunctional Atmospheric Organics by Selecting Conformers in COSMOtherm. *J. Phys. Chem. A*, 124, 23, 4801–4812, 2020.
51. Morris, G. and Huey, R., AutoDock4 and AutoDockTools4: Automated Docking with Selective Receptor Flexibility. *J. Comput. Chem.*, *48*, 443–453, 2009.
52. Fu, Q., Y.Han, Y., Gong, N., Guo, F., Carbamazepine Cocrystals with Several Aromatic Carboxylic Acids in Different Stoichiometries: Structures and Solid State Characterization. *J. Mol. Struct.*, *1168*, 145–152, 2018.
53. Aitipamula, S. and Vangala, V.R., X-Ray Crystallography and its Role in Understanding the Physicochemical Properties of Pharmaceutical Cocrystals. *J. Indian Inst. Sci.*, 97, 2, 227–243, 2017.
54. Ouyang, J., Xing, X., Zhou, L., Zhang, C., Heng, J.Y.Y., Cocrystal design of vanillin with amide drugs: Crystal structure determination, solubility enhancement, DFT calculation. *Chem. Eng. Res. Des.*, 183, 170–180, 2022.
55. Ali, A., Kuznetsov, A., Ibrahim, M., Abbas, A., Akram, N., Maqbool, T., Ushna, Chemistry and Modern Techniques of Characterization of Co-Crystals, *Drug Formulation Design*, IntechOpen, London, UK, 2022.
56. Bhattacharya, S., Peraka, K.S., Zaworotko, M.J., The Role of Hydrogen Bonding in Cocrystals. *Monograph Supramol. Chem.*, 33–79, 2018.
57. Turek, M., Różycka-Sokołowska, E., Koprowski, M., Marciniak, B., Bałczewski, P., Role of Hydrogen Bonds in Formation of Co-amorphous Valsartan/Nicotinamide Compositions of High Solubility and Durability with Anti-hypertension and Anti-COVID-19 Potential. *Mol. Pharm.*, 18, 5, 1970–1984, 2021.
58. Ilma, N., Agiana, P.E., Dwichandra, P.O., theophylline–na–sacharrine single crystal isolation for its structure determination. *Int. J. Pharm. Pharm. Sci.*, 7, 12, 15–24, 2015.
59. Ovchinnikov, O.V., Evtukhova, A.V., Kondratenko, T.S., Smirnov, M.S., Khokhlov, V.Y., Erina, O.V., Manifestation of intermolecular interactions in FTIR spectra of methylene blue molecules. *Vib. Spec.*, 86, 181–189, 2016.
60. Garbacz, P. and Wesolowski, M., DSC, FTIR and Raman Spectroscopy Coupled with Multivariate Analysis in a Study of Cocrystals of Pharmaceutical Interest. *Mol.*, 23, 9, 2136, 2018.
61. Garbacz, P., Paukszta, D., Sikorski, A., Wesolowski, M., Structural Characterization of Cocrystals of Chlordiazepoxide with p-Aminobenzoic

Acid and Lorazepam with Nicotinamide by DSC, X-ray Diffraction, FTIR and Raman Spectroscopy. *Pharmaceu.*, 12, 7, 648, 2020.

62. Maruthamuthu, M.K., Rudge, S.R., Ardekani, A.M., Ladisch, M.R., Verma, M.S,. Process Analytical Technologies and Data Analytics for the Manufacture of Monoclonal Antibodies, *Trends Biotechnol*, 38, 10, 1169–1186, 2020.
63. Kumar Bandaru, R., Rout, S.R., Kenguva, G., Gorain, B., Alhakamy, N.A., Kesharwani, P., Dandela, R., Recent Advances in Pharmaceutical Cocrystals: From Bench to Market. *Front. Pharmacol.*, 12, 780582, 2021.
64. Sun, C.C., Cocrystallization for successful drug delivery, *Expert Opin Drug Deliv*, 10, 2, 201–13, 2013.
65. Gadade, D.D. and Pekamwar, S.S., Pharmaceutical Cocrystals: Regulatory and Strategic Aspects, Design and Development. *Adv. Pharm. Bull.*, 6, 4, 479–494, 2016.
66. Sonali, M. and Amruta Suryakant, S., Regulatory and strategic aspects in preparation of cocrystals. *Int. J. Crea. Res. Thou.*, 9, 8, d845–d853, 2021.
67. Devi, P., Budhwar, V., Kakkar, S., Kumar, A., Pharmaceutical Co-Crystallization: Strategies for Co-Crystal Design. *Int. J. Life Sci. Pharma. Res.*, 13, 6, P87–P105, 2023.
68. Trzeciak, K.., Dudek, M.K., Potrzebowski, M.J., Mechanochemical Transformations of Pharmaceutical Cocrystals: Polymorphs and Coformer Exchange. *Chem. A Eur. J.*, e202402683, 2024.
69. Kumar, A., Kumar, S., Nanda, A., A Review about Regulatory Status and Recent Patents of Pharmaceutical Cocrystals. *Adv. Pharm. Bull.*, 8, 3, 355, 2018.
70. Liu, L., Wang, J.R., Mei, X., Enhancing the stability of active pharmaceutical ingredients by the cocrystal strategy. *Cryst. Eng. Comm.*, 24, 11, 2002–2022, 2022.
71. Li, S., Yu, T., Tian, Y., Lagan, C., Jones, D.S., Andrews, G.P., Mechanochemical Synthesis of Pharmaceutical Cocrystal Suspensions via Hot Melt Extrusion: Enhancing Cocrystal Yield. *Mol. Pharm.*, 15, 9, 3741–3754, 2018.
72. Chavan, R.B., Thipparaboina, R., Yadav, B., Shastri, N.R., Continuous manufacturing of cocrystals: challenges and prospects. *Drug Deliv. Trans. Res.*, 8, 6, 1726–1739, 2018.

3

Advancements in Computational Screening Methods for Pharmaceutical Cocrystals

Maan Singh[1], Abhishek Mali[1], Madhukiran Dhondale[1], Ashish Kumar Agrawal[1], Amrit Paudel[2*] and Dinesh Kumar[1†]

[1]Department of Pharmaceutical Engineering and Technology, Indian Institute of Technology (Banaras Hindu University), Varanasi, Uttar Pradesh, India
[2]Research Centre Pharmaceutical Engineering GmbH (RCPE), University of Technology, Institute of Process and Particle Engineering, Graz, Austria

Abstract

In crystal engineering, pharmaceutical cocrystals are designed as co-systems composed of an active pharmaceutical ingredient (API) and a coformer held together by noncovalent interaction. Cocrystallization is a commonly employed technique to enhance the solubility and bioavailability of drugs. The selection of a suitable coformer has a paramount role in cocrystal development. A vast range of coformers is currently available for cocrystallization experiments. Coformer selection is either done by experimental- or computational-based methods. Experimental methods predominantly rely on the traditional empirical approach and are thus extremely tedious. Computational methods can boost the cocrystal research capabilities by providing information about the API-coformer complex even before experimentation by considering the structural features of the components. Computational-based screening methods can screen libraries of compounds in a relatively short time. Molecular electrostatic potential surfaces (MEPS), Hirshfeld surface analysis (HSA), Cambridge structure database (CSD), COSMO-RS (conductor-like screening model for real solvents), and Crystal structure prediction (CSP) are the prominent cocrystal screening tools employed by the researchers. These methods can also save significant time and resources. Apart from assisting the screening, the physical and chemical properties of formed cocrystals can also

**Corresponding author*: amrit.paudel@rcpe.at
†Corresponding author: dinesh.phe@itbhu.ac.in

Inderbir Singh, Subrata Mallick and Ritu Rathi (eds.) Cocrystals in Pharmaceutical Sciences: Design to Applications, (55–92) © 2025 Scrivener Publishing LLC

be approximated by these computational methods. This review provides insights into the latest findings and advances in computational-based cocrystal screening methodologies.

Keywords: Cocrystal screening, computational, molecular electrostatic potential surface, Cambridge structural database, COSMO-RS, crystal structure prediction, Hirshfeld surface analysis

3.1 Introduction

Crystal engineering (CE) is the understanding of intermolecular interaction between the API to API and API to coformer for synthesizing new crystalline species or modifying the existing crystal structure. CE is an innovative way to improve the physicochemical and pharmaceutical properties of pharmaceutical solids and has become a preferred method for several pharmaceutical companies [1]. Most of the newly discovered APIs (about 60% to 70%) belong to biopharmaceutical classification system (BCS) class II and IV having poor solubility and permeability which causes a remarkable reduction in their solubility and finally oral bioavailability [2, 3].

Solvates, polymorphs, cocrystals, hydrates, and salts are examples of different crystalline forms of the API possessing altered solubility, dissolution rate, flowability, compressibility, tabletability, and compactability properties [4, 5]. Among these, salt formation has long been an effective method to enhance drug solubility, but it has limitations, such as molecules with an ionizable group can only form salts [6, 7]. Hydrates and solvates have stability issues as there is a tendency for the loss of entrapped water or solvent over time [8]. Amorphous solid dispersions are helpful to improve the solubility of poorly aqueous soluble APIs but these have stability issues due to the presence of moisture-absorbing polymers. These polymers result in phase separation from amorphous to crystalline [9–12].

Pharmaceutical cocrystals are thermodynamically stable moieties having a definite stoichiometry which is maintained throughout the storage period [4]. Cocrystal effectively improves the physicochemical properties such as the solubility of API while maintaining the stability of drug molecules [13, 14]. Cocrystals are formed by several noncovalent interactions (i.e., hydrogen bonds, van der Waals forces, and p stacking), between the API and coformer. Coformer act as building blocks containing a few binding sites for forming cocrystals. The presence of good hydrogen bond acceptors and donors in both the API and coformer increases the possibility of cocrystal formation. Functional groups such as alcohols, carboxylic acids, and amides are commonly used to create supramolecular synthons [15]. Supramolecular synthons are the core of structural units that are capable of forming large crystal structures. Coformers act as homosynthons or heterosynthons to form the cocrystals. Supramolecular synthons are classified as (a) homosynthons, which have identical self-complimentary functions, and (b) heterosynthons, which have distinct but complementary functionalities. Supramolecular heterosynthons are frequently chosen over homosynthons due to the presence of carboxylic acids with pyridine and carboxylic acid with amide [16]. These heterosynthons are more likely to interact with the APIs by forming hydrogen bonds compared to homosynthons.

However, to successfully synthesize a cocrystal with desired physicochemical characteristics, the choice of a specific coformer plays a significant role [17]. Broadly, experimental and computational approaches are used for the screening of suitable coformers. Experimental screening is also referred to as the traditional screening method and has long been employed in the pharmaceutical sector. Here, selected APIs are cocrystallized with an array of coformers and then individually screened by solid-state characterization techniques to confirm the successful formation of cocrystals [17]. Some useful solid-state characterization techniques are scanning electron microscopy (SEM), Fourier transformation infrared (FTIR), powder x-ray diffraction (PXRD), differential scanning calorimetry (DSC), and thermogravimetry analysis (TGA). Other models that can help in the experimental screening of coformers are the Hansen solubility parameter, hydrogen bonding, lattice energy calculation, pKa-based models, etc. [15]. These approaches are currently used for screening while hundreds of trials are required which are time-consuming and expensive.

Alternatively, computational approaches can be more suitable, efficient, and time-saving [18]. Over the years, scientists have developed a range of computational cocrystal prediction algorithms for comparing the original crystal structures and interactions between APIs and coformers to provide a useful model [19]. Molecular electrostatic potential surface (MEPS) measurement, Cambridge structural database (CSD) analysis, COSMO-RS (conductor-like screening model for real solvents) approach, crystal structure prediction (CSP) method, and Hirshfeld surface (HS) analysis are prominent computational techniques employed for the prediction of cocrystal formation as well as describe a mechanistic explanation of the cocrystallization process (Figure 3.1). MEPS effectively represents electrostatic interactions between the different functional groups of interacting

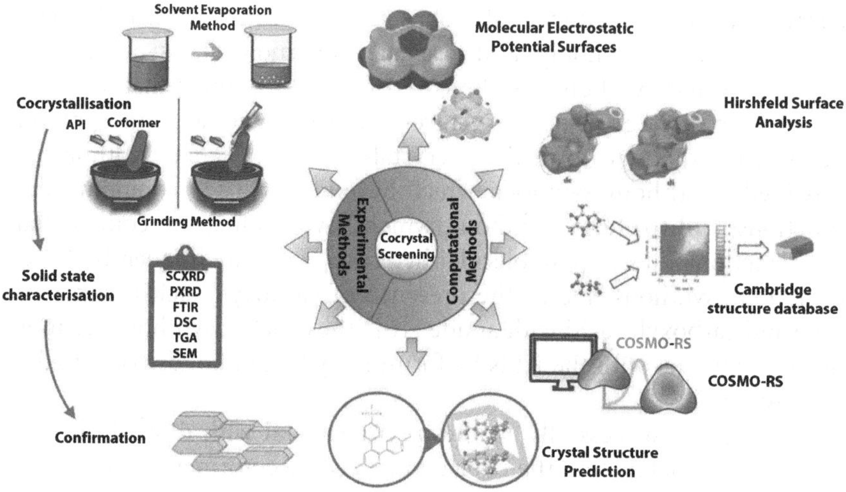

Figure 3.1 Depiction of various computational methods for the screening of cocrystals.

molecules and assesses molecular complementarity [20]. To gain an additional understanding of the system's stability, Hirshfeld surface (HS) analysis can be employed. It provides a piece of information related to crystal packing behavior [21]. Yu *et al.* [22] used both MEPS and HS analysis to predict the formation of 5-fluorouracil (5-FU) and L-proline cocrystal which supported the experimental results. The COSMO-RS approach is a unique blend of quantum chemical knowledge and liquid-phase thermodynamics and when coupled with COSMOtherm software makes it an appropriate toolset for cocrystal screening. Chapman *et al.* [23] used this approach in their study to predict the cocrystal formation of CL-20 material. The CSD system along with software like mercury, ORTEP (Oak Ridge Thermal Ellipsoid Plot Program), and WinGX is used to solve, refine, analyze, and visualize crystal structures and estimate hydrogen bond formation and their bond distance. CSP is generally used to predict the crystal structure using the predefined stored data [24]. CSP technology considers the contribution of crystallinity in solubility enhancement and is based on the lattice energy approach [25].

A detailed explanation of the applicability of computational screening methods in predicting cocrystal formation and how artificial intelligence (AI) and machine learning (ML), in the future will advance the cocrystal screening process are discussed in the upcoming section [26, 27].

3.2 Experimental Screening of Cocrystals

Experimental screening methodologies include the preparation of cocrystals followed by solid-state characterization to confirm the formation of cocrystal complexes [28, 29]. The experimental screening method is a tedious approach for identifying potential coformers for a given API. For example, Przybyłek *et al.* [30] could identify the successful coformer pair for methylxanthines only after cocrystallizing the APIs with eight coformers and analyzing each product by solid-state characterization techniques for confirming cocrystal formation. The said example suggests the tediousness of the experimental screening approach. Figure 3.2 depicts the methodology of experimental cocrystal screening.

The first step in experimental screening is the cocrystallization experiment with probable coformer pairs. Several reported methods such as solvent evaporation and mechanochemical- and slurry-based processes can be used to prepare cocrystals. In the solvent evaporation technique, the drug and coformer are dissolved in a solvent, which is then allowed to slowly evaporate to obtain a cocrystal [31]. The solubility of the API and coformer in the selected solvent should be nearly equal otherwise either API or coformer is precipitated out. Other methods such as cooling [32] and antisolvent addition [33] are also suitable for cocrystal formation. The mechanochemical methods are based on the mechanical activation of coformer and API by grinding with/without solvent. Accordingly, grinding of API and coformer without solvent is called the neat grinding method [34], and grinding of API and coformer with the solvent is called

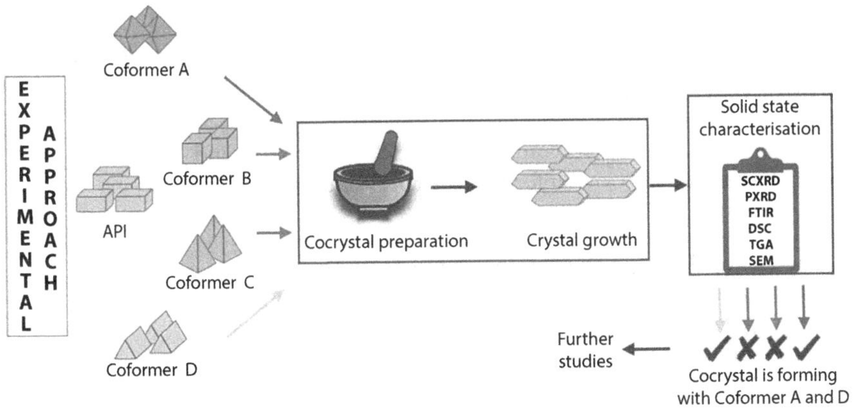

Figure 3.2 Depiction of experimental approach for screening cocrystals.

the liquid-assisted grinding method [35, 36]. The cocrystal preparation methods are represented in Figure 3.3.

In the second step after the preparation of the new product, it is mandatory to confirm whether the formed new product is cocrystals or not. The product is subjected to a variety of spectroscopic, microscopic, thermal analyses, and diffraction methods as listed in Table 3.1 to confirm the cocrystallization. Only the diffraction methods (single-crystal x-ray diffraction, in particular) can confirm the formation of cocrystals [37].

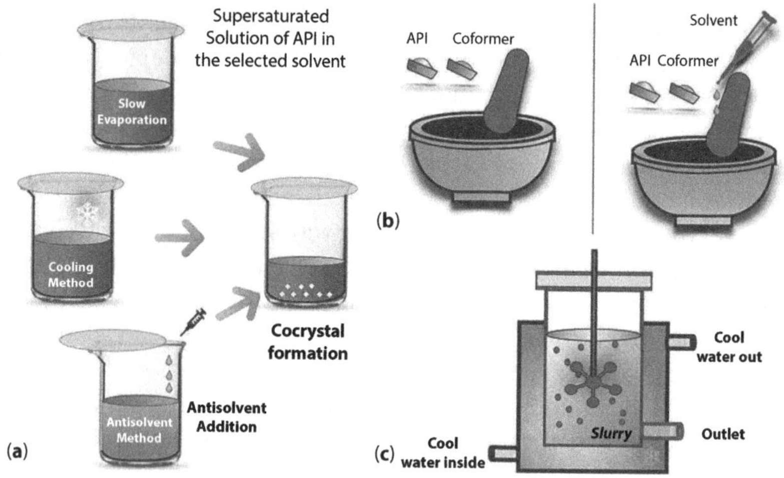

Figure 3.3 Depiction of cocrystal preparation by (a). Solvent evaporation technique (b). Neat grinding and liquid assisted grinding (c). Slurry method.

Table 3.1 Different techniques employed for the characterization of cocrystals.

Technique	Characteristic features indicative of cocrystal formation	References
Diffraction methods		
Powder X-ray Diffraction (PXRD)	As compared to peaks of pure API and coformer, the cocrystal shows altered or new peaks.	[38, 39]
Single-crystal X-ray Diffraction (SCXRD)	It gives the complete crystal structure information such as atomic/molecular arrangement, stoichiometry, and hydrogen bonding of the formed crystal.	[37]

(Continued)

Table 3.1 Different techniques employed for the characterization of cocrystals. (*Continued*)

Technique	Characteristic features indicative of cocrystal formation	References
Micro Electron Diffraction (MicroED)	Determines the crystal structure of the cocrystal and the position of atoms/molecules in the cocrystal accurately due to high resolution and sensitivity, which helps in understanding the interaction between APIs and coformers	[40]
Spectroscopy		
Fourier Transform Infrared Spectroscopy (FTIR)	Changes in the vibrational frequencies of the product are compared with APIs and coformers. Intermolecular interactions such as hydrogen bonding and π-π stacking can be effectively studied.	[41]
X-ray Photoelectron Spectroscopy (XPS)	The surface chemistry study of individual components and their mixture provides information about intermolecular interactions (binding energy) between API and coformer at the surface level.	[42]
Thermal analysis		
Differential Scanning Calorimetry (DSC)	Changes in the thermal behavior of the product leading to an altered melting point, change in peak pattern, or shift in the peak temperature relative to the individual components indicate cocrystal formation.	[43, 44]
Thermogravimetric Analysis (TGA)	The determination of differences in melting point and thermal decomposition behavior of cocrystals compared to their individual components.	[39, 45]

(*Continued*)

Table 3.1 Different techniques employed for the characterization of cocrystals. (*Continued*)

Technique	Characteristic features indicative of cocrystal formation	References
Microscopic methods		
Scanning Electron Microscopy (SEM)	Does not exactly confirm the formation of cocrystals. SEM analysis reveals the change in the crystal shape and size of the crystals formed.	[46]

Other methods are useful in characterizing the formed cocrystal with respect to its physical properties.

In conclusion, the experimental cocrystal screening approach is a time, facility, and resource-extensive operation. The success rate of producing a cocrystal for any given API is very low. There is also a huge requirement for organic solvents especially in solvent evaporation, cooling crystallization, and antisolvent addition methods during the cocrystallization experiments. These aspects increase the cost incurred in the research and development of cocrystal-based products [47].

3.3 Computational Method for Screening Cocrystals

By utilizing literature from previous studies on APIs and coformers to form cocrystals available on several structural database systems, the interactions between APIs and numerous coformers can be studied within a short duration. The advanced computational techniques offer us the possibility of screening libraries of structures and choosing the best pair of API and coformer to form cocrystals. These computational tools offer means for the deliberate design of cocrystals with required properties. A few widely used computational tools and software for predicting cocrystallization are briefly discussed in the following sections [26].

3.3.1 Molecular Electrostatic Potential Surfaces (MEPS) Measurement

A MEPS is a three-dimensional representation of charge distribution in a molecule. MEPS effectively represents electrostatic interactions between the different functional groups of interacting molecules. These electrostatic interactions then play a major part in the creation of hydrogen bonds (non-covalent bonds). The charge densities, including regions with an excess of electrons or a deficiency of electrons within a molecule, can be visualized. Thus, the complementarity of two potential cocrystallization partners can be determined using MEPS [26]. The color spectrum from red to blue shows the rising order of electrostatic potential [48]. The order of electrostatic potential's intensity grows from darkest red < orange < yellow < green < cyan < deepest blue. The most electrostatically negative potential, zero electrostatic potential, and most electrostatically positive regions are represented by the colors red-orange (location for electrophilic reaction), green, and blue (location for nucleophilic reaction), respectively [49]. The neutral atoms are involved in hydrophobic interactions and are shown by green-yellow color [50]. The darker the color, the greater the absolute value and the stronger the electrostatic interactions [51, 52]. Figure 3.4 shows the scale range from positive (blue) to negative (red) electrostatic potential.

There are various computational tools available to generate MEPS. One of them is "MolView," a crucial web-based computational tool that enables us to construct MEP surfaces [53]. Other software that can be used to determine MEPS is Multiwfn. To discover surface site interaction points (SSIPs),

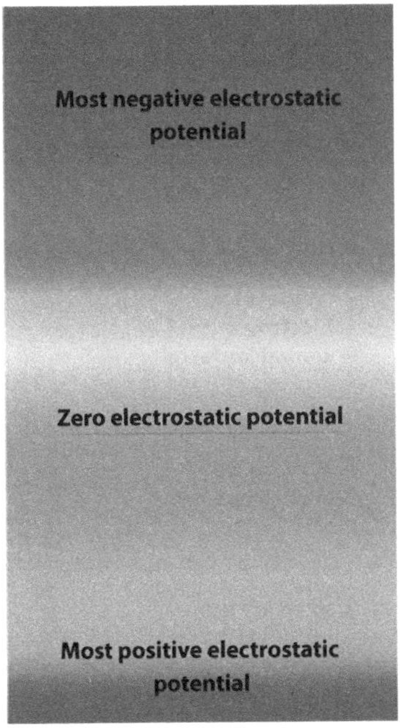

Figure 3.4 Electrostatic potential scale.

the computational technique makes use of estimated MEPS generated by software. The molecular recognition properties of the entire molecule's surface can be evaluated by utilizing the H-bond donor and acceptor parameters [54, 55]. Then, an interaction parameter is assigned to each SSIP (H-bond donor with positive property and H-bond acceptor with negative property). The difference between pure component interaction site pairing energies and cocrystal corresponding energy offers an approximation of cocrystal formation [56]. Furthermore, the MEPS converted into SSIPs predicts cocrystal formation chance for a given compound and a large library of potential coformers. Based on the results, the most promising candidates are selected for experimental testing [29, 57]. Figure 3.5 depicts the process used for predicting new cocrystals using MEPS and SSIP pathways.

Zhou *et al.* [58] used MEPS values projected on the van der Waals surface to visualize electrophilic and nucleophilic charge areas. Considering the structural features and MEPS values, the molecular recognition and subsequent packing behavior of the cocrystal during crystallization were explained. Krueger *et al.* [59] used the MEPS approach to study electrostatic potential at different hydrogen bonding sites formed in various alternative ligand conformations produced during the study. MEPS and the total energy of the molecules were computed using Spartan'04 software. This study provides considerable insight into how conformational changes produce an impact on the formation of cocrystals. The electrostatic

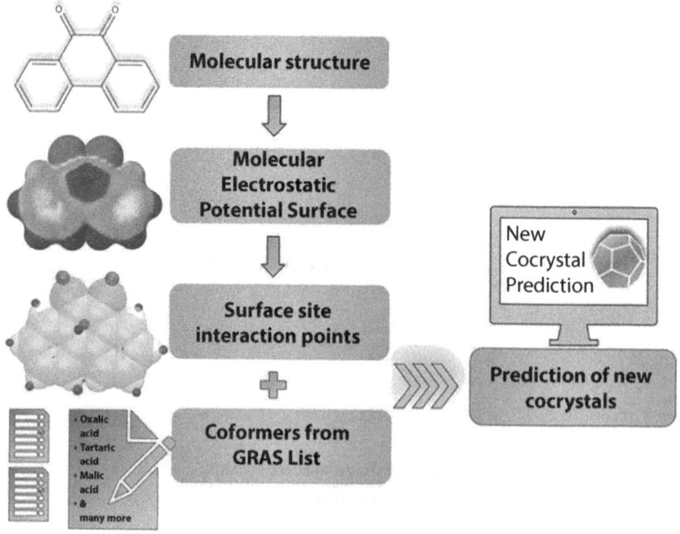

Figure 3.5 Molecular electrostatic potential surfaces measurement process.

potential of different conformations of the ligands was calculated, and the most probable hydrogen bonding sites were recognized from their negative electrostatic potential. Recently, Huang *et al.* [60] used the MEPS approach to confirm the intermolecular interactions within the chlorbipram-m-hydroxy benzoic acid cocrystal hydrate. The potential hydrogen bond interactions on the surface were determined by measuring the MEPS of all individual molecules using the density functional theory (DFT) approach using the GAUSSIAN 16 software. Local maxima and minima indicating hydrogen bond donors and acceptors, respectively, were generated from the MEPS and then transferred onto the isosurface using the Multiwfn 3.6 application. Using Mercury software having a hydrate analyzer feature, the

Table 3.2 Recent publications on cocrystal screening using molecular electrostatic potential surface measurement methods.

API	Coformer	Reference
Chlorbipram	M-hydroxybenzoic acid	[60]
Etoricoxib	Oxalic acid, fumaric acid, malic acid, hydroquinone, resorcinol, and catechol	[51]
5-fluorouracil	L-proline	[22]
Acridine	1,2,4,5-tetracyano-benzene	[62]
Antipyrine	Amino phenazone	[19]
Sulfathiazole	Theophylline and sulfanilamide	[63]
2,2':6',2"-terpyridine	1,2,4,5-tetracyanobenzene	[58]
Palbociclib	Kaempferol	[64]
Luteolin	Caffeine and isoniazid	[65]
Voriconazole	Tartaric acid, protocatechuic, gallic acid, and 3,5-dinitrobenzoic acids	[66]
Rhein	Matrine	[67]
5-fluorocytosine	Ferulic acid	[68]
CL-20	2,4-DNI	[69]
2-hydroxy pyridine	Oxalic acid	[70]
Rhein	Berberine	[71]

water molecule contact surfaces of the under-investigation cocrystals were examined. Another approach based on MEPS was used by Yu *et al.* [22] while designing cocrystals of 5-FU with proline. The polarity changes in the cocrystal compared to pure 5-FU were studied by measuring the MEPS values of the cocrystal and API. It was found that the global maximum (G_{max}) and global minimum (G_{min}) for the cocrystal had increased due to the formation of new hydrogen bonds, resulting in a polarity increment.

Huang *et al.* [61] employed the MEPS to investigate the formation of hydrogen bonds. As hydrogen bond formation is the primary step driven by electrostatic interactions, MEPS can help demonstrate the electrostatic interaction by visualizing the potential hydrogen acceptor and donor sites. A total of eight coformers were screened based on MEPS, and experimentally, it was found that four coformers out of eight were able to form the cocrystals with the API. The differences in the ratio of screened coformer by MEPS and finally obtained potential coformer experimentally were 50:50.

Even though this MEPS method is used widely for cocrystal screening (Table 3.2), it is dependent on one general or simplified rule for cocrystallization and focuses on isolated molecular features, i.e., electrostatic charge on the surface.

3.3.2 Hirshfeld Surface Analysis (HSA)

All information regarding intermolecular interactions is encoded on the crystal surface. HSA is used to visualize the molecule's space and establish a structure-property relationship, particularly in relation to various kinds of noncovalent interactions in the lattice. The normalized contact distance (d_{norm}) of the HSA is described by two distances: d_i—the distance inside the surface from the nearest nucleus—and d_e—the distance outside the surface to the nearest nucleus. The descriptor d_{norm} is utilized to map the directions and intensities of intermolecular interactions inside the molecular crystal onto HS and is expressed in terms of Equation 3.1 as given below:

$$d_{norm} = (d_i - r_i^{vdw})/r_i^{vdw} + (d_e - r_e^{vdw})/r_e^{vdw} \quad (3.1)$$

where r_i^{vdw} is the internal van der Waals radii of the atom and r_e^{vdw} is the exterior van der Waals radii of the atom [72–74]. Negative and positive d_{norm} values indicate shorter and longer intermolecular interactions, respectively. The d_{norm} HSA is displayed with red emphasizing shorter connections, whereas blue emphasizing longer contacts. Moreover, owing to the symmetry in the d_{norm} equation between d_e and d_i, the intersection location of Hirshfeld surfaces (HS) in both locations where two HS intersect will be marked with a red spot that has equivalent color intensity, size, and shape [75, 76].

For the cocrystals and salts, Abidi *et al.* [77] constructed HS mapped on d_{norm} using the CrystalExplorer 3.1 software. The complexities in intermolecular interactions and packing of cocrystal and salt packing were investigated using HS analysis. HS and fingerprint plots were produced with the aid of the application CrystalExplorer. The main objective of CrystalExplorer is to facilitate the exploration and visualization of molecular crystal structures, particularly *via* the embellished HS and its accompanying two-dimensional imprint, as well as visualizing quantitative intermolecular interactions.

The molecular surface alone provides the required visual information but the addition of scalar surface properties decorating the surface of a crystal further increases the understanding of crystal packing [78]. The CrystalExplorer now has the capacity to generate fingerprint plots for certain molecular interactions, enabling more detailed comparison [79]. Novena *et al.* [80] utilized HS analysis and CrystalExplorer 3.1 software to examine theophylline cocrystals with resorcinol monohydrate (THR) and Pyrogallol monohydrate (THPY). To get information about the properties of the intermolecular interactions within crystal structures, the research team utilized a 2D finger plot to assess the strength of these interactions in the crystal packing modes. The analysis of finger plots reveals N···H/H···N interactions, with 1 molecule acting as a donor and the other as an acceptor. In THR and THPY, the N···H/H···N contacts account for 6.5 and 8.1 %, respectively. THPY exhibits an imbalance due to the presence of N···H···N interaction, while THR shows C···H···N interaction. The identification of intermolecular interactions through color-coding distances from the HS provides evidence of cocrystal formation.

Another example is the study involving seselin by Hussain *et al.* [81] using HS analysis to screen different coformers and their interactions with seselin. This study underscores the significance of computational cocrystal prediction. The impact on the stability of the crystal lattice configurations was assessed. The shifting of the contacts made with the HS can be known by comparing the surface imprinted on the d_{norm} values with that of d_i and d_e. The contacts between the surfaces of di and de are plainly visible in the highlighted region over those surfaces; bright red dots change as a result of interacting atoms on both inside and outside of a surface. The close connections between molecules are indicated by bright red regions on the Hirshfeld surface plotted on the compound's d_{norm} value. Figure 3.6 shows the HS analysis of seselin-thiourea cocrystals. The red patches represent tight interactions with the Hirshfeld surface caused by two hydrogen bonds.

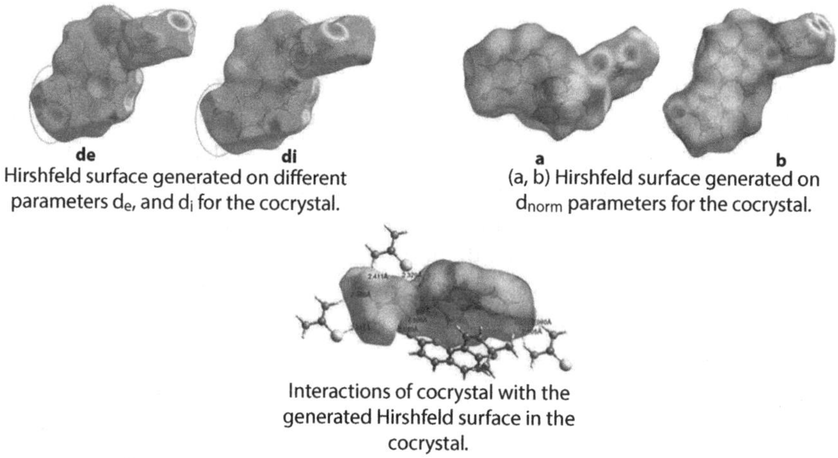

Figure 3.6 Hirshfeld surface analysis of seselin-thiourea cocrystal in 1:1 stoichiometry. Reprinted (adapted) with permission from [81]. Copyright 2018 American Chemical Society.

Recently, Yu *et al.* [82] used Hirshfeld surface analysis for cocrystal analysis of fluorouracil-phenylalanine that shows distinct distributions of red spots, indicating different intermolecular bonding modes compared to 5-FU. These observations are supported by quantitative fingerprint plots, which demonstrate the contributions of various interactions. While O···H interactions are significant in both Fu and the cocrystal, they constitute less than forty percent of all contacts, emphasizing the importance of other interactions. The cocrystal exhibits a higher proportion (91.7%) of strong interactions, including F···H, N···H, O···H, C···C, C···H, and H···H, compared to Fu alone (68.8%), indicating a more stable crystal lattice in the cocrystal. Similarly, Yu *et al.* [66] analyzed molecular Hirshfeld surfaces of voriconazole (VZL) multicomponent crystals using CrystalExplorer 3.1. The surfaces showed strong (red) and weak (blue) contacts. Two-dimensional (2D) fingerprints revealed dominant N···H, F···H, and O···H interactions matching the Hirshfeld surfaces. Triazole N_3 and pyrimidine N_5 acted as the main hydrogen bond acceptors. SCXRD and MEPS calculations supported the findings, showing stronger N···H and H···N interactions in VZL cocrystals. As this method also focuses on isolated molecular features, i.e., surface study, it will show some limitations in the prediction accuracy. So often, both MEPS and HSA are used to predict cocrystal formation. Table 3.3 enlists some examples of cocrystal screening by HSA, which are based on recent publications.

Table 3.3 Recent publications of cocrystal screening from Hirshfeld surface analysis methods.

API	Coformer	Reference
Theophylline	Resorcinol	[80]
Indole-3-acetic acid	Indole-3-butyric acid	[77]
Creatinine	2,3-pyridine dicarboxylic acid	[83]
Metformin	Citric acid	[7]
Urea	Furan tetracarboxylic acid	[49]
4-(1-napthylvinyl) pyridine	Terephthalic acid	[72]
Sulfamethoxazole	Succinimide	[84]
5-fluorouracil	L-phenylalanine	[82]
Magnolol	Nicotinamide, isonicotinamide and picolinamide	[85]
Voriconazole	Tartaric acid, protocatechuic, gallic acid, and 3,5-dinitrobenzoic acids	[66]
Berberine	Malonic acid, tartaric acid, citric acid	[86]
Myricetin	Piracetam	[87]
Genistein	Caffeine	[88]
Scoparone	Succinimide, urea, pyrimethamine	[89]
Zinc phenylacetate	Isonicotinamide	[90]

3.3.3 Cambridge Structural Database (CSD)

Each small-molecule crystal structure of a drug molecule that has been approved for use has been published and accessible through the CSD [24]. As per the Cambridge Crystallographic Data Centre (CCDC) website, unique curated structures in CSD are more than 1,100,000. Currently, it contains over 1.2 million crystal structures [91].

This database stores crystallographic information and is freely accessible to the research fraternity. The stored data can be used to better understand intermolecular interactions between molecules and design suitable supramolecular synthons (in the form of cocrystals). The solid-form informatics tools created by the CCDC have continuously shown the prediction capability of such a sizable structural data collection [24]. The CSD system offers various software components, including the Mercury program (Version 2020.1). The researchers use the Mercury program for visualizing molecule structures and estimating bond distances. Additionally, the data analysis module of the CSD system is employed to perform statistical analyses on the data obtained from these software components. The CSD program ConQuest.1 is used for searching the sub-structure. Some examples based on recent publications to screen cocrystals by CSD are enlisted in Table 3.4 [92, 93]. Fornari *et al.* [94] used the ConQuest program for searching, analyzing, and retrieving crystal structure from the CCDC, and then Mercury

Table 3.4 Recent publications of cocrystal screening from Cambridge structural database methods.

API	**Coformer**	**Reference**
Lenalidomide	Nicotinamide	[100]
Orcinol	Dipyridine	[101]
Diacerein	Isonicotinamide	[98]
Daidzein	4,4′-bipyridine	[99]
Captopril	l-proline, Sarcosine	[27]
Levetiracetam	Tartaric acid	[102]
Ammonium nitrate	20 molecular coformers	[103]
Quinine, d-mannitol, barbituric acid, l-histidine, theophylline.	2-aminopyridine, benzoic acid, benzenesulfonic acid, phenol, and phenylboronic acid	[104]
Diclofenac	Isonicotinamide, pyrazole, 2-chloropyrimidine, etc.	[57]
Piracetam	Hydroquinone, gentisic acid, fumaric acid, etc.	
Tegafur	Nicotinamide, pyrogallol, etc.	[105]
Pymetrozine	Oxalic acid, p-hydroxybenzoic acid, 2,3-dihydroxy benzoic acid, etc.	[106]

was used for visualizing and identifying sets of molecules that have formed cocrystals with each other and for which a cocrystal structure has been previously published in the written material and is included in the CSD.

It is a known fact that cocrystal formation takes place only when molecules have the ability to form intermolecular interactions. To identify such molecules, researchers often need the structure of the target molecules, which can be obtained from the CSD. Some physicochemical parameters, such as pKa, have also been reported to guide cocrystal design [27]. Also, examining the polarity of molecules can offer strong correlations between molecules that are prone to form cocrystals with each other. However, it is important to note that these parameters may not be always accurate [95]. Devogelaer *et al.* [96] used data mining and network science to evaluate coformers having a tendency to form cocrystals from the CSD system. A set of 9222 cocrystals, formed by 7188 unique coformers, was successfully extracted from CSD using the classifier algorithm. A group of coformers that have a common tendency to form cocrystals were isolated from CSD using this network.

Cocrystal structures and molecular descriptors, which are frequently used in quantitative structure-activity relationship analyses of compounds, were gathered by Fabian *et al.* [95] from the CSD. Based on their expected molecular features, the generated database describes pairs of compounds that can cocrystallize together. Figure 3.7 shows the molecular complementarity analysis (MCA) of cocrystals using the CSD. In this analysis, we can see a density plot showing regions with lighter colors indicating more molecular pairing. This MCA study concludes cocrystal formation.

Kamal *et al.* [97] analyzed the cyano-halogen interactions using CSD version 5.36 and ConQuest software. After scanning the database, 383

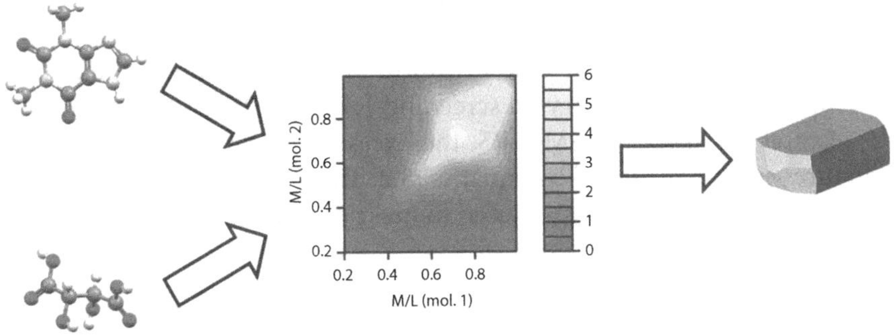

Figure 3.7 Molecular complementarity analysis in cocrystals using the Cambridge structural database ("reprinted (adapted) with permission from [95]. Copyright 2009 American Chemical Society").

unique hits were retrieved which would favorably form halogen bonds with the selected molecules. Tomar *et al.* [98] used ConQuest software to study the probability of the generation of supramolecular synthon of coformers from a CSD system with diacerein. With an intent to improve the aqueous solubility, CSD was scanned for possible hits to form H-bonding with the pyridine nitrogen and amide groups of diacerein. Through the search, it was revealed that carboxylic and carbonyl functional groups possessing compounds could be best-suited coformers for diacerein. From the search results, three cocrystals of diacerein were prepared. The CSD was utilized by Bhalla *et al.* [99] to gather statistics on existing supramolecular synthons involving hydroxyl and carbonyl groups in relation to other functional groups. This information guided the selection of coformers for cocrystal formation of daidzein based on their generally recognized safe (GRAS) status, structural/functional properties, and high-throughput results from the CSD. The CSD statistics highlighted the strong preference for supramolecular heterosynthons (-OH-----COOH and -OH-----N) over homosynthons. This CSD study served as a valuable resource for the molecular design approach in developing new crystalline forms of daidzein, a well-known isoflavone. From these examples, we can say that the CSD method relies on the availability of crystal structures for cocrystals. While the CSD contains an extensive collection of crystal structures, it may not cover all possible cocrystal combinations. This limitation restricts the application of the CSD method to only those cocrystals with known crystal structures.

3.3.4 COSMO-RS Approach

The COSMO-RS approach, which Klamt and his colleague created during their 12 years at Bayer, is presently the most precise approach for molecular free energy determination in solution. It is a unique blend of quantum chemical knowledge and liquid-phase thermodynamics. The COSMO-RS theory covers a wide range of key features of solubilization, making it an appropriate toolset for cocrystal screening Table 3.5 [107, 108]. According to researchers, using COSMO-RS, the excess enthalpy of stoichiometric mixtures can be calculated by Equation 3.2. The stoichiometric ratio of 1:1 of pure components A and B is used to make this combination.

$$H_{ex} = H_{AB} - mH_{pureA} - nH_{pureB} \tag{3.2}$$

The enthalpies in the m:n mixture and the pure reference condition, respectively, are represented by the letters $H_{AB\ \text{and}}\ H_{pure}$ [109, 110]. Despite

Table 3.5 Recent publications of cocrystal screening from COSMO-RS approach methods.

API	Coformer	Reference
Carbamazepine	Saccharin	[15]
Theophylline/Caffeine	4-amino-2-hydroxybenzoic acid, 4-nitrophenol	[30]
Minoxidil	Aromatic Carboxylic Acids	[110]
2-hydroxybenzylamine	Succinic acid, p-aminobenzoic acid, p-nitrobenzoic acid, o-nitrobenzoic acid, p-toluic acid, 2,3-dihydroxybenzoic acid, 3,4-dihydroxybenzoic acid, p-nitrophenol, and 5-hydroxyisophthalic acid	[114]
Abiraterone acetate	Succinic acid	[115]
Carbamazepine	DL-mandelic acid and DL-tartaric acid	[116]
SHR0302	Oxalic acid, phloroglucinol, gallic acid, etc.	[113]
Ciprofloxacin	Acetanilide, acetaminophen, isonicotinamide, etc.	[117]

the fact that COSMOtherm can extract hydrogen bonding interactions from the total enthalpy, H_{ex} contains all enthalpic contributions. Compared to the pure hydrogen bonding interaction, the excess enthalpy, H_{ex}, is a better description. As a result of their strong interactions in solution, compounds with $H_{ex} < 0$ prefer the combination enthalpically over their pure liquids (corresponding to a negative deviation from Raoult's law). An enhanced likelihood of cocrystal formation is strongly correlated with excess enthalpy, H_{ex}. Based on an assumption of enthalpic preference similarities in a liquid phase and cocrystals, it is appropriate to utilize the excess enthalpy in the liquid phase as a reference for cocrystal screening [16, 78].

COSMOtherm software is used to predict the thermodynamic properties of liquid systems which helps in precise and effective cocrystallization screening [111]. By evaluating a huge collection of molecules based on their excess enthalpy with the API, we may use COSMOtherm to discover potential coformers. In order to determine a drug's likelihood for cocrystallization, COSMOquick utilizes the excess enthalpy of an undercooled coformer melt. This has been proven to be a reliable approach for swiftly screening a vast array of possible coformers [112]. Figure 3.8 shows an example of how the COSMO-RS approach predicts cocrystal formation.

Deng *et al.* [110] used the COSMO-RS approach to estimate the likelihood of cocrystallization between minoxidil and 27 coformers (results shown in Figure 3.9). A lower H_{ex} value indicates more miscibility of coformers and minoxidil. The cutoff value for H_{ex} is also obvious because when H_{ex} is below 2.00 kcal mol^{-1}, coformers can always build multicomponent crystals with minoxidil. Otherwise, the coformers will be unsuccessful. The COSMO-RS approach performed significantly better than other screening methods used. They also used the Hansen solubility parameters method of the COSMO-RS approach for identifying suitable coformers for cocrystal formation.

Wu *et al.* [106] used the COSMO-RS approach to screen the cocrystal formation of pymetrozine (PMZ). The COSMO-RS approach employs the distribution of polarization charge density (σ) on the molecular surface to generate a σ-profile, reflecting the polarity distribution. The ensemble of interacting surface segments is addressed using statistical thermodynamics, which leads to the determination of the σ-potential. The integration of the σ-potential over the surface of a molecule yields the chemical potential. The mixing enthalpy (ΔHex) serves as a parameter for assessing

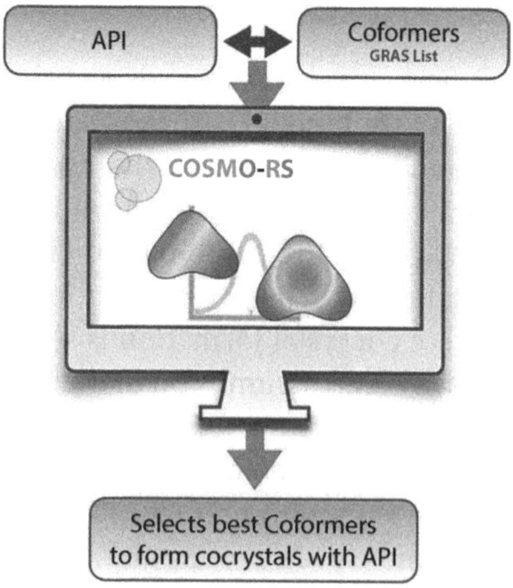

Figure 3.8 COSMO-RS screening method.

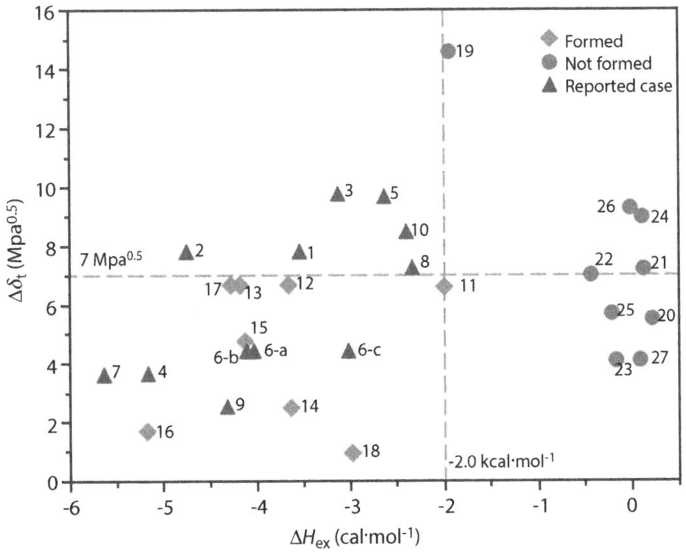

Figure 3.9 Results of the COSMO-RS approach and another method of minoxidil with aromatic carboxylic acids as coformers (HSP method) ("Reprinted (adapted) with permission from [110]. Copyright 2009 American Chemical Society").

the propensity of compounds to form cocrystals or salts. It is utilized as an indicator, with more negative values indicating a greater likelihood of cocrystal or salt formation. In this study, ΔH_{ex} values were calculated for 39 PMZ-coformer mixtures using COSMOthermXver2019 software. Results showed the formation of cocrystals with 3,4-dinitrobenzoic acid (3,4NBA) and 3-hydroxybenzoic acid which was further confirmed by PXRD and DSC pattern. Xie *et al.* [113] conducted calculations of excess enthalpy (ΔH_{ex}) between SHR0302 (JAK inhibitor) and various coformers using the COSMOthermX_3.0 program. In the study, COSMO files of 42 coformers and the active pharmaceutical ingredient (API) were utilized. These files were obtained from a DMol3-COSMO-PBE database or generated using the DMol3 package from Accelrys. The calculations were performed at the DFT level using the Perdew-Burke_Ernzerhof (PBE) functional and double numeric plus polarization (DNP) basis sets. Only stable conformations were taken into account for the calculations.

This method is one of the widely used methods for cocrystal screening but it predicts cocrystal formation from excess enthalpy alone while there is a need to consider intermolecular interactions, molecular packing, and geometrical factors for more accuracy in prediction.

3.3.5 Crystal Structure Prediction (CSP)

The virtual coformer screening method known as crystal structure prediction technology considers the contribution of crystallinity in solubility enhancement. This technique is utilized in the pharmaceutical sector during drug development to optimize lead compounds. The pharmaceutical industry's use of CSP technology for stable solid-form drug selection has been proven [25]. CSP and the FLEXCRYST program in combination can provide the free lattice energies of crystal structures. Using the free lattice energies, the cocrystal formation can be predicted based on simple thermodynamics. Cocrystal formation is favored only if the compound's free lattice energy is lower than the sum of the free lattice energies of the pure components. A lower free lattice energy indicates that the formed compound is thermodynamically more stable than the individual components used [118, 119]. FLEXCRYST is a feasible tool for computational screening of the stability of pharmaceutical cocrystals. Crystal structure predictions can be made using the FLEXCRYST program based on the crystal structure's free energy. In FLEXCRYST, all interaction locations are relocated to a single location on a grid, making it possible to discretize configuration space and easily compute and look up the scoring function for each interaction [120]. The calculations are carried out using FLEXCRYST software components (Figure 3.10). The SCORE module is used to measure and reduce the energies of the experimental structure of a crystal, the PREDICT module is used to anticipate crystal structures, and the module COMPARE is used to analyze the prediction outcomes [121].

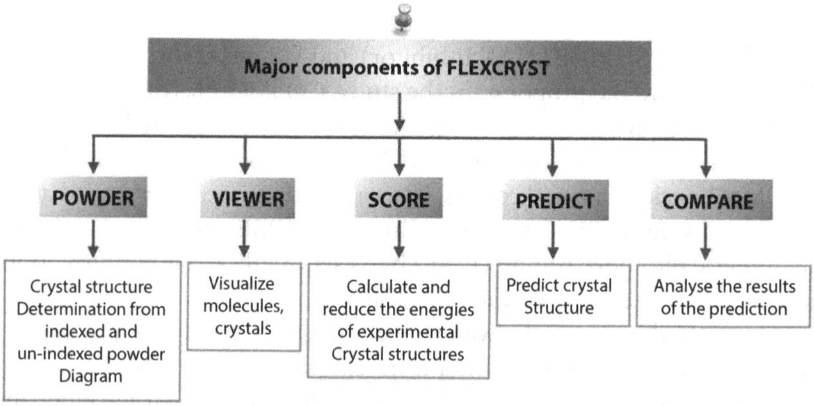

Figure 3.10 Flowchart of the major component of FLEXCRYST software used for crystal structure prediction.

Lattice Energy Calculations

As mentioned before, if the cocrystal's lattice energy is less than the total of the lattice energies of its constituent parts, a cocrystal will form.

Consider the following Equation 3.3

$$xA(solid) + yB(solid) \rightarrow A_xB_y(solid) + \Delta H \tag{3.3}$$

Here, in a stoichiometric ratio of x:y, molecules A and B form a cocrystal, and the resulting change in enthalpy is denoted by the symbol ΔH. Furthermore,

$$\Delta H \approx E_{AB} - (xE_A + yE_B) \tag{3.4}$$

Here, A_xB_y(solid), A(solid), and B(solid) have calculated lattice energies E_{AB}, E_A, and E_B, respectively. If entropic influences are disregarded, the cocrystal (or salt) for an exothermic reaction should be stable in relation to its constituents because ΔH for exothermic reactions is negative. The cocrystal is unlikely to form if ΔH is positive. In general, experimental crystal structures for single-component crystals or cocrystals may not be accessible, in that case, the proper lattice energies must be estimated using CSP analysis [122]. These calculations can be used to assess the chance of cocrystal formation and determine which cocrystal candidates offer the best chances. The stability and energy of the resulting cocrystal in cocrystal screening can be predicted using lattice energy estimates based on the interactions between the API and coformer. Table 3.6 shows the evidence from the set of examples that the CSP is a useful tool for cocrystal screening.

Issa *et al.* [123] applied the lattice energy calculation approach to predict the success of cocrystallization. The lattice energies were compared using the CrystalPredictor program. In the later part of this study, using cocrystals of 4-aminobenzoic acid, 2,2′-bipyridine, and 4-nitrophenylacetic acid, Karamertzanis *et al.* [124] established a multistage lattice energy minimization approach to produce stable packing configurations of cocrystals. They discovered that kinetic parameters are critical to successful cocrystallization and that using this method to identify cocrystal structures may be advantageous. Although lattice energy calculations are highly accurate, they have some limitations compared to other computational methodologies, and their efficiency decreases as the system's complexity grows. This method is not widely used because it has various problems and expensive calculations that are time-consuming. Computational crystal structure prediction (CSP) problems involve two parts: sampling and ranking.

Table 3.6 Recent publications of cocrystal screening from crystal structure prediction methods.

API	Coformer	Reference
2-methyl benzoic acid	2-amino-4-methyl pyrimidine	[126]
Paracetamol	Anthranilic acid	[122]
4-aminobenzoic acid	2,2′-bipyridine	[124]
1,3,5,7-tetranitro-1,3,5,7-tetrazocane	1,3,5-triamino-2,4,6-trinitrobenzene	[127]
Nicotinamide, isonicotinamide, and picolinamide	Carbamazepine, ibuprofen, and celecoxib	[122]

Sampling explores different ways a molecule can form a crystal while ranking identifies the most likely structures experimentally [125].

3.4 Recent Trends in Cocrystal Screening and Future Approach

Our early data search published in the past 15 years found a consistent rise in the publications related to the subject of the study. The following keywords were entered into Science Direct: "Molecular Electrostatic Potential Surfaces-cocrystal, Hirshfeld Surface Analysis-cocrystal, Cambridge structure database-cocrystal, COSMO-RS-cocrystal, Crystal Structure Prediction-cocrystal." The search results are shown in Figure 3.11. The enormous volume of publications and increasing trends in recent years show that the utility of these methods is growing in modern pharmaceutical research. Table 3.2 enlists some of the recent publications in computational cocrystal screening, which provide insight into new cocrystal candidates, and the latest computational approach used in cocrystal screening. The graphs of MEPS and HSA show a drastic increase in the use of these methods for computational cocrystal prediction. The simple calculation-based approach of COSMO-RS has also increased its use in prediction. The other methods CSD and CSP are seldom used for cocrystal prediction, as there is still a need to develop straightforward software to ease the use of these methods.

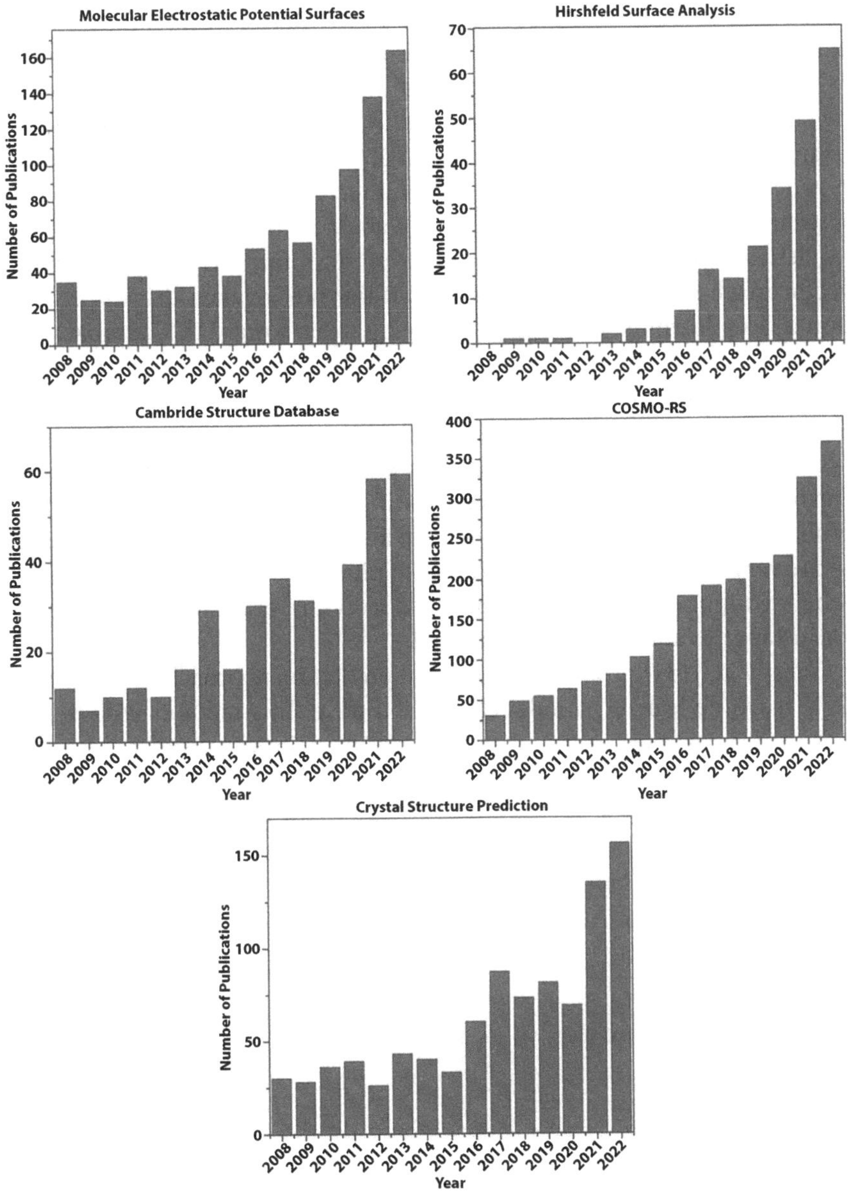

Figure 3.11 Recent trends in computational cocrystal screening in the scientific database (data taken from Science Direct).

Even though the scope of the above computational screening methods is increasing, the use in cocrystal screening faces limitations due to their reliance on assumptions and approximations, and their inability to access enough data for predicting the formation of cocrystals for new compounds. However, these limitations can be circumvented by utilizing ML and AI. AI and ML techniques can improve the accuracy of predictive models by leveraging large datasets and enabling learning from them. Examples of AI include deep learning, ML, data mining, and data analysis. A crucial use of AI is ML as it is a useful technique for a more advanced screening of cocrystals [128]. The popular AI methods are artificial neural network (ANN), logistic regression, random forest (RF), and support vector machine (SVM). Another promising tool for data-driven prediction is deep learning. Some steps are being taken toward this approach by some scientists. Wicker *et al.* [129] in 2017 first tried to predict cocrystal formation with a data-driven, machine learning algorithm approach. In this approach, results showed 2.6-fold enrichment in cocrystal formation from a set of coformers as compared to the experimental approach. Devogelaer *et al.* [130] used a deep learning–based ANN to determine the probability that two molecules will form a cocrystal. This method is useful at the very beginning of the therapeutic pipeline since prediction can be generated immediately as the molecular structure of an active ingredient, for instance, is anticipated or identified.

Yang *et al.* [131] proposed a cocrystal prediction approach that is based on a machine learning model called XGBoost. This method utilizes the

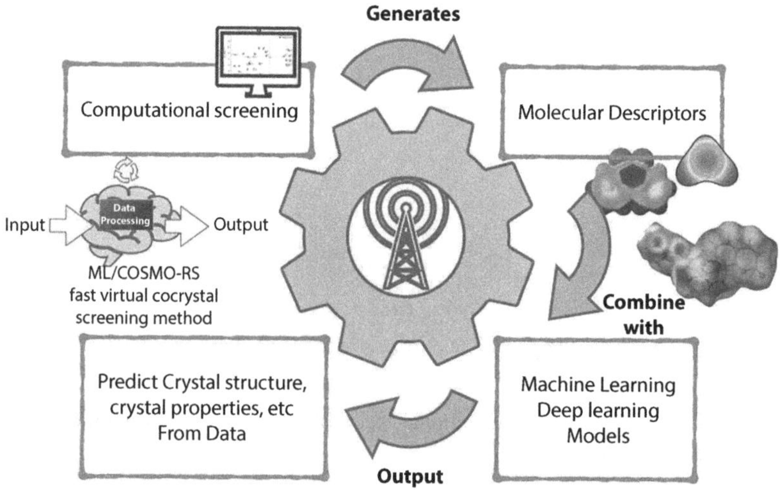

Figure 3.12 Data-driven machine learning algorithm for advanced cocrystal screening.

SMILES notation of two compounds as input to confirm cocrystal formation. The analysis is done using data that includes information on cocrystal formation extracted from the CSD, as well as previous research and experimental data on instances where cocrystal formation did not occur. This approach based on data aims to accurately predict the formation of cocrystal based on a set of input parameters. Hao *et al.* [132] also developed two ML models—COSMO-SVM and 3D-CNN ML on 3D spatial charge distribution descriptors. Yuan *et al.* [133] used physics-based and ML models and conducted a virtual screening of cocrystals. They offer a faster virtual cocrystal screening method that combines ML and COSMO-RS. Figure 3.12 explains how AI and ML approaches can be used in cocrystal screening.

3.5 Conclusion

Cocrystal engineering has proven to be a successful strategy for enhancing the efficacy of drugs with solubility and dissolving problems. However, choosing the right coformer is crucial due to the plethora of options available. Proper screening is necessary to successfully design cocrystals and can be effectively achieved through computational screening methods. These methods are advantageous in terms of time and resources in comparison to the experimental screening approach.

Nowadays, with the increasing demand for improved specificity, sensitivity, and accuracy in cocrystal screening, predictive approaches have evolved into more sophisticated methodologies that assist mechanistic understanding of cocrystal formation. When it comes to screening a large library of coformers, computational approaches have proven to be more effective than empirical screening techniques. These approaches involve measuring molecular complementarity, intermolecular interactions, and lattice energy calculations using MEPS, HSA, CSP, etc. Open-access databases such as CSD have been a boon for cocrystal research as these databases supply data essential to run computational analysis. Comparatively speaking, the use of the HSA and MEPS approach is more prevalent when compared to other methods. In the coming future, the use of ML and AI along with a computational approach will boost the screening method of cocrystals. Although the effectiveness of these computational methods has been established in the literature, these methods may not always accurately predict the actual stability of a cocrystal and may require significant computational resources. In the future, these limits in computational approaches can be overcome by using AI and ML approaches.

Acknowledgments

The authors are grateful to the Indian Institute of Technology (Banaras Hindu University), Varanasi, India, and the University of Technology, Institute of Process and Particle Engineering, Austria, for providing infrastructure facilities.

References

1. Gao, Z.G., Rohani, S., Gong, J.B., Wang, J.K., Recent Developments in the Crystallization Process: Toward the Pharmaceutical Industry. *Engineering*, 3, 3, 343–353, 2017.
2. Rathi, R., Kushwaha, R., Goyal, A., Singh, I., Oxaliplatin-flavone pharmaceutical co-crystal-CN111205332A: patent spotlight. *Pharm. Pat. Anal.*, 11, 5, 147–154, 2022.
3. Singh, M., Dhondale, M.R., Kumar, D., Sonocrystallization Approach to Modify Crystal Habit for Improving Powder Processability, in: *International Conference on Separation and Purification Technologies 2023 (ICSPT 2023)*, p. 21, 2023.
4. Gong, W., Mondal, P.K., Ahmadi, S., Wu, Y., Rohani, S., Cocrystals, Salts, and Salt-Solvates of olanzapine; selection of coformers and improved solubility. *Int. J. Pharm.*, 608, 121063, 2021.
5. Datir, R.S., Kumar, D., Bele, H.M., Modified crystal habits of glimepiride to improve manufacturing processability. *J. Cryst. Growth*, 592, 126711, 2022.
6. Gupta, D., Bhatia, D., Dave, V., Sutariya, V., Varghese Gupta, S., Salts of Therapeutic Agents: Chemical, Physicochemical, and Biological Considerations. *Molecules*, 23, 7, 1719–1734, 2018.
7. Yan, Y., Wang, L., Si, Z., Zhang, X., Yuan, W., A novel cocrystal of metformin hydrochloride with citric acid: Systematic synthesis and computational simulation. *Eur. J. Pharm. Biopharm.*, 179, 37–46, 2022.
8. Zhang, Y.N., Zhang, Y.F., Chang, L., Ji, Y.Y., Liu, L.X., Feng, Y.R., Wu, L.L., Zhang, L., Zhang, Y.J., Zou, D.Y., Liu, Y.L., Su, X., Crystalline palmatine saccharinate pharmaceutical salt without reducing solubility and improving its hygroscopic stability with regard to palmatine chloride. *J. Mol. Struct.*, 1230, 129631, 2021.
9. Yang, R., Zhang, G.G.Z., Kjoller, K., Dillon, E., Purohit, H.S., Taylor, L.S., Phase separation in surfactant-containing amorphous solid dispersions: Orthogonal analytical methods to probe the effects of surfactants on morphology and phase composition. *Int. J. Pharm.*, 619, 121708, 2022.
10. Knopp, M.M., Wendelboe, J., Holm, R., Rades, T., Effect of amorphous phase separation and crystallization on the *in vitro* and *in vivo* performance of an amorphous solid dispersion. *Eur. J. Pharm. Biopharm.*, 130, 290–295, 2018.

11. Rumondor, A.C., Wikstrom, H., Van Eerdenbrugh, B., Taylor, L.S., Understanding the tendency of amorphous solid dispersions to undergo amorphous-amorphous phase separation in the presence of absorbed moisture. *AAPS PharmSciTech.*, 12, 4, 1209–1219, 2011.
12. Li, N., Cape, J.L., Mankani, B.R., Zemlyanov, D.Y., Shepard, K.B., Morgen, M.M., Taylor, L.S., Water-Induced Phase Separation of Spray-Dried Amorphous Solid Dispersions. *Mol. Pharm.*, 17, 10, 4004–4017, 2020.
13. Teng, R.G., Wang, L.P., Chen, M.S., Fang, W., Gao, Z.G., Chai, Y.H., Zhao, P., Bao, Y., Amino acid based pharmaceutical cocrystals and hydrate cocrystals of the chlorothiazide: Structural studies and physicochemical properties. *J. Mol. Struct.*, 1217, 128432, 2020.
14. Dhondale, M.R., Thakor, P., Nambiar, A.G., Singh, M., Agrawal, A.K., Shastri, N.R., Kumar, D., Co-Crystallization Approach to Enhance the Stability of Moisture-Sensitive Drugs. *Pharmaceutics*, 15, 1, 189–215, 2023.
15. Banerjee, M., Nimkar, K., Naik, S., Patravale, V., Unlocking the potential of drug-drug cocrystals - A comprehensive review. *J. Control. Release*, 348, 456–469, 2022.
16. Qiao, N., Li, M., Schlindwein, W., Malek, N., Davies, A., Trappitt, G., Pharmaceutical cocrystals: an overview. *Int. J. Pharm.*, 419, 1-2, 1–11, 2011.
17. Malamatari, M., Ross, S.A., Douroumis, D., Velaga, S.P., Experimental cocrystal screening and solution based scale-up cocrystallization methods. *Adv. Drug Deliv. Rev.*, 117, 162–177, 2017.
18. Zhou, J.H., Chen, M.B., Chen, W.M., Shi, L.W., Zhang, C.Y., Li, H.Z., Virtual screening of cocrystal formers for CL-20. *J. Mol. Struct.*, 1072, 179–186, 2014.
19. Yang, D.Z., Wang, R.N., Jin, G.M., Zhang, B.X., Zhang, L., Lu, Y., Du, G.H., Structural and Computational Insights into Cocrystal Interactions: A Case on Cocrystals of Antipyrine and Aminophenazone. *Cryst. Growth Des.*, 19, 11, 6175–6183, 2019.
20. Aakeröy, C.B., Wijethunga, T.K., Desper, J., Practical crystal engineering using halogen bonding: A hierarchy based on calculated molecular electrostatic potential surfaces. *J. Mol. Struct.*, 1072, 20–27, 2014.
21. Nasaruddin, N.H., Ahmad, S.N., Sirat, S.S., Tan, K.W., Zakaria, N.A., Mohamad Nazam, S.S., Rahman, N., Mohd Yusof, N.S., Bahron, H., Synthesis, Structural Characterization, Hirshfeld Surface Analysis, and Antibacterial Study of Pd(II) and Ni(II) Schiff Base Complexes Derived from Aliphatic Diamine. *ACS Omega*, 7, 47, 42809–42818, 2022.
22. Yu, Y.M., Liu, L., Bu, F.Z., Li, Y.T., Yan, C.W., Wu, Z.Y., A novice cocrystal nanomicelle formulation of 5-fluorouracil with proline: The design, self-assembly and *in vitro/vivo* biopharmaceutical characteristics. *Int. J. Pharm.*, 617, 121635, 2022.
23. Chapman, C.J. and Groven, L.J., Evaluation of solvate and co-crystal screening methods for CL-20 containing energetic materials. *J. Energ. Mater.*, 40, 3, 258–272, 2022.

24. Bryant, M.J., Black, S.N., Blade, H., Docherty, R., Maloney, A.G.P., Taylor, S.C., The CSD Drug Subset: The Changing Chemistry and Crystallography of Small Molecule Pharmaceuticals. *J. Pharm. Sci.*, 108, 5, 1655–1662, 2019.
25. Sun, G., Jin, Y., Li, S., Yang, Z., Shi, B., Chang, C., Abramov, Y.A., Virtual Coformer Screening by Crystal Structure Predictions: Crucial Role of Crystallinity in Pharmaceutical Cocrystallization. *J. Phys. Chem. Lett.*, 11, 20, 8832–8838, 2020.
26. Musumeci, D., Hunter, C.A., Prohens, R., Scuderi, S., McCabe, J.F., Virtual cocrystal screening. *Chem. Sci.*, 2, 5, 883–890, 2011.
27. Wang, D.Y., Yang, Z., Zhu, B.Q., Mei, X.F., Luo, X.M., Machine-Learning-Guided Cocrystal Prediction Based on Large Data Base. *Cryst. Growth Des.*, 20, 10, 6610–6621, 2020.
28. Shaikh, R., Singh, R., Walker, G.M., Croker, D.M., Pharmaceutical Cocrystal Drug Products: An Outlook on Product Development. *Trends Pharmacol. Sci.*, 39, 12, 1033–1048, 2018.
29. Kumar, A. and Nanda, A., *In-silico* methods of cocrystal screening: A review on tools for rational design of pharmaceutical cocrystals. *J. Drug Deliv. Sci. Technol.*, 63, 102527, 2021.
30. Przybyłek, M., Ziółkowska, D., Mroczyńska, K., Cysewski, P., Applicability of Phenolic Acids as Effective Enhancers of Cocrystal Solubility of Methylxanthines. *Cryst. Growth Des.*, 17, 4, 2186–2193, 2017.
31. Rathi, R., Guggal, Y., Twinkle, V., Singh, I., Exploring the Patent Landscape and Regulatory Prospective on Pharmaceutical Cocrystals. *J. Pharm. Technol. Res. Manag.*, 11, 49–60, 2023.
32. Zhao, W., Li, B., Liu, S., Deng, Y., Zhang, R., Jiang, Y., Kinetic study of complicated anti-solvent and cooling crystallization of disodium 5′-ribonucleotide. *Particuology*, 73, 103–112, 2023.
33. Ober, C.A. and Gupta, R.B., Formation of itraconazole-succinic acid cocrystals by gas antisolvent cocrystallization. *AAPS PharmSciTech.*, 13, 4, 1396–1406, 2012.
34. Zhang, G.C., Lin, H.L., Lin, S.Y., Thermal analysis and FTIR spectral curve-fitting investigation of formation mechanism and stability of indomethacin-saccharin cocrystals via solid-state grinding process. *J. Pharm. Biomed. Anal.*, 66, 162–169, 2012.
35. Ying, P., Yu, J.B., Su, W.K., Liquid-Assisted Grinding Mechanochemistry in the Synthesis of Pharmaceuticals. *Adv. Synth. Catal.*, 363, 5, 1246–1271, 2021.
36. Rahman, F., Winantari, A.N., Setyawan, D., and Siswandono. Comparison Study of Grinding and Slurry Method on Physicochemical Characteristic of Acyclovir – Succinic Acid Cocrystal. *Asian J. Pharm. Clin. Res.*, 10, 3, 153–158, 2017.
37. Hong, M.H., Li, S.J., Ji, W.J., Qi, M.H., Ren, G.B., Cocrystals of Lenvatinib with Sulfamerazine and Salicylic Acid: Crystal Structure, Equilibrium

Solubility, Stability Study, and Anti-Hepatoma Activity. *Cryst. Growth Des.*, 21, 7, 3714–3727, 2021.

38. Liang, F., Tan, X., Hao, S., Liu, W., Duan, C., Zhang, G., Zhuang, T., Chen, Y., Hao, C., Synthesis and structural characterization of two novel olanzapine cocrystals with decreased or enhanced dissolution rate. *J. Mol. Struct.*, 1255, 132340, 2022.
39. Meng, Y., Tan, F., Yao, J., Cui, Y., Feng, Y., Li, Z., Wang, Y., Yang, Y., Gong, W., Yang, M., Kong, X., Gao, C., Preparation, characterization, and pharmacokinetics of rivaroxaban cocrystals with enhanced *in vitro* and *in vivo* properties in beagle dogs. *Int. J. Pharm. X.*, 4, 100119, 2022.
40. Sasaki, T., Nakane, T., Kawamoto, A., Nishizawa, T., Kurisu, G., Microcrystal electron diffraction (MicroED) structure determination of a mechanochemically synthesized co-crystal not affordable from solution crystallization. *CrystEngComm.*, 25, 3, 352–356, 2023.
41. Wasim, M., Mannan, A., Asad, M., Amirzada, M.I., Shafique, M., Hussain, I., *In Vitro* and Comparative *In Vivo* Evaluation. *Biomed Res. Int.*, 2021, 6685806, 2021.
42. Stevens, J.S., Byard, S.J., Schroeder, S.L.M., Characterization of Proton Transfer in Co-Crystals by X-ray Photoelectron Spectroscopy (XPS). *Cryst. Growth Des.*, 10, 3, 1435–1442, 2010.
43. Patel, A.M. and Patel, S.R., Membrane assisted crystallization of active pharmaceutical ingredients (APIs) by forward osmosis and its effect on crystal size and morphology. *Mater. Today Proc.*, 57, 2406–2411, 2022.
44. Mashhadi, S.M.A., Yufit, D., Liu, H.Y., Hodgkinson, P., Yunus, U., Synthesis and structural characterization of cocrystals of isoniazid and cinnamic acid derivatives. *J. Mol. Struct.*, 1219, 128621, 2020.
45. Li, M., Yang, Z., Zhang, Y., Xu, H., S.-h. Zhang, Sun, J., Wang, J., Synthesis, crystal structure, and solubility study of a supramolecular assembly cocrystal formed by levofloxacin and nicotinic acid. *J. Mol. Struct.*, 1249, 131549, 2022.
46. Wang, L., Li, S.Y., Xu, X.Z., Xu, X.Y., Wang, Q., Li, D.X., Zhang, H.L., Drug-drug cocrystals of theophylline with quercetin. *J. Drug Deliv. Sci. Technol.*, 70, 103228, 2022.
47. Rathi, R., Kaur, S., Singh, I., A review on co-crystals of Herbal Bioactives for solubility enhancement: preparation methods and characterization techniques. *Cryst. Growth Des.*, 22, 3, 2023–2042, 2022.
48. Guin, M., Halder, S., Chatterjee, S., Konar, S., Synthesis, X-ray crystal structure of Cu(II) 1D coordination Polymer: In View of Hirshfeld surface, FMO, Molecular electrostatic potential (MEP) and Natural Bond orbital (NBO) analyses. *J. Mol. Struct.*, 1270, 133949, 2022.
49. Azim, Y. and Garg, U., Experimental and computational analyses of the cocrystal of tetrahydrofuran-2,3,4,5-tetracarboxylic acid and urea. *J. Mol. Struct.*, 1269, 133820, 2022.

50. Eissa, I.H., Khalifa, M.M., Elkaeed, E.B., Hafez, E.E., Alsfouk, A.A., Metwaly, A.M., *In Silico* Exploration of Potential Natural Inhibitors against SARS-Cov-2 nsp10. *Molecules*, 26, 20, 6151, 2021.
51. Wang, Y.Y., Wang, L.P., Zhang, F., Wang, N., Gao, Y.T., Xiao, Y.T., Wang, Z., Bao, Y., Structure analysis and insight into hydrogen bond and van der waals interactions of etoricoxib cocrystals and cocrystal solvate. *J. Mol. Struct.*, 1258, 132665, 2022.
52. Kumar, R., Kamal, R., Kumar, V., Parkash, J., Bifunctionalization of α,β-unsaturated diaryl ketones into α-aryl-β,β-ditosyloxy ketones: Single crystal XRD, DFT, FMOs, molecular electrostatic potential, hirshfeld surface analysis, and 3D-energy frameworks. *J. Mol. Struct.*, 1250, 131754, 2022.
53. Lakshminarayanan, S., Jeyasingh, V., Murugesan, K., Selvapalam, N., Dass, G., Molecular electrostatic potential (MEP) surface analysis of chemo sensors: An extra supporting hand for strength, selectivity & non-traditional interactions. *J. Photochem. Photobiol.*, 6, 100022, 2021.
54. Grecu, T., Prohens, R., McCabe, J.F., Carrington, E.J., Wright, J.S., Brammer, L., Hunter, C.A., Cocrystals of spironolactone and griseofulvin based on an screening method. *Crystengcomm.*, 19, 26, 3592–3599, 2017.
55. Oliver, A., Hunter, C.A., Prohens, R., Rossello, J.L., An improved methodology to compute surface site interaction points using high density molecular electrostatic potential surfaces. *J. Comput. Chem.*, 39, 28, 2371–2377, 2018.
56. Grecu, T., Adams, H., Hunter, C.A., McCabe, J.F., Portell, A., Prohens, R., Virtual Screening Identifies New Cocrystals of Nalidixic Acid. *Cryst. Growth Des.*, 14, 4, 1749–1755, 2014.
57. Grecu, T., Hunter, C.A., Gardiner, E.J., McCabe, J.F., Validation of a Computational Cocrystal Prediction Tool: Comparison of Virtual and Experimental Cocrystal Screening Results. *Cryst. Growth Des.*, 14, 1, 165–171, 2014.
58. Zhou, Q., Yu, Y., Du, H., Liu, C., Ren, J., Han, S., Pang, Z., Preparation, structure and optical properties of mixed stacked cocrystals of 2,2':6',2"-Terpyridine and 1,2,4,5-tetracyanobenzene. *J. Lumin.*, 247, 118903, 2022.
59. Krueger, E.L., Sinha, A.S., Desper, J., Aakeröy, C.B., Exploring binding preferences in co-crystals of conformationally flexible multitopic ligands. *Crystengcomm.*, 19, 31, 4605–4614, 2017.
60. Huang, D.D., Yang, D.Z., Lv, Y., Zhou, J.L., Li, L., Xu, J.P., Yang, X.M., Zhou, Z.Z., Lattice water provides hydrogen atom donor to form hydrate: A case study of chlorbipram: m-hydroxybenzoic acid (1:1) cocrystal. *J. Mol. Struct.*, 1250, 131891, 2022.
61. Huang, S., Cheemarla, V.K.R., Tiana, D., Lawrence, S.E., Experimental and Theoretical Investigation of Hydrogen-Bonding Interactions in Cocrystals of Sulfaguanidine. *Cryst. Growth Des.*, 23, 4, 2306–2320, 2023.
62. Zhou, Q., Du, H., Chen, Z., Yu, Y., Ren, J., Han, S., Fan, J., Pang, Z., Two-photon excited fluorescence from ground-state charge-transfer cocrystals of acridine and 1,2,4,5-tetracyano-benzene. *Dyes Pigm.*, 198, 109965, 2022.

63. Mary, Y.S., Mary, Y.S., Analysis, D.F.T., and Molecular Docking Studies of the Cocrystals of Sulfathiazole-Theophylline and Sulfathiazole-Sulfanilamide. *Polycyc. Aromat. Comp.*, 42, 6, 3809–3820, 2022.
64. Zhou, H.L., Duan, C.X., Qin, H.M., Huang, C.A., Hou, J.X., Chen, Y.M., Zhu, J., Xu, C.C., Jin, J., Zhuang, T., Synthesis and structural characterization of a novel palbociclib-kaempferol cocrystal with improved tabletability and synergistic antitumor activity. *J. Mol. Struct.*, 1281, 135101, 2023.
65. Luo, Y.M., Chen, S.Q., Zhou, J.L., Chen, J.M., Tian, L.W., Gao, W., Zhang, Y.Y., Ma, A.D., Li, L., Zhou, Z.Z., Luteolin cocrystals: Characterization, evaluation of solubility, oral bioavailability and theoretical calculation. *J. Drug Deliv. Sci. Technol.*, 50, 248–254, 2019.
66. Yu, H.M., Zhang, B.X., Xing, W.H., Liu, M.J., Wang, F.F., Gong, N.B., Zhang, L., Lu, Y., Du, G.H., Investigation into the structures and physicochemical properties of multi-component crystals of voriconazole. *Chin. Chem. Lett.*, 34, 6, 107668, 2023.
67. Wang, H., Yang, D., Zhang, W., Song, J., Gong, N., Yu, M., Yang, S., Zhang, B., Liu, Q., Du, G., Lu, Y., An innovative rhein-matrine cocrystal: Synthesis, characterization, formation mechanism and pharmacokinetic study. *Chin. Chem. Lett.*, 34, 2, 107258, 2023.
68. Bu, F.-Z., Yu, Y.-M., Shen, Y.-L., Wu, Z.-Y., Li, Y.-T., Cocrystallization with nutrient ferulic acid towards reducing the dissolubility behaviors of antifungal drug 5-fluorocytosine: An integrated theoretical and experimental case research. *J. Mol. Struct.*, 1275, 134601, 2023.
69. Li, H.J., Liu, J.C., Yang, L., Yan, Z.Z., Lu, Y.W., Han, J.M., Ren, X.T., Li, W., Theoretical predict structure and property of the novel CL-20/2,4-DNI cocrystal by systematic search approach. *Def. Technol.*, 18, 6, 907–917, 2022.
70. Khan, I.M., Shakya, S., Akhtar, R., Alam, K., Islam, M., Alam, N., Exploring interaction dynamics of designed organic cocrystal charge transfer complex of 2-hydroxypyridine and oxalic acid with human serum albumin: Single crystal, spectrophotometric, theoretical and antimicrobial studies. *Bioorg. Chem.*, 100, 103872, 2020.
71. Yang, D.Z., Wang, H.J., Liu, Q.W., Yuan, P.H., Chen, T., Zhang, L., Yang, S.Y., Zhou, Z.Z., Lu, Y., Du, G.H., Structural landscape on a series of rhein: Berberine cocrystal salt solvates: The formation, dissolution elucidation from experimental and theoretical investigations. *Chin. Chem. Lett.*, 33, 6, 3207–3211, 2022.
72. Lakshmipathi, M., Emmerling, F., Bhattacharya, B., Ghosh, S., Structure-mechanical property correlation of a series of 4-(1-Napthylvinyl) pyridine based cocrystals. *J. Mol. Struct.*, 1268, 133670, 2022.
73. Prabhu, S.R., Upadhyaya, V., Jayarama, A., Synthesis, crystal structure and Hirshfeld surface analysis of a novel chalcone derivative: (2 E)-3-(2,3-dimethoxyphenyl)-1-(3-nitrophenyl)prop-2-en-1-one. *Chem. Data Collect.*, 11-12, 199–210, 2017.

74. Spackman, P.R., Turner, M.J., McKinnon, J.J., Wolff, S.K., Grimwood, D.J., Jayatilaka, D., Spackman, M.A., CrystalExplorer: a program for Hirshfeld surface analysis, visualization and quantitative analysis of molecular crystals. *J. Appl. Crystallogr.*, 54, Pt 3, 1006–1011, 2021.
75. Luo, Y.H., Zhang, C.G., Xu, B., Sun, B.W., A cocrystal strategy for the precipitation of liquid 2,3-dimethyl pyrazine with hydroxyl substituted benzoic acid and a Hirshfeld surfaces analysis of them. *Crystengcomm.*, 14, 20, 6860–6868, 2012.
76. Chethan, B.S. and Lokanath, N.K., Study of the crystal structure, H-bonding and noncovalent interactions of novel cocrystal by systematic computational search approach. *J. Mol. Struct.*, 1251, 131936, 2022.
77. Abidi, S.S.A., Azim, Y., Gupta, A.K., Pradeep, C.P., Cocrystals of indole-3-acetic acid and indole-3-butyric acid: Synthesis, structural characterization and Hirshfeld surface analysis. *J. Mol. Struct.*, 1166, 202–213, 2018.
78. Loschen, C. and Klamt, A., New Developments in Prediction of Solid-State Solubility and Cocrystallization Using COSMO-RS Theory. in: *Computational Pharmaceutical Solid State Chemistry*, 211–233, 2016.
79. Jayatilaka, D., Wolff, S.K., Grimwood, D.J., McKinnon, J.J., Spackman, M.A., CrystalExplorer: A Tool for Displaying Hirshfeld Surfaces and Visualising Intermolecular Interactions in Molecular Crystals. *Acta Cryst. a-Foundation Adv.*, 62, a1, S90–S90, 2006.
80. Novena, L.M., Athimoolam, S., Anitha, R., Bahadur, S.A., Synthesis, crystal structure, hirshfeld surface analysis, spectral and quantum chemical studies of pharmaceutical cocrystals of a bronchodilator drug (Theophylline). *J. Mol. Struct.*, 1249, 131585, 2022.
81. Hussain, E., Kumar, R., Choudhary, M.I., Yousuf, S., Crystal Engineering of Naturally Occurring Seselin To Obtain Cocrystal with Enhanced Anti-Leishmanial Activity, Hirshfeld Surface Analysis, and Computational Insight. *Cryst. Growth Des.*, 18, 8, 4628–4636, 2018.
82. Yu, Y.M., Bu, F.Z., Liu, L., Yan, C.W., Wu, Z.Y., Li, Y.T., A novel sustained-release formulation of 5-fluorouracil-phenylalanine cocrystal self-assembled by cocrystal-entrapped micelle strategy displays enhanced antitumor efficacy. *J. Mol. Liq.*, 368, 120665, 2022.
83. Matheswari, P.P., Asha, R.N., Bhuvanesh, N., Nayagam, B.R.D., Hirshfeld surface, D.F.T., and molecular docking studies of a new organic cocrystal: creatinine:2,3-pyridinedicarboxylic acid. *J. Phys. Org. Chem.*, 34, 9, e4247, 2021.
84. Roy, P., Pandey, N., Kumari, N., Baidya, R., Mary, Y.S., Mary, Y.S., Ghosh, A., Development of sulfamethoxazole-succinimide cocrystal by mechanochemical cocrystallization - An insight into spectroscopic, electronic, chemical conformation and physicochemical properties. *Chem. Eng. Res. Des.*, 185, 446–457, 2022.

85. Qu, H., Li, Z., Wu, S., Gong, J., Mechanosynthesis of magnolol multicomponent crystalline solids for improved natural antibiotics and customizable release profiles. *Int. J. Pharm.*, 632, 122530, 2023.
86. Zhang, G., Yang, X., Shang, X., Han, W., Wang, F., Ban, S., Zhang, S., Novel multi-component crystals of berberine with improved pharmaceutical properties. *IUCrJ*, 10, Pt 1, 66–76, 2023.
87. Sowa, M., Slepokura, K., Matczak-Jon, E., A 1:1 pharmaceutical cocrystal of myricetin in combination with uncommon piracetam conformer: X-ray single crystal analysis and mechanochemical synthesis. *J. Mol. Struct.*, 1058, 114–121, 2014.
88. Sowa, M., Slepokura, K., Matczak-Jon, E., Solid-state characterization and solubility of a genistein-caffeine cocrystal. *J. Mol. Struct.*, 1076, 80–88, 2014.
89. Chen, J.Y., Wang, X.Y., Hong, M.H., Yi, D.X., Qi, M.H., Ren, G.B., Structure properties of scoparone: Polymorphs and cocrystals. *J. Mol. Struct.*, 1191, 323–336, 2019.
90. Ahmadi, S. and Rohani, S., Overcoming the Hydrophobic Nature of Zinc Phenylacetate Through Co-Crystallization with Isonicotinamide. *J. Pharm. Sci.*, 112, 7, 1929–1938, 2023.
91. Singh, M., Barua, H., Jyothi, V., Dhondale, M.R., Nambiar, A.G., Agrawal, A.K., Kumar, P., Shastri, N.R., Kumar, D., Cocrystals by Design: A Rational Coformer Selection Approach for Tackling the API Problems. *Pharmaceutics*, 15, 4, 1161, 2023.
92. Tetteh, S., Halide effect on the electron distribution in oxovanadium complexes bearing side-on (η2) dioxygen ligands: A combined crystallographic and theoretical study. *J. Mol. Struct.*, 1223, 129187, 2021.
93. Bruno, I.J., Cole, J.C., Edgington, P.R., Kessler, M., Macrae, C.F., McCabe, P., Pearson, J., Taylor, R., New software for searching the Cambridge Structural Database and visualizing crystal structures. *Acta Crystallogr. B.*, 58, Pt 3 Pt 1, 389–397, 2002.
94. Fornari, F., Montisci, F., Bianchi, F., Cocchi, M., Carraro, C., Cavaliere, F., Cozzini, P., Peccati, F., Mazzeo, P.P., Riboni, N., Careri, M., Bacchi, A., Chemometric-assisted cocrystallization: supervised pattern recognition for predicting the formation of new functional cocrystals. *Chemom. Intellig. Lab. Syst.*, 226, 104580, 2022.
95. Fábián, L., Cambridge Structural Database Analysis of Molecular Complementarity in Cocrystals. *Cryst. Growth Des.*, 9, 3, 1436–1443, 2009.
96. Devogelaer, J.J., Meekes, H., Vlieg, E., de Gelder, R., Cocrystals in the Cambridge Structural Database: a network approach. *Acta Crystallogr. B Struct. Sci. Cryst. Eng. Mater.*, 75, Pt 3, 371–383, 2019.
97. Kamal, K., Maurya, H.K., Gupta, A., Vasudev, P.G., Crystal structures of four delta-keto esters and a Cambridge Structural Database analysis of cyano-halogen interactions. *Acta Crystallogr. C Struct. Chem.*, 71, Pt 10, 921–928, 2015.

98. Tomar, S., Chakraborti, S., Jindal, A., Grewal, M.K., Chadha, R., Cocrystals of diacerein: Towards the development of improved biopharmaceutical parameters. *Int. J. Pharm.*, 574, 118942, 2020.
99. Bhalla, Y., Chadha, K., Chadha, R., Karan, M., Daidzein cocrystals: An opportunity to improve its biopharmaceutical parameters. *Heliyon*, 5, 11, e02669, 2019.
100. Wang, L., Yan, Y., Zhang, X., Zhou, X., Novel pharmaceutical cocrystal of lenalidomide with nicotinamide: Structural design, evaluation, and thermal phase transition study. *Int. J. Pharm.*, 613, 121394, 2022.
101. Rautenberg, M., Bhattacharya, B., Akhmetova, I., Emmerling, F., Mechanochemical and solution syntheses of two novel cocrystals of orcinol with two N,N 0-Dipyridines: Structural diversity with varying ligand flexibility. *J. Mol. Struct.*, 1217, 128303, 2020.
102. Springuel, G., Robeyns, K., Norberg, B., Wouters, J., Leyssens, T., Cocrystal Formation between Chiral Compounds: How Cocrystals Differ from Salts. *Cryst. Growth Des.*, 14, 8, 3996–4004, 2014.
103. Bennett, A.J. and Matzger, A.J., Progress in Predicting Ionic Cocrystal Formation: The Case of Ammonium Nitrate. *Chem. (Easton)*, 29, 27, e202300076, 2023.
104. Cappuccino, C., Cusack, D., Flanagan, J., Harrison, C., Holohan, C., Lestari, M., Walsh, G., Lusi, M., How Many Cocrystals Are We Missing? Assessing Two Crystal Engineering Approaches to Pharmaceutical Cocrystal Screening. *Cryst. Growth Des.*, 22, 2, 1390–1397, 2022.
105. Aitipamula, S., Chow, P.S., Tan, R.B.H., Crystal Engineering of Tegafur Cocrystals: Structural Analysis and Physicochemical Properties. *Cryst. Growth Des.*, 14, 12, 6557–6569, 2014.
106. Wu, D., Li, J.L., Xiao, Y.T., Ji, X., Li, C., Zhang, B., Hou, B.H., Zhou, L.N., Xie, C., Gong, J.B., Chen, W., New Salts and Cocrystals of Pymetrozine with Improvements on Solubility and Humidity Stability: Experimental and Theoretical Study. *Cryst. Growth Des.*, 21, 4, 2371–2388, 2021.
107. Klamt, A., Solvent-screening and co-crystal screening for drug development with COSMO-RS. *J. Cheminform.*, 4, S1, O14, 2012.
108. Loschen, C. and Klamt, A., Solubility prediction, solvate and cocrystal screening as tools for rational crystal engineering. *J. Pharm. Pharmacol.*, 67, 6, 803–811, 2015.
109. Abramov, Y.A., Loschen, C., Klamt, A., Rational coformer or solvent selection for pharmaceutical cocrystallization or desolvation. *J. Pharm. Sci.*, 101, 10, 3687–3697, 2012.
110. Deng, Y.H., Chen, Y.C., Xie, F., Tang, J.W., Zhang, R., Yang, H.Y., Jiang, Y.B., Liu, S.Y., Minoxidil Multi-Component Crystals with Aromatic Carboxylic Acids: Theoretical Calculation and Structural Analysis. *Cryst. Growth Des.*, 22, 6, 3941–3953, 2022.
111. Loschen, C. and Klamt, A., COSMOtherm as a Valuable Tool for Cocrystal Screening and Development.

112. Loschen, C. and Klamt, A., COSMO: A Novel Interface for Fast σ-Profile Composition and Its Application to COSMO-RS Solvent Screening Using Multiple Reference Solvents. *Ind. Eng. Chem. Res.*, 51, 43, 14303–14308, 2012.
113. Xie, Y.J., Shi, G.P., Sun, J., Li, S., Gao, W., Hu, Y.M., Zu, C., Tang, W.W., Gong, J.B., Computational Screening and Experimental Validation on Multicomponent Crystals of a New Class of Janus Kinase (JAK) Inhibitor Drug with Improved Solubility. *Crystals*, 12, 12, 1722, 2022.
114. Li, J., Wu, D., Xiao, Y., Li, C., Ji, X., Sun, Q., Chang, D., Zhou, L., Jing, D., Gong, J., Chen, W., Salts of 2-hydroxybenzylamine with improvements on solubility and stability: Virtual and experimental screening. *Eur. J. Pharm. Sci.*, 169, 106091, 2022.
115. Chennuru, R., Devarapalli, R., Rengaraj, P., Srinivas, P.L., Dey, S., Reddy, C.M., Improving Solubility of Poorly Soluble Abiraterone Acetate by Cocrystal Design Aided by *In Silico* Screening. *Cryst. Growth Des.*, 20, 8, 5018–5030, 2020.
116. Roca-Paixão, L., Correia, N.T., Affouard, F., Affinity prediction computations and mechanosynthesis of carbamazepine based cocrystals. *CrystEngComm.*, 21, 45, 6991–7001, 2019.
117. Zheng, Z.Y., Deng, Y.H., Xie, W.N., Chen, S.R., Liang, X.X., Liu, S.Y., Jiang, Y.B., Yang, H.Y., Co-Former Screening Method for Multicomponent Crystals Based on Partial Least Squares Regression: A Case Study of Ciprofloxacin. *Cryst. Growth Des.*, 23, 5, 3244–3257, 2023.
118. Kuleshova, L.N., Hofmann, D.W.M., Boese, R., Lattice energy calculation - A quick tool for screening of cocrystals and estimation of relative solubility. Case of flavonoids. *Chem. Phys. Lett.*, 564, 26–32, 2013.
119. Karagianni, A., Malamatari, M., Kachrimanis, K., Pharmaceutical Cocrystals: New Solid Phase Modification Approaches for the Formulation of APIs. *Pharmaceutics*, 10, 1, 18, 2018.
120. Day, G.M., Current approaches to predicting molecular organic crystal structures. *Crystallogr. Rev.*, 17, 1, 3–52, 2011.
121. Stepanovs, D., Jure, M., Kuleshova, L.N., Hofmann, D.W.M., Mishnev, A., Cocrystals of Pentoxifylline: *In Silico* and Experimental Screening. *Cryst. Growth Des.*, 15, 8, 3652–3660, 2015.
122. Chan, H.C.S., Kendrick, J., Neumann, M.A., Leusen, F.J.J., Towards screening of co-crystal formation through lattice energy calculations and crystal structure prediction of nicotinamide, isonicotinamide, picolinamide and paracetamol multi-component crystals. *Crystengcomm.*, 15, 19, 3799–3807, 2013.
123. Issa, N., Karamertzanis, P.G., Welch, G.W.A., Price, S.L., Can the Formation of Pharmaceutical Cocrystals Be Computationally Predicted? I. Comparison of Lattice Energies. *Cryst. Growth Des.*, 9, 1, 442–453, 2008.
124. Karamertzanis, P.G., Kazantsev, A.V., Issa, N., Welch, G.W., Adjiman, C.S., Pantelides, C.C., Price, S.L., Can the Formation of Pharmaceutical Cocrystals

Be Computationally Predicted? 2. Crystal Structure Prediction. *J. Chem. Theory Comput.*, 5, 5, 1432–1448, 2009.

125. Cruz-Cabeza, A.J., Crystal structure prediction: are we there yet? *Acta Crystallogr. B Struct. Sci. Cryst. Eng. Mater.*, 72, Pt 4, 437–438, 2016.
126. Thakur, T.S. and Desiraju, G.R., Crystal Structure Prediction of a Co-Crystal Using a Supramolecular Synthon Approach: 2-Methylbenzoic Acid-2-Amino-4-methylpyrimidine. *Cryst. Growth Des.*, 8, 11, 4031–4044, 2008.
127. Wei, C.X., Huang, H., Duan, X.H., Pei, C.H., Structures and Properties Prediction of HMX/TATB Co-Crystal. *Propellants Explos. Pyrotech.*, 36, 5, 416–423, 2011.
128. Heng, T., Yang, D., Wang, R., Zhang, L., Lu, Y., Du, G., Progress in Research on Artificial Intelligence Applied to Polymorphism and Cocrystal Prediction. *ACS Omega*, 6, 24, 15543–15550, 2021.
129. Wicker, J.G.P., Crowley, L.M., Robshaw, O., Little, E.J., Stokes, S.P., Cooper, R.I., Lawrence, S.E., Will they co-crystallize? *CrystEngComm.*, 19, 36, 5336–5340, 2017.
130. Devogelaer, J.J., Meekes, H., Tinnemans, P., Vlieg, E., de Gelder, R., Co-crystal Prediction by Artificial Neural Networks*. *Angew. Chem. Int. Ed. Engl.*, 59, 48, 21711–21718, 2020.
131. Yang, D.Z., Wang, L., Yuan, P.H., An, Q., Su, B., Yu, M.C., Chen, T., Hu, K., Zhang, L., Lu, Y., Du, G.H., Cocrystal virtual screening based on the XGBoost machine learning model. *Chin. Chem. Lett.*, 34, 8, 107964, 2023.
132. Hao, Y.Q., Hung, Y.C., Shimoyama, Y., Investigating Spatial Charge Descriptors for Prediction of Cocrystal Formation Using Machine Learning Algorithms. *Cryst. Growth Des.*, 22, 11, 6608–6615, 2022.
133. Yuan, J.C., Liu, X.T., Wang, S.M., Chang, C., Zeng, Q., Song, Z.T., Jin, Y.D., Zeng, Q., Sun, G.X., Ruan, S.G., Greenwell, C., Abramov, Y.A., Virtual coformer screening by a combined machine learning and physics-based approach. *Crystengcomm.*, 23, 35, 6039–6044, 2021.

4

Preparation Techniques of Pharmaceutical Cocrystals

Nikita Thakur[1], Mansi Prashar[2], Sukhvir Kaur[1], Neelima Dhingra[2], Manish Kumar[3], Ritu Rathi[3], Inderbir Singh[3] and Poonam Arora[3*]

[1]Chitkara University School of Pharmacy, Chitkara University, Himachal Pradesh, India
[2]University Institute of Pharmaceutical Sciences, Panjab University, Chandigarh, India
[3]Chitkara College of Pharmacy, Chitkara University, Punjab, India

Abstract
Academic and industrial scientists are becoming more concerned with novel solid forms, i.e., pharmaceutical cocrystals, emerging as an innovative class of solid pharmaceuticals with enhanced physicochemical properties. By incorporating APIs with pharmaceutically acceptable coformers, cocrystals offer improved solubility, stability, dissolution rates, and bioavailability. The preparation of cocrystals can be achieved using various techniques, broadly categorized into solid-based and solution-based methods. Solid-based methods, such as dry grinding, wet grinding, and hot-melt extrusion, involve minimal solvents and are advantageous for solvent-sensitive compounds. On the other hand, solution-based methods—including solvent evaporation, cooling crystallization, antisolvent crystallization, and supercritical fluids—provide a versatile approach to producing cocrystals, allowing better control over crystal morphology and purity. Optimization of these techniques requires a systematic understanding of several critical parameters. A quality-by-design framework is essential for identifying critical quality attributes and critical process parameters, ensuring consistency and reproducibility in cocrystal production. The kinetics of crystallization, particularly nucleation and crystal growth, are crucial for regulating particle size distribution, crystal habit, and yield. Moreover, the selection of an appropriate coformer is of paramount importance; it should be based on molecular complementarity, hydrogen bonding potential, and compatibility with the API to achieve the desired physicochemical

**Corresponding author*: poonamarora0803@gmail.com

Inderbir Singh, Subrata Mallick and Ritu Rathi (eds.) Cocrystals in Pharmaceutical Sciences: Design to Applications, (93–120) © 2025 Scrivener Publishing LLC

enhancements. This chapter provides a comprehensive overview of the preparation techniques and the key optimization parameters that influence cocrystal formation, emphasizing their significance in developing cocrystals with enhanced therapeutic efficacy for pharmaceutical applications.

Keywords: Optimization, quality-by-design, green chemistry, efficiency, PAT, innovation

4.1 Introduction

Over the past two decades, cocrystals have gained noteworthy interest from academia and industry for their ability to enhance API properties without changing pharmacological effects. Cocrystals, which contain more than two coformer in a fixed stoichiometric ratio, offer a strategic approach to improving the chemical and physical properties, such as stability, apparent bioavailability, and solubility studies, of active pharmaceutical ingredients (APIs) [1]. Research has highlighted the importance of crystal engineering and supramolecular synthons in designing pharmaceutical cocrystals, which advanced the use of cocrystal strategies to enhance drug properties. Various vigorous supramolecular synthons are key factors in driving pharmaceutical cocrystal formation. Functional groups like carboxylic acids, amides, and alcohols are particularly suitable for forming these synthons [2]. Supramolecular synthons are classified into two types: homosynthons and heterosynthons. Homosynthons are formed by self-complementary collections, such as amide dimers or carboxylic acid, while heterosynthons arise from different but complementary groups, like hydrogen bonding between pyridines and carboxylic acids or alcohols and aromatic nitrogen ring [3].

Historically, cocrystal preparation methods were limited and often relied on trial-and-error approaches. Early techniques primarily involved solvent evaporation or mechanical grinding. While effective in some cases, lacked consistency and efficiency. As research progressed, solid-state methods, such as ball milling and extrusion were introduced. These techniques generally use little or no solvent, making them more environmentally friendly and simpler to implement. However, they often required extensive experimentation to fine-tune the process and achieve desirable cocrystal qualities. Moreover, recent research has emphasized optimizing various parameters, including temperature, pressure, and solvent choice, to achieve the desired cocrystal characteristics. The ability to manipulate these factors has led to the development of new cocrystals with tailored properties, further expanding their potential applications in pharmaceuticals and beyond [4]. This chapter explores various techniques commonly used for cocrystallization, including liquid-assisted and solid-state approaches. It also evaluates several novel approaches, such as spray drying and supercritical CO_2-assisted and ultrasound-assisted cocrystallization, for their potential in cocrystal preparation [5]. It also provides a comprehensive conversation on continuous manufacturing and scale-up efforts undertaken over the past 15 years. Finally, it addresses the regulatory aspects of solid forms (pharmaceutical cocrystals), focusing on the perspectives and classifications provided by regulatory agencies regarding their manufacturing.

4.2 Cocrystal Preparation Techniques

Cocrystal preparation approaches can be categorized into three main types: solid-state approach, solution-based approach, and miscellaneous techniques. Solid-state approaches, such as hot-melt extrusion, solvent-assisted grinding, and dry grinding, use little to no solvent and rely on mechanical forces to promote cocrystal formation [6]. Solution-based methods, involving cooling crystallization, solvent drop grinding, antisolvent crystallization, and solvent

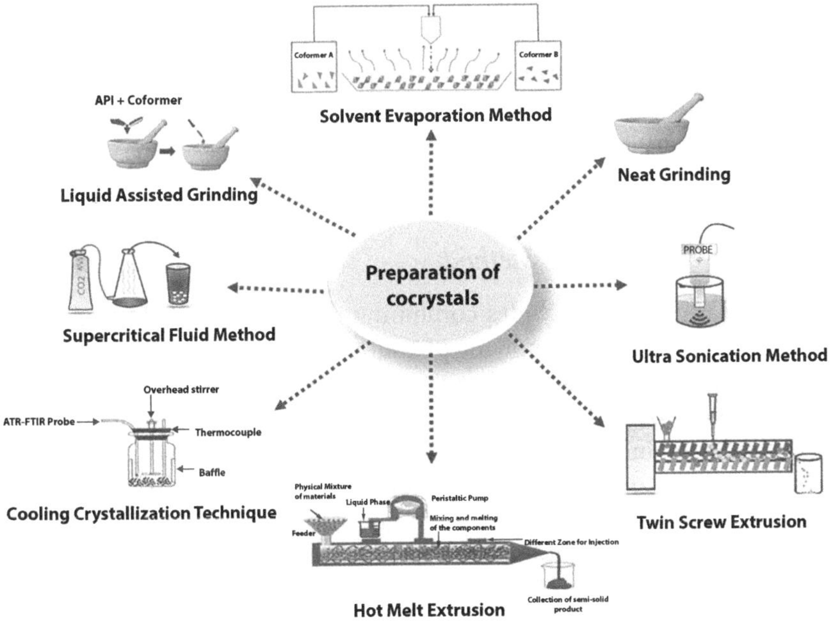

Figure 4.1 Various techniques for the preparation of cocrystals.

evaporation, include using solvents to dissolve and recrystallize the coformer and API (Figure 4.1) [7]. Miscellaneous methods, like supercritical CO_2 crystallization, ultrasound-assisted cocrystallization, spray drying, and twin-screw extrusion, utilize specialized technologies that do not fit strictly into solid or solution-based approaches. These techniques are selected based on the properties of the coformer and API, as well as the desired characteristics and scalability of the novel solid forms [8]. The different techniques used for cocrystal development and various molecules engaged in cocrystal formation in literature have been shown in Figure 4.1 and Table 4.1.

4.2.1 Solid-State Approach

The spontaneous creation of novel solid form (cocrystals) occurs through the combining of pure drug and coformer in a regulated environment, without the application of mechanical forces [4]. However, a short-term pre-milling of the individual constituent before mixing has been employed. Rodriguez-Hornedo and the research team studied that the cocrystallization rate of nicotinamide and carbamazepine was significantly faster when pre-milled reactants were used compared to unmilled reactants [9]. Furthermore, it was observed that elevated temperatures and higher relative humidity significantly enhanced the cocrystallization rate, in the absence of mechanical activation. Similarly in another application, Ibrahim and the team learned the effect of particle sizes on spontaneous cocrystallization of 2-methoxybenzamide and urea, showing that smaller particle sizes resulted in earlier cocrystal formation. In particular, an increment in the cocrystallization rate was observed with particles sized between 20 and 45μm, where no amorphous or eutectic intermediate phase was distinguished [6].

4.2.1.1 Neat-Grinding Method

Dry or neat grinding includes combining the pure API and coformer in their dry solid forms and applying pressure either mechanically (automated equipment like a vibratory mill or a ball mill) or manually with a pestle and mortar. The process typically takes 30 to 60 min, during which particle size is reduced, increasing surface area and enhancing molecular interactions [10]. This simple method allows for the rapid, uniform formation of novel solid forms with higher yields and improved quality. Unlike melt crystallization, neat grinding does not involve melting the solid starting materials [11]. For example, two cocrystals of carboxylic acid and sulfathiazole were successfully prepared by grinding stoichiometric equivalents in a mill (Retsch mixer) for 90 min at a frequency of 25 Hz, ensuring the temperature remained below the body temperature (37°C). The advantage of mechanical grinding over solvent-based approaches is that it eliminates the need for solvent, so there is no loss of yield due to solubility issues [12]. However, challenges include incomplete generation of amorphous content, cocrystal formation, and crystalline defects. Inadequate conversion, leading to a mix of cocrystals and unreacted precursors, is undesirable, as it necessitates additional purification steps [13]. Prolonging the grinding time may sometimes address this issue, but mixed products may also indicate nonstoichiometric cocrystal formation, where excess starting material remains. Typically, neat grinding uses stoichiometric ratios of solids, which requires a stoichiometric cocrystal to fully consume the starting materials [14]. If a nonstoichiometric cocrystal forms, excess material will remain. Therefore, neat grinding should ideally be performed with molar equivalents and excess starting material to facilitate the discovery of alternative cocrystals within the system [15]. The process of neat grinding can be explained through various mechanisms, including the formation of eutectics, amorphous, and molecular diffusion, where liquid, amorphous, or gas, the three intermediate phases, exhibit higher energy or mobility than the initial crystalline forms, aiding the reaction [16].

4.2.1.2 Liquid-Assisted Grinding Method

The second technique for the preparation of cocrystals through grinding is liquid-assisted grinding (LAG), termed wet co-grinding, kneading, or solvent-drop grinding method. This technique involves adding a minor amount of solvent to the dry solid components before the grinding approach initiates [17]. The medium plays a catalyst in the production of cocrystal and remains present throughout the grinding procedure. LAG is generally more efficient than neat grinding, with some evidence indicating that the rate of novel solid forms (cocrystals) increases with the amount of solvent added, although this has not yet been conclusively proven. The solvent accelerates reaction kinetics by wetting the solid surface, thereby enhancing the rate of cocrystallization [18]. Additionally, LAG can influence the polymorphic outcome of cocrystallization, with the choice of solvent playing a significant role in the mechanochemical reaction. Therefore, LAG proves to be an environmentally friendly, cost-effective, and reliable process for both preparing existing ones and discovering new solid forms (cocrystals) [19]. Advantages of this technique include better control over the process, increased production rates, and improved quality of the final cocrystals. However, drawbacks of this method include the small scale of production, high energy consumption, and relatively low performance. Trask *et al.* applied LAG to form maleic acid and caffeine cocrystals, demonstrating that 1:1 or 2:1 cocrystals could be produced within 30 to 60 min, depending on the solvent used. For poorly soluble nutraceuticals like hesperetin, LAG was employed to generate highly water-soluble cocrystals. Chadha *et al.* have used numerous coformers, such as nicotinamide, caffeine, and picolinic acid, to optimize the pharmacokinetic properties including relative bioavailability. The intrinsic dissolution rate studies of the cocrystals in an aqueous buffer observed that the extreme concentration of hesperetin was nearly four- to fivefold higher than that of the drug [20].

4.2.1.3 Hot-Melt Extrusion Approach

HME, hot-melt extrusion, is an operative approach for synthesizing novel solid forms due to its ability to efficiently mix components and enhance surface interactions, all without the need for solvents. In this process, drug molecules and coformers are miscellaneous in a specific stoichiometric ratio and fed into a heated screw extruder. As the materials melt, they are thoroughly mixed, enabling cocrystal nucleation to take place directly within the molten state [21]. The resulting pure cocrystal is continuously extruded and collected. HME is particularly suitable for poorly soluble drugs with high thermodynamic stability. Ideal drug candidates for HME possess low glass transition temperatures, low viscosity, and low melting points, enabling rapid solidification [22]. The technique offers several advantages, including continuous processing, ease of scale-up, and the elimination of organic solvents. It also boasts fast handling times, enhanced conversion efficiency associated with solvent-based methods, and diminished waste, making it well-suited for continuous pharmaceutical production. As a result, HME has to be situated widely used in the manufacturing of pharmaceutical products delivered through transdermal, transmucosal routes, and oral formulations, including capsules, tablets, sustained-release formulations, implants, vaginal rings, and transdermal films [23]. Liu *et al.* selected nicotinamide and carbamazepine as a coformer and drug, respectively, to produce an amorphous nicotinamide and carbamazepine cocrystal solid dispersion using the approach HME. The primary goal was to demonstrate that cocrystallization can be an effective strategy to avoid thermal degradation of heat-sensitive medications. The newly formed cocrystal melted at approximately 160°C, which is 30°C lower than the melting point of carbamazepine. This allowed for the preparation of a nicotinamide and carbamazepine polymer solid dispersion, effectively protecting the heat-sensitive API from thermal degradation *via* the HME method [24]. HME, a continuous manufacturing technique, was employed to produce saccharin with indomethacin cocrystals. The study demonstrated that three critical process parameters (CPP)—screw rotation speed, feeding rate, and temperature settings—are essential for engineering good-quality cocrystals. Additionally, the researchers found that temperature had no noteworthy result on the dissolution of the cocrystals [25].

4.2.2 Solution-Based Method

Solution-based cocrystallization technique follows two main approaches: (1) using solvent mixtures or solvents in which the cocrystal constituents have like solubilities, leading to congruent saturation, or (2) using unequal to modify reactant concentrations to attain the desired results the cocrystal constancy in noncongruently saturating solvents, as confirmed by isothermal TPDs [26]. When the two concentrations exhibit comparable solubilities in a solvent, solution cocrystallization with equimolar quantities will lead to the production of a stoichiometric new solid form *via* liquid evaporation. When two concentrations have unequal solubilities, evaporation of an equimolar solution may produce a single-component crystal instead of a cocrystal [27]. Solution crystallization is a well-established method for single-component systems, making it an appealing option for scaling up cocrystal production without relying on specialized techniques. Additionally, screening for potential cocrystal systems through a combination of solution cocrystallization and grinding methods can significantly broaden the range of explored cocrystal systems. For example, the cocrystallization potential of dicarboxylic acids with caffeine was examined through the solution cocrystallization of L-malic acid used as a coformer and caffeine [28].

4.2.2.1 Solvent Evaporation Approach

Solvent evaporation is a broadly used method intended for formulating the novel solid form (cocrystals), especially for producing high-quality single crystals appropriate for analysis (structural) *via* single-crystal X-ray diffraction [29]. This technique is essential for cocrystal formation, as the functional groups present in the coformer and drug undergo intermolecular interactions, resulting in the creation of novel solid forms. It includes the nucleation and growth of cocrystals from a solution containing both coformers, with supersaturation achieved by gradually evaporating the solvent [30]. The choice of solvent is crucial, as it influences the cocrystallization procedure and affects the reactant's solubility. Ideally, cocrystal constituents should be congruently solvent-soluble. However, if the molecules have incongruent solubilities, the least soluble constituent may precipitate first, resulting in a solid mixture of cocrystals and unreacted materials, or even preventing cocrystal formation altogether. A slow evaporation rate is usually preferred to encourage the growth of larger, higher-quality crystals instead of numerous smaller crystals. In lab-scale applications, evaporative cocrystallization generally arises from solutions (very dilute). Although this method is typically used on a small scale, there is no inherent reason why it could not be applied to larger-scale production of bulk cocrystals. In such cases, the focus would shift from single-crystal formation to optimizing the overall crystallization process [31]. The primary objective is to evaporate the solvent through a solvent evaporation approach, leaving behind crystals with unique physical and chemical properties compared to the pure starting materials. This technique offers advantages like simplicity, ease of scale-up, and high yield. However, its main drawback is the need for large amounts of solvent. A prominent instance of evaporative cocrystallization involves the creation of a block-shaped single crystal of a stoichiometric ratio piroxicam and febuxostat cocrystal, which was formed through interactions involving azole and carboxylic acid synthon. This cocrystal was generated by the gradual evaporation of solvent, i.e., acetonitrile for three to five days at room temperature [37]. The resultant cocrystal demonstrated enhanced solubility and tabletability compared to its individual components. Similarly, Basavoju *et al.* designed the production of single crystals of a cocrystal (isonicotinamide as coformer and norfloxacin) from the solvent (chloroform). In their method, 0.1 mmol of isonicotinamide and norfloxacin was dissolved in 8 mL of chloroform, and slowly at room temperature, it evaporated and produced rod-shaped single crystals. This approach, using stoichiometric solutions (1:1 molar ratio), is common in evaporative cocrystallization. However, relying exclusively on stoichiometric solutions can limit the discovery of cocrystals with unequal API/coformer ratios. Ideally, evaporative cocrystallization should be performed using three different solutions: with a stoichiometric ratio (1:1), excess coformer, and excess target molecule. This strategy not only increases the chances of forming a cocrystal but also helps to identify cocrystals with nonstoichiometric API/coformer ratios when such ratios are possible [32].

4.2.2.2 Slurry Conversion Approach

The slurry conversion technique is a phase change mediated by a solution method in which excess cocrystal constituents are introduced into a solvent. During this process, each constituent slowly dissolves and produces a complex, which aids in the growth and nucleation of cocrystals. As nucleation and growth occur, the reactants become undersaturated (with respect to the reactants), further promoting the dissolution of the components and continuous cocrystallization [29]. However, there are few kinetic studies on isothermal slurry conversion for any cocrystal system. As cocrystals form, the concentration of the reactants decreases, leading to further undersaturation, and allowing the cocrystal components to dissolve progressively. The concentration and temperature ranges for the components are governed by a TPD, which guides the generation of supersaturation needed for cocrystallization. The advantage of this method is its simplicity and minimal solvent requirement [33]. Recently, the slurry conversion method has evolved into a multistep procedure, offering benefits such as faster cocrystal synthesis, increased surface interaction range between the components, and higher molecular diffusivity.

This method was used to patent celecoxib-venlafaxine cocrystals, a multidrug cocrystal that combines the anti-inflammatory properties of celecoxib, beneficial for individuals suffering from long-term inflammatory musculoskeletal disorders, with the antidepressant effect of venlafaxine. Huang *et al.* discovered that the rate of novel solid form (theophylline-benzoic acid cocrystal) was considerably influenced by the initial focus on the molecules and then the effective temperature, as monitored through Raman spectroscopy [34]. A higher concentration enhanced the likelihood of collisions between the molecules and increased the total surface area, whereas elevated temperatures permitted the reactants to quickly reach an activated state, thus accelerating the formation of cocrystals. Similarly, Croker and Rasmuson demonstrated that isothermal slurry conversion could achieve high purity and process efficiency in the toluenesulfonamide:triphenylphosphine oxide cocrystal system. Cheney *et al.* also achieved a 96% cocrystal yield by applying slurry conversion to aspirin and meloxicam in tetrahydrofuran (THF) under ambient conditions. Although this method typically requires more starting materials and can lead to significant material loss *via* residual solubility, its great efficiency makes it one of the most promising screening methods [35].

4.2.2.3 *Cooling Crystallization*

In the crystallization process, the effective province is determined by the stoichiometric ratios of the cocrystal and the thermodynamic stability at both the start and end temperatures. Typically, great quantities of solvent and reactants are dissolved in a jacketed vessel, where the solution is heated to ensure all solutes are completely dissolved, followed by a cooling phase. As the temperature is lowered, the solution develops supersaturated, which causes novel solid forms to precipitate. The cooling crystallization method can be used in conjunction with TPD, i.e., TPDs to depict thermodynamic stability in multicomponent systems and predict cocrystal formation [36]. This approach is widely used for preparing large-scale, purified crystals, where crystal properties such as purity, polymorphism, morphology, and size distribution are established by the controlled local supersaturation by method parameters like heat and mass transfer. For example, novel solid forms (p-hydroxybenzoic acid and caffeine cocrystal) were obtained through this approach. Studies on their intermolecular interactions at various concentration ratios in methanol have shown that understanding these interactions not only enhances cocrystal screening but also acts as a predictive tool and qualitative for the final products, i.e., in crystalline nature. Much research has demonstrated that this technique is effective for scaling up cocrystal production. For instance, a new solid form of nicotinamide with α-lipoic acid was produced on a kilogram scale with a purity of 99% and even particle sizes, employing a continuous oscillatory baffled crystallizer (330 g/h). A 10% w/w suspension for seeding was used to induce cocrystal nucleation. Solvent selection, thermodynamically stable cocrystal operating range, and desupersaturation kinetics were considered in designing the process, which was established on a 1-L scale with a 90% yield and a throughput of 14 L/kg. A similar method was applied by Holan and the group to prepare agomelatine and citric acid cocrystals, where they evaluated the impact of seed quantity and cooling rate on the ultimate crystal size distribution [37].

4.2.2.4 Antisolvent Method

Antisolvent crystallization is an effective approach for regulating the quality, particle size, and properties of cocrystals and can be employed in either semibatch or continuous manufacturing processes. In this technique, the solubility of the cocrystal is reduced by adding an antisolvent, which causes the solution to become supersaturated and results in the precipitation of cocrystals. Selecting the appropriate miscible solvent combination is crucial, as the cocrystal ought to be poorly soluble. The stoichiometric ratio of cosolvent can notably impact the yield of cocrystals because it affects the solubility of both the cocrystal and its constituents [38]. However, this method has some drawbacks, including lower performance compared to solvent-based grinding methods and the requirement for large volumes of solvent. The yield of carbamazepine/saccharin cocrystals was found to be maximized when the volume ratio of water to methanol was 2:1; at ratios below this, carbamazepine hydrates tended to form instead. Additionally, the ratio of coformer to the drug was identified as a critical factor influencing the purity of cocrystal and solid yield. In solutions with lower saccharin concentrations, carbamazepine hydrates were more likely to form as impurities during the antisolvent cocrystallization process. This occurred because saccharin has a higher affinity toward water than carbamazepine, so a lower saccharin content resulted in more water molecules coordinating with carbamazepine, thereby promoting the formation of carbamazepine hydrates [39].

4.2.3 Miscellaneous Approaches

4.2.3.1 Spray Drying and Supercritical CO_2-Assisted Spray Drying

Spray drying is an individual, incessant step procedure that transforms solvents, such as solutions, slurries, and suspensions, into solid powders. This method can yield pure cocrystals from separately saturated systems, unlike the mixes often achieved through the solvent evaporation approach. The process allows for kinetic monitoring and can be intermediated by the material's glassy state; it is highly controllable and efficient. Although the spray drying approach is commonly utilized to create amorphous solid dispersions due to its rapid solidification, it is also effective for cocrystal synthesis. For instance, the spray drying approach has been successfully applied to formulate the novel solid form (carbamazepine with nicotinamide cocrystal) in stoichiometric ratio, and the technique has demonstrated feasibility in industrial settings. Additionally, spray drying can produce cocrystals integrated into a matrix of excipients with better rheological characteristics. Walsh *et al.* showed that spray drying could generate cocrystals of 4-aminosalicylic acid (a coformer) and sulfadimidine when used with a third component excipient [40]. They observed that a larger disparity between the excipient and the cocrystal's Hansen solubility parameters enhances the production of cocrystals. This occurs because the cocrystal components are not miscible with the excipient but still interact and cocrystallize with each other. As a result, the cocrystal phase forms within the matrix excipient. Compared to spray-dried cocrystal systems, the compaction characteristics of these co-spray–dried cocrystal systems were noticeably improved cocrystals without the excipient, due to reduced sticking characteristics [41].

Using supercritical CO_2 as a unique property of supercritical fluids to improve the disintegration of liquid depressurizes and shoots into tiny droplets. This process involves spraying a solution with dissolved cocrystal constituents through a supercritical CO2 nozzle into an atmospheric pressure drying chamber. Theophylline micro- to nanosized cocrystals were created using the supercritical enhanced atomization (SEA) technique with several coformers, effectively improving the cocrystal particle shape and dissolution analysis. The approach was also effectively applied to create micro-composites of novel solid form (theophylline-saccharin cocrystal) dispersed in hydrogenated palm oil for controlled-release formulations [42].

4.2.3.2 Twin-Screw Extruder

Extrusion is a technique used to convert raw materials or blends into uniformly shaped and dense products by forcing them through controlled die parameters. Unlike HME, which occurs at temperatures above the melting points of the materials, twin-screw extrusion functions at temperatures lower than the melting points of the initial materials. This process employs a twin-screw extruder, a specialized machine equipped with two screws rotating in the same or opposite directions within a single barrel. The screw action simultaneously mixes and moves the material along the barrel. In twin-screw extrusion, a mixture of the excipients, active substances, auxiliary agents (such as antioxidants and plasticizers), and thermoplastic polymer carrier, the extruder is heated [43]. At the end of the process, the output is typically a solid form, such as granules. The matrix-assisted cocrystallization approach leverages this technique to formulate cocrystals embedded inside a matrix for formulation. Equimolar quantities of the coformer and API are mixed in solid form with a matrix material before being fed into the extruder. The extruder is established to a temperature at which only the matrix is liquefied, either by forming a soft material or by melting it. The cocrystal particles formed are then incorporated into the matrix, remaining in a fluidized state until they exit the extruder [44]. In this process, the matrix material serves a twin role: (1) during extrusion, it functions similarly to a catalyst by improving mixing and reducing excessive shear stress that could damage the crystalline structure and (2) in the final product, it provides additional functionality, such as enhanced compressibility, flowability, and release kinetics of the drug. The selection of the matrix material is crucial for achieving these benefits and ensuring good-quality cocrystals. The matrix-assisted cocrystallization approach offers simultaneous formulation and production of novel solid forms, i.e., pharmaceutical cocrystals. For example, the twin-screw extruder was used for the cocrystallization of L-tartaric acid with L-malic acid. Daurio *et al.* demonstrated the formation of four model cocrystals theophylline with citric acid, nicotinamide with cinnamic acid, oxalic acid with caffeine, and carbamazepine with saccharin, utilizing a 16-mm twin-screw extruder featuring four temperature-controllable zones. They found that temperature impacts on cocrystal formation were specific to each cocrystal system; for instance, carbamazepine/saccharin showed no temperature dependence, whereas solid form (cocrystal of nicotinamide with cinnamic acid) exhibited a strong temperature dependence. In a different study, a similar research group compared saccharin with AMG-157 cocrystals produced by twin-screw extrusion with those produced by solution crystallization [45].

Table 4.1 Various techniques employed for cocrystal synthesis.

Methods	Drugs	Coformer	Conclusion	Reference
Solvent evaporation	Artemisinin	Orcinol	26-fold increase in solubility	[57]
	Rosuvastatin	L-asparagine L-glutamine	2.17-fold increase in solubility	[58]
	Valsartan	Sacubitril	Improved pharmacokinetics and bioavailability of valsartan	[59]
	Ibuprofen	4,4-dipyridyl nicotinamide	Solubility was increased by 62 times in suspension form	[60]
	Azithromycin	Neotame	<2-fold increase in solubility	[61]
	Acyclovir	Tartaric acid	Higher saturated solubility	
	Lornoxicam	1,3-dimethyl urea	2.52-fold increase in solubility	[62]
	Fluoxetine HCl	Benzoic acid	Enhance dissolution rate	[63]
	Itraconazole	Malic acid Tartaric acid Succinic acid	Enhanced dissolution and solubility rate as amorphous form	[64]
	Piroxicam	Saccharin	Improved dissolution rate and bioavailability	[65]
	Acyclovir	Nicotinamide	Increase in dissolution rate	[66]

(*Continued*)

Table 4.1 Various techniques employed for cocrystal synthesis. (*Continued*)

Methods	Drugs	Coformer	Conclusion	Reference
Liquid-assisting grinding	Indomethacin	Nicotinamide	Enhance solubility	[67]
	5-Fluorouracil	Phenolic acid Nnutraceutical ferulic acid + water	1.64-fold increase in solubility	[68]
	Acyclovir	Tartaric acid	Higher saturated solubility	[68]
	Aceclofenac	Dimethyl urea	3.43-fold increase in solubility	[69]
	Naproxen	L-alanine	1.4-fold increase in solubility	[70]
	Didanosin	Benzoic acid and salicylic acid	Didanosine-benzoic acid cocrystal and didanosine-salicylic acid cocrystal successfully formed	[71]
	Rebamipide	Oxalic acid Citric acid 3,4-dihydroxybenzoic acid	7.29-fold increase in solubility	[72]
	Telmisartan	3,5-dihydroxy benzoic acid	Threefold increase in solubility	[72]
	Bergenin	Isonicotinamide	1.57- to 2.87-fold increase in solubility	[73]
	Axitinib	Fumaric acid Suberic acid Trans-cinnamic acid	11.7-, 5.5-, 1.9-fold increase in solubility respectively	[74]
Neat grinding	Aceclofenac	Dimethyl urea	Acyclovir-glutaric acid cocrystal showed a similar solubility with pure drug, meanwhile acyclovir-fumaric acid cocrystal has a higher solubility compared to pure acyclovir	[69]
	Ipragliflozin	L-prolin	Stability against hydrate formation	[75]

(*Continued*)

Table 4.1 Various techniques employed for cocrystal synthesis. (*Continued*)

Methods	Drugs	Coformer	Conclusion	Reference
	Aceclofenac	Salicylic acid	Sevenfold increase in solubility	[76]
	Acyclovir	Fumaric Acid and glutaric Acid	Cocrystals of acyclovir-fumaric acid have a higher solubility as compared to pure drug	[77]
	Piroxicam	Piroxicam Sodium acetate	Improved the solubility and dissolution rate	[78]
Antisolvent crystallization	Diacerein	β-Resorcylic acid	2.3- to 2.8-fold increase in solubility	[79]
	Paracetamol	Caffeine	Enhance dissolution rate	[80]
Supercritical fluid	Indomethacin	Saccharin	Differences in morphology and particle size distribution	[81]
Hot-melt extrusion	Aripiprazole	Adipic acid	Enhanced solubility and dissolution rate	[82]
Mechanochemical method	5-Fluorouracil	3,4-dihydroxybenzoic 4-aminopyridine and gentisic acid	Improve permeability and bioavailability	[83]
	Dimethylglyoxime	Acridine, phenanthroline monohydrate, and acetamide	Improve solubility and bioavailability	[84]
	Chrysin	Cytosine and thiamine hydrochloride	Enhanced dissolution and solubility profile with improved bioavailability parameters	[85]
Solvent-assisted grinding and Solvent ultrasonic method	Dipfluzine	Benzoic acid	Enhance apparent solubility and dissolution rate	[86]
Slurry method	Quercetin	Caffeine:methanol Isonicotinamide Theobromine dihydrate	Improved solubility and oral bioavailability	[87]

(*Continued*)

Table 4.1 Various techniques employed for cocrystal synthesis. (*Continued*)

Methods	Drugs	Coformer	Conclusion	Reference
	Naringenin	Isonicotinamide, Picolinic acid Betaine	Improved dissolution rate	[88]
	Simvastatin	Aspartame	Improved dissolution rate	[89]
	Tadalafil	Malonic acid	Enhance solubility and dissolution rate	[90]

4.3 Optimization Parameters for Cocrystal Formation

4.3.1 Choice of Coformer

Selecting a coformer consistent with an API is a vital phase in novel solid form design, though it presents a significant challenge during the cocrystal formation process. The USFDA has compiled an inventory of thousands of substances that might potentially serve as coformers for novel solid forms. The coformer, which is the non-API constituent, must be non-toxic and free of side effects. Ideally, non-API should hold generally regarded as safe (GRAS) status. Various techniques, such as lattice energy calculations, thermal analysis, the Kofler contact method, the Cambridge structure database (CSD), and synthon engineering, can be employed to choose an appropriate coformer for synthesizing the novel solid form (cocrystals). When a coformer possesses complementary functional groups, it can bond with the API through supramolecular synthons using intermolecular interactions identified *via* the CSD [46]. This strategy helps prevent undesirable polymorphs, which can have adverse effects on drug performance. Strong hydrogen bond cocrystals have been shown by the Cambridge structure database. These synthons serve as structural units within supermolecules and are assembled through known or potential synthetic processes involving intermolecular interactions. While hydrogen bonds are commonly involved in other intermolecular forces, including halogen bonds, van der Waals forces, π-π interactions in aromatic compounds, and others, because of their intensity and directionality, stacking interactions also contribute significantly to the crystallization process and play an important role in newly solid forms. Strong hydrogen bonds commonly observed in solid materials like cocrystals include N–H•••O, O–H•••N and O–H•••O. Weaker hydrogen bonds involve interactions such as N–H•••π, C–H•••N and C–H•••O. van der Waals forces, which arise from inductive and dispersive attractions between molecules, contribute to the production of solid dispersions [47]. Halogen-containing molecules, i.e., those with fluorine, tend to create halogen connections throughout the process of self-assembly. Desiraju and the research team describe a halogen bond as a net attractive connection that occurs between a nucleophilic region in one molecule or within the same molecule with the electrophilic portion of a halogen atom in another molecule. The therapeutic effects of the drugs are of interest to solid-state chemists, as cocrystallizing them with APIs can improve solubility, physical and chemical properties, stability, and bioavailability. This approach produces valuable, reasonably priced, and easily obtainable synergistic hybrids. Together with their complementary benefits, nutraceutical-API combos may help connect APIs within a multidrug crystal lattice. Flavonoids and vitamins have been explored as potential nutraceutical candidates for cocrystals to improve their physicochemical properties and bioavailability. The selection of coformers typically follows a supramolecular strategy [48].

4.3.2 Kinetics of Cocrystallization

The exploration of nucleation and growth in cocrystals formed through the combination of different stoichiometric ratios of coformers is a fundamental characteristic of cocrystallization. When coformers in solution reach supersaturation, the nucleation of numerous cocrystals competes with the formation of the novel solid form, i.e., cocrystals. The solubility of coformers and cocrystals, along with their metastable regions, influences the relative nucleation rates and growth, which can shift the order of nucleation based on how fast supersaturation occurs. Although TPDs offer data about stable crystal forms, the cocrystallization process is primarily governed by the kinetics of nucleation and growth [27]. Understanding these factors is crucial for controlling cocrystal production and achieving the desired dispersion of sizes. The bare minimum of time needed for batch or residence for pure cocrystal production depends on both growth and dissolution kinetics. Furthermore, adjusting coformer concentrations can affect the growing rate of different crystal faces, which in turn influences crystal morphology. Techniques like heteronuclei or ultrasound can be integrated into cocrystallization processes due to their impact on nucleation kinetics [49]. Understanding the kinetics of both nucleation and growth of cocrystal and conformer crystal is crucial for efficient large-scale production. Although small-scale manufacturing of novel cocrystals can be achieved through a variety of solid-state or solution-based methods, large-scale production calls for a thorough grasp of these kinetics to optimize the process. For example, producing a pure 1:2 cocrystal (such as maleic acid with caffeine) in many solvents is challenging, as both 1:1 and 1:2 forms tend to nucleate simultaneously, resulting in an impure mixture that eventually transitions to the 1:1 form. As a result, most attempts to obtain a pure 1:2 cocrystal focus on rapid kinetic methods. However, recognizing the thermodynamic and kinetic hurdles has led to the development of specialized conditions or steps to produce the metastable cocrystal form. Croker *et al.* investigated the nucleation behavior of p-toluenesulfonamide with triphenylphosphine oxide cocrystals at different stoichiometric proportions throughout a cooling cocrystallization process. By cooling mixes in the area of the stable cocrystal, as predicted by a TPD, they successfully produced a novel solid form. However, in the mixed-phase region, one cocrystal form nucleated first, followed by the other. The proportions of cocrystals in the mixture changed over time, suggesting that either the nucleation did not occur simultaneously or their growth rates varied depending on the coformer combination. Thus, the nucleation sequence is determined by kinetic factors rather than the stability as shown by the phase diagram [50].

4.3.3 Application of Modeling and Process Analytical Technology

The quality-by-design (QbD) technique emphasizes the development of formulations and manufacturing processes to confirm that products meet predefined quality standards. QbD emphasizes the creation of a highly efficient manufacturing process, enabling constant, instantaneous assurance of product worth through comprehensive process comprehension and management. This methodology utilizes thorough knowledge of the process and its variables to ensure that batches consistently meet specifications, minimizing the need for extensive post-production testing and allowing for real-time batch release. Formulations that combine drugs or drugs with coformers to create cocrystals, solid-state co-amorphous formations, crystalline solid dispersions, or amorphous solid dispersions, are increasingly being explored [51]. While these combination strategies offer several benefits, they also present challenges related to physical stability, requiring more rigorous process controls to ensure that the drug remains in its intended solid-state form. Recent advancements have been made in process analytical technology (PAT) modeling, simulation, and application for innovative cocrystallization methods. These advancements allow for the real-time monitoring of essential quality parameters throughout production, promoting consistency and improved product quality. Using PAT modeling also facilitates the application of QbD principles in industrial crystallization. Yu *et al.* created the method of moments that was used to solve a mathematical model for the batch-cooling crystallization of a cocrystal of caffeine and glutaric acid. Three basic equations that addressed the cooling trajectory, one-dimensional population balance, and mass balance were included in the model [52]. Among the presumptions were a growth rate that is independent of size, agglomeration, no nucleation or breakage, and ideal mixing conditions. The model aimed to maintain polymorphic purity by controlling supersaturation. Experimental data were used to estimate growth rates and crystal growth order. Simulations and experiments optimized seeding parameters such as timing, seed loading, size, and cooling profile to prevent unwanted polymorph formation. PATs, like ATR-FTIR for solute concentration and videography for monitoring cocrystal morphology, were employed. Monte Carlo simulations were used to assess model parameter uncertainty and process variability, and experimental data were used to validate the model. It was also scaled from a 1-L vessel to a 10-L crystallizer, demonstrating successful validation at both scales [53].

While modeling aids in understanding and designing processes, ongoing observation of crucial process parameters and crucial quality traits is necessary for effective quality control, which is fundamental to the implementation of QbD. In crystallization, the most frequently monitored attributes include solution composition, solid form morphology, and particle size [54]. Tools such as near-infrared FTIR measurement, ATR-UV/Vis spectroscopy, and ATR-FTIR are utilized for monitoring the composition of the solution. Raman analysis, turbidity measurements, particle vision microscopy (PVM), and FBRM help observe the production and characteristics of the solid form. In a study by Powel *et al.*, batch and semibatch crystallizers were used to study the p-toluenesulfonamide with triphenylphosphine oxide cocrystal in acetonitrile. Using PAT technologies to support a control plan for real-time process monitoring resulted in the successful production of pure 1:1 and 3:2 novel solid forms. Even in mixed-phase regions, the desired crystal form was obtained through the use of thermal cycling and Raman spectroscopy. This demonstrates the critical role of PAT tools in monitoring and controlling crystal transformations during the cocrystallization process. Huang *et al.* utilized Raman spectroscopy to investigate the effects of coformer concentration and temperature on the cocrystallization of theophylline with benzoic acid cocrystals through cooling and slurry crystallization methods. The real-time data obtained from the study also enabled the identification of the nucleation temperature and the cooling process endpoint [55].

Due to its resemblance to conventional crystallization, solution cocrystallization can employ the same PAT tools typically used for single-component crystallization. However, cocrystallization introduces the challenge of differentiating and managing the production of various concurrent solid forms. This means the models of processes and control techniques for cocrystallization may differ significantly. Further research is needed to develop strong models for cocrystallization, focusing on characteristics of the product and process factors that can be monitored using the current PAT tools. Continuous cocrystallization techniques, with close monitoring and real-time adjustments through PAT tools, are expected to be more effective than batch processes in maintaining the quality of the goods. With the creation of accurate models, active control, and real-time monitoring, ongoing cocrystallization could progress toward a quality-by-control (QbC) approach, offering enhanced process efficiency and reliability [56].

4.4 Future Perspectives and Challenges

Investigating cocrystals represents a sustainable and eco-friendly approach, as it avoids complex reactions and minimizes the need for excessive solvent use. Cocrystals, also referred to as multi-component crystals, can be developed using simple methods like grinding, aligning with the green chemistry approach. These structures, consisting of two or more molecular constituents bound by non-covalent interactions, offer unique properties and superior performance compared to their individual constituents. Cocrystals hold significant promise across various sectors, though they also present certain challenges. In the pharmaceutical industry, cocrystals have the potential to enhance drug stability, solubility, and bioavailability, leading to more effective medications with fewer side effects. They can also aid in extending patent protection for drugs by offering novel formulations. In materials science, cocrystals exhibit enhanced physicochemical properties, opening novel possibilities in areas like catalysis, electronics, and energy storage. They could significantly impact technologies like batteries, where better performance and stability are critical.

The preparation and optimization of novel solid forms involve specialized approaches and skills, and scaling up production for industrial use remains a key hurdle. Moreover, the regulatory framework for novel solid forms in the pharmaceutical sector is still evolving, necessitating clear evaluation and approval guidelines. Further research is also needed to understand the principles governing novel solid forms' development and stability. Predictive models and computational tools could assist in designing cocrystals with desirable properties. In conclusion, novel solid forms hold vast potential in industries like materials science and pharmaceuticals. Overcoming obstacles related to preparation, production, regulation, and fundamental understanding will be essential to harness their capabilities fully. Continued research and cooperation between engineers, scientists, and regulatory bodies will drive the development and practical application of novel solid forms in the future. The future of novel solid form development is set to advance through the adoption of cutting-edge technologies that improve both the efficiency and functionality of cocrystallization. Green chemistry techniques, such as using supercritical CO_2 or eco-friendly solvents, offer more sustainable approaches to cocrystal synthesis. Innovations like high-throughput screening and automation are expected to accelerate the discovery of new cocrystals by streamlining the process of identifying suitable coformers and crystallization conditions. The artificial intelligence and integration of machine learning could also revolutionize cocrystal design, predicting molecular interactions and crystallization outcomes, thus reducing reliance on experimental trial and error. There is increasing interest in combining cocrystals with advanced drug delivery systems, such as nanoparticles and polymers, to improve the bioavailability and therapeutic effects of poorly soluble drugs. Regulatory pathways for cocrystals are becoming more defined, facilitating smoother transitions from discovery to market. These advancements promise to expand the use of novel solid forms in pharmaceutical formulations, paving the way for more efficient, targeted, and patient-friendly treatments.

Acknowledgments

I would like to acknowledge the support and resource guidance provided by Chitkara College of Pharmacy, Chitkara University, Rajpura/Chitkara University School of Pharmacy, Chitkara University, Himachal Pradesh, and University Institute of Pharmaceutical Sciences, Panjab University, Chandigarh by for the completion of this book chapter.

References

1. Rathi, R., Guggal, Y., Twinkle, V., Sandhu, Singh, I., Exploring the Patent Landscape and Regulatory Prospective on Pharmaceutical Cocrystals. *J. Pharm. Technol. Res. Manag.*, 11, 49–60, 2023.
2. Singh, M., Barua, H., Jyothi, V.G., Dhondale, M.R., Nambiar, A.G., Agrawal, A.K., Kumar, P., Shastri, N.R., Kumar, D., Cocrystals by design: a rational coformer selection approach for tackling the API problems. *Pharm.*, 15, 4, 1161, 2023.
3. Kansiz, S., Azam, M., Dege, N., Ermis, N., Al-Resayes, S.I., Alam, M., Supramolecular assembly in designing cocrystals of fumaric acid and pyrimidine/picolinate derivatives. *Green Chem. Lett. Rev.*, 15, 3, 825–36, 2022.
4. Patel, D.J. and Puranik, P.K., Pharmaceutical Cocrystal: An Emerging Technique to enhance Physicochemical properties of drugs. *Int. J. ChemTech Res.*, 13, 283–90, 2020.
5. Panzade, P.S. and Shendarkar, G.R., Pharmaceutical cocrystal: an antique and multifaceted approach. *Curr. Drug Deliv.*, 14, 8, 1097–105, 2017.
6. Kumar Bandaru, R., Rout, S.R., Kenguva, G., Gorain, B., Alhakamy, N.A., Kesharwani, P., Dandela, R., Recent advances in pharmaceutical cocrystals: From bench to market. *Front. Pharmacol.*, 12, 780582, 2021.
7. Manchanda, D., Kumar, A., Nanda, A., Recent advancements in pharmaceutical cocrystals, preparation methods, and their applications. *Curr. Pharm. Des.*, 27, 44, 4477–95, 2021.
8. Douroumis, D., Ross, S.A., Nokhodchi, A., Advanced methodologies for cocrystal synthesis. *Adv. Drug Deliv. Rev.*, 117, 178–95, 2017.
9. Karimi-Jafari, M., Padrela, L., Walker, G.M., Croker, D.M., Creating cocrystals: A review of pharmaceutical cocrystal preparation routes and applications. *Cryst. Growth Des.*, 18, 10, 6370–87, 2018.
10. Ibrahim, A.Y., Forbes, R.T., Blagden, N., Spontaneous crystal growth of cocrystals: the contribution of particle size reduction and convection mixing of the coformers. *Cryst.Eng. Comm.*, 13, 4, 1141–52, 2011.
11. Hussain, A., A study of the mechanisms of milling-induced enhancement of solubility and dissolution rate of poorly soluble drugs, 2015.
12. Gajda, M., Nartowski, K.P., Pluta, J., Karolewicz, B., Continuous, one-step synthesis of pharmaceutical cocrystals via hot melt extrusion from neat to matrix-assisted processing–State of the art. *Int. J. Pharm.*, 558, 426–40, 2019.
13. Rathi, R., Kushwaha, R., Goyal, A., Singh, I., Oxaliplatin-flavone pharmaceutical cocrystal-CN111205332A: patent spotlight. *Pharm. Pat. Anal.*, 11, 5, 147–54, 2022.
14. Baláž, P., Achimovičová, M., Baláž, M., Billik, P., Cherkezova-Zheleva, Z., Criado, J.M., Delogu, F., Dutková, E., Gaffet, E., Gotor, F.J., Kumar, R., Hallmarks of mechanochemistry: from nanoparticles to technology. *Chem. Soc. Rev.*, 18, 7571–637, 2013.

15. Niu, X., Yang, R., Zhang, H., Yang, J., Crystal engineering in the development of improved pesticide products. *Adv. Agrochem*, 1, 1, 39–60, 2022.
16. Khan, S., Zahoor, M., Rahman, M.U., Gul, Z., Cocrystals; basic concepts, properties and formation strategies. *Z. Phys. Chem.*, 237, 3, 273–332, 2023.
17. Hasa, D. and Jones, W., Screening for new pharmaceutical solid forms using mechanochemistry: A practical guide. *Adv. Drug Deliv. Rev.*, 117, 147–61, 2017.
18. Rathi, R. and Singh, I., Multicomponent crystal compromising dasatinib and selected cocrystals formers: a patent evaluation of EP2861589B1. *Pharm. Pat. Anal.*, 11, 1, 15–21, 2021.
19. Sakhiya, D.C. and Borkhataria, C.H., A review on advancement of cocrystallization approach and a brief on screening, formulation and characterization of the same. *Heliyon*, 10, 7, e29057 2024.
20. Solares-Briones, M., Coyote-Dotor, G., Páez-Franco, J.C., Zermeño-Ortega, M.R., de la O Contreras, C.M., Canseco-González, D., Avila-Sorrosa, A., Morales-Morales, D., Germán-Acacio, J.M., Mechanochemistry: A green approach in the preparation of pharmaceutical cocrystals. *Pharm.*, 13, 6, 790, 2021.
21. Kumar, P. and Singh, I., Formulation and characterization of tramadol-loaded IPN microgels of alginate and gelatin: Optimization using response surface methodology. *Acta Pharm.*, 60, 3, 295–310, 2010 Sep 1.
22. Rathi, R. and Singh, I., Hot Melt Extrusion Technique for Developing Pharmaceutical Cocrystals A Review. *Drug Deliv. Lett.*, 14, 2, 97–108, 2024.
23. Rathi, R., Kaur, S., Singh, I., A review on cocrystals of Herbal Bioactives for solubility enhancement: preparation methods and characterization techniques. *Cryst. Growth Des.*, 22, 3, 2023–42, 2022.
24. Shen, L., Pang, S., Zhong, M., Sun, Y., Qayum, A., Liu, Y., Rashid, A., Xu, B., Liang, Q., Ma, H., Ren, X., A comprehensive review of ultrasonic assisted extraction (UAE) for bioactive components: Principles, advantages, equipment, and combined technologies. *Ultrason. Sonochem.*, 106646, 2023.
25. Zhou, H., Wang, Y., Li, S., Lu, M., Improving chemical stability of resveratrol in hot melt extrusion based on formation of eutectic with nicotinamide. *Int. J. Pharm.*, 607, 25, 121042, 2021.
26. Rajadhyax, A.R., Shinde, U.J., Desai, H.A., Mane, S.H., Hot melt extrusion in engineering of drug cocrystals: a review. *Asian J. Pharm. Clin. Res.*, 14, 8, 10–9, 2021.
27. Tupe, S.A., Khandagale, S.P., Jadhav, A.B., Pharmaceutical cocrystals: An emerging approach to modulate physicochemical properties of active pharmaceutical ingredients. *J. Drug Deliv. Ther.*, 13, 4, 101–12, 2023.
28. Zalte, A.G. and Saudagar, R.B., Advanced techniques in preparation of cocrystals. *Int. J. Sci. Progr. Res.*, 12, 32–5, 2015.
29. MacEachern, L., Kermanshahi-pour, A., Mirmehrabi, M., Ajiboye, L., Trivedi, V., Rohani, S., He, Q., Cocrystal formation of niclosamide and urea in supercritical CO2 and impact of cosolvent. *J. Supercrit. Fluids*, 201, 106029, 2023.

30. Pando, C., Cabañas, A., Cuadra, I.A., Preparation of pharmaceutical cocrystals through sustainable processes using supercritical carbon dioxide: a review. *RSC Adv.*, 6, 75, 71134–50, 2016.
31. Al Rahal, O., Majumder, M., Spillman, M.J., van de Streek, J., Shankland, K., Cocrystal structures of furosemide: Urea and carbamazepine: Indomethacin determined from powder X-ray diffraction data. *Crys.*, 10, 1, 42, 2020.
32. Zhang, X., Xing, H., Zhao, Y., Ma., Z., Pharmaceutical dispersion techniques for dissolution and bioavailability enhancement of poorly water-soluble drugs. *Pharm.*, 10, 3, 74, 2018.
33. Sekhon, B.S., Surfactants pharmaceutical and medicinal aspects. *J. Pharm. Technol. Res.*, 1, 1, 43–68, 2013.
34. Hiendrawan, S., Veriansyah, B., Tjandrawinata, R.R., Solid-state properties and solubility studies of novel pharmaceutical cocrystal of itraconazole. *Int. J. Appl. Pharm.*, 10, 97–104, 2018.
35. Emami, S., Siahi-Shadbad, M., Adibkia, K., Barzegar-Jalali, M., Recent advances in improving oral drug bioavailability by cocrystals. *BI*, 8, 4, 305, 2018.
36. Jayasankar, A., Reddy, L.S., Bethune, S.J., Rodríguez-Hornedo, N., Role of cocrystal and solution chemistry on the formation and stability of cocrystals with different stoichiometry. *Cryst. Growth Des.*, 9, 2, 889–97, 2009.
37. Biscaia, I.F., Gomes, S.N., Bernardi, L.S., Oliveira, P.R., Obtaining cocrystals by reaction crystallization method Pharmaceutical applications. *Pharma.*, 13, 6, 898, 2021.
38. Saha, A., Ahangar, A.A., Dar, A.A., Thirunahari, S., Parambil, J.V., Pharmaceutical cocrystals: a perspective on development and scale-up of solution cocrystallization. *Cryst. Growth Des.*, 23, 11, 7558–81, 2023.
39. Dhondale, M.R., Thakor, P., Nambiar, A.G., Singh, M., Agrawal, A.K., Shastri, N.R., Kumar, D., Cocrystallization approach to enhance the stability of moisture-sensitive drugs. *Pharma*, 15, 1, 189, 2023.
40. Mohammad, K.A., Wei, M.H., Zainudin, E.N.E., Abd Rahim, S., Impact of the SAC: CBZ ratio on polymorphic transformation and morphology of carbamazepine-saccharin cocrystals using fast cooling crystallization. *J. Therm. Anal. Calorim.*, 15, 1–3, 2024.
41. Bhatia, M. and Devi, S., Cocrystallization: a green approach for the solubility enhancement of poorly soluble drugs. *Cryst. Eng. Comm.*, 26, 3, 293–311, 2024.
42. Shao, S., Stocker, M.W., Zarrella, S., Korter, T.M., Singh, A., Healy, A.M., *In Situ* Cocrystallization via Spray Drying with Polymer as a Strategy to Prevent Cocrystal Dissociation. *Mol. Pharm.*, 20, 9, 4770–85, 2023.
43. Dias, J.L., Rebelatto, E.A., Lanza, M., Ferreira, S.R., Production of quercetin-proline cocrystals by means of supercritical CO2 antisolvent. *Adv. Powder Technol.*, 34, 11, 104222, 2023.

44. Madanayake, S.N., Manipura, A., Thakuria, R., Adassooriya., N.M., Opportunities and challenges in mechanochemical cocrystallization toward scaled-up pharmaceutical manufacturing. *Org. Lett.*, 27, 3, 409–22, 2023.
45. Vasoya, J.M., Lee, H.L., Lee, T., Serajuddin, A.T., Continuous Synthesis of Cinnarizine Salt with Malic Acid by Applying Green Chemistry Using Water-Assisted Twin Screw Extrusion. *Mol. Pharm.*, 20, 10, 5160–72, 2023.
46. Patil, H., SK. Vemula, Lakkala, P., Munnangi, S.R., Narala, N., Jara, M.O., Williams III, R.O., Terefe, H., Repka, M.A., Hot-melt extrusion: from theory to application in pharmaceutical formulation—where are we now? *AAPS PharmSciTech.*, 25, 2, 37, 2024.
47. Chettri, A., Subba, A., Singh, G.P., Bag, P.P., Pharmaceutical cocrystals: A green way to enhance drug stability and solubility for improved therapeutic efficacy. *J. Pharm. Pharmacol.*, 76, 1, 1–2, 2024.
48. Chen, K., Wu, W., Zhao, H., Wei, Y., Zhu, S., Huang, X., Wang, N., Wang, T., Hao, H., Designing Dynamic Crystals with Multiple Flexibilities through a Cocrystal Strategy for Flexible Photo-controlled Switches. *Cryst. Growth Des.*, 23, 12, 8637–45, 2023.
49. Dhibar, M., Chakraborty, S., Basak, S., Pattanayak, P., Chatterjee, T., Ghosh, B., Raafat, M., Abourehab., M.A., Critical analysis and optimization of stoichiometric ratio of drug-coformer on cocrystal design: molecular docking, *in vitro* and *in vivo* assessment. *Pharma*, 16, 2, 284, 2023.
50. Sarangi, S., Remya, P.N., Damodharan, N., Advances in solvent based cocrystallization: Bridging the gap between theory and practice. *J. Drug Deliv. Sci. Technol.*, 28, 105619, 2024.
51. M .Aucamp, and Milne, M., The physical stability of drugs linked to quality-by-design (QbD) and in-process technology (PAT) perspectives. *Eur. J. Pharm. Sci.*, 1, 139, 105057, 2019.
52. Ng, L.H., Ling, J.K., Hadinoto, K., Formulation strategies to improve the stability and handling of oral solid dosage forms of highly hygroscopic pharmaceuticals and nutraceuticals. *Pharma.*, 14, 10, 2015, 2022.
53. Liendo, F., Arduino, M., Deorsola, F.A., Bensaid, S., Nucleation and growth kinetics of CaCO3 crystals in the presence of foreign monovalent ions. *J. Cryst. Growth*, 578, 126406, 2022.
54. Bondi Jr., R.W. and Drennen III, J.K., Quality by Design and the Importance of PAT in QbD. *J. Sep. Sci.*, 10, 195–224, 2011.
55. He, G., Hermanto, M.W., Tjahjono, M., Chow, P.S., Tan, R.B., Garland, M., Comparison of dielectric constant meter with turbidity meter and focused beam reflectance measurement for metastable zone width determination. *Chem. Eng. Res. Des.*, 90, 2, 259–65, 2012.
56. Haskins, M.M. and Zaworotko, M.J., Screening and preparation of cocrystals: A comparative study of mechanochemistry vs slurry methods. *Cryst. Growth Des.*, 9;21, 7, 4141–50, 2021.

57. Kaur, M., Yardley, V., Wang, K., Masania, J., Botana, A., Arroo, R.R., Li, M., Artemisinin cocrystals for bioavailability enhancement. Part 1: Formulation design and role of the polymeric excipient. *Mol. Pharm.*, 18, 12, 4256–71, 2021.
58. Vemuri, V.D. and Lankalapalli, S., Rosuvastatin cocrystals: an attempt to modulate physicochemical parameters. *Futur. J. Pharm. Sci.*, 7, 1–2, 2021.
59. Zhang, Y., Du, X., Wang, H., He, Z., Liu, H., Sacubitril-Valsartan Cocrystal revisited: role of polymer excipients in the formulation. *Drug. Deliv.*, 18, 4, 515–26, 2021.
60. Soares, F.L. and Carneiro, R.L., Green synthesis of ibuprofen–nicotinamide cocrystals and in-line evaluation by Raman spectroscopy. *Cryst. Growth Des.*, 13, 4, 1510–7, 2013.
61. Upadhye, K.P., Dhakate, C.S., Dixit, G.R., Bakhle, S.S., *Development and Evaluation of Taste Masked Azithromycin by CrystEngComm*, .
62. Fatima, K., Bukhari, N.I., Latif, S., Afzal, H., Hussain, A., Shamim, R., Abbas, N., Amelioration of physicochemical, pharmaceutical, and pharmacokinetic properties of lornoxicam by cocrystallization with a novel coformer. *Drug Dev. Ind. Pharm.*, 47, 3, 498–508, 2021.
63. Childs, S.L., Chyall, L.J., Dunlap, J.T., Smolenskaya, V.N., Stahly, B.C., Stahly, G.P., Crystal engineering approach to forming cocrystals of amine hydrochlorides with organic acids. Molecular complexes of fluoxetine hydrochloride with benzoic, succinic, and fumaric acids. *J. Am. Chem. Soc.*, 126, 41, 13335–42, 2004.
64. Machado Cruz, R., Boleslavská, T., Beránek, J., Tieger, E., Twamley, B., Santos-Martinez, M.J., Dammer, O., Tajber, L., Identification and pharmaceutical characterization of a new itraconazole terephthalic acid cocrystal. *Pharma.*, 12, 8, 741, 2020.
65. Chattoraj, S., Shi, L., Chen, M., Alhalaweh, A., Velaga, S., Sun, C.C., Origin of deteriorated crystal plasticity and compaction properties of a 1: 1 cocrystal between piroxicam and saccharin. *Cryst. Growth Des.*, 14, 8, 3864–74, 2014.
66. Nawatila, R., Nuniek, W.A., Siswodihardjo, S., Setyawan, D., Preparation of acyclovir-nicotinamide cocrystal by solvent evaporation technique with variation of solvent. *Asian J. Pharm. Anal.*, 10, 3, 283–7, 2017.
67. Lin, H.L., Zhang, G.C., Huang, Y.T., Lin, S.Y., An investigation of indomethacin–nicotinamide cocrystal formation induced by thermal stress in the solid or liquid state. *J. Pharm. Sci.*, 103, 8, 2386–95, 2014.
68. Kumbhar, P., Kolekar, K., Khot, C., Dabhole, S., Salawi, A., Sabei, F.Y., A . Mohite, K., Mhatre, S., Jha, N.K., Manjappa, A., Cocrystal nanoarchitectonics as an emerging strategy in attenuating cancer: Fundamentals and applications. *J. Control. Release*, 353, 1150–70, 2023.
69. Afzal, H., Abbas, N., Hussain, A., Latif, S., Fatima, K., Arshad, M.S., Bukhari. Physicomechanical, N.I., stability, and pharmacokinetic evaluation of aceclofenac dimethyl urea cocrystals. *AAPS PharmSciTech.*, 22, 1–3, 2021.

70. Tumanova, N., Tumanov, N., Robeyns, K., Filinchuk, Y., Wouters, J., Leyssens, T., Structural insight into cocrystallization with zwitterionic coformers: cocrystals of S-naproxen. *CrystEngComm.*, 16, 35, 8185–96, 2014.
71. Manin, A.N., Voronin, A.P., Drozd, K.V., Manin, N.G., Bauer-Brandl, A., Perlovich, G.L., Cocrystal screening of hydroxybenzamides with benzoic acid derivatives: a comparative study of thermal and solution-based methods. *Eur. J. Pharm. Sci.*, 65, 56–64, 2014.
72. O'Sullivan, A., Long, B., Verma, V., Ryan, K.M., Padrela, L., Solid-state and particle size control of pharmaceutical cocrystals using atomization-based techniques. *Int. J. Pharm.*, 621, 121798, 2022.
73. Liu, H., Lin, H., Zhou, Z., Li, L., Bergenin-isonicotinamide (1: 1) cocrystal with enhanced solubility and investigation of its solubility behavior. *Drug Deliv. Sci. Technol.*, 64, 102556, 2021.
74. Ren, B.Y., Dai, X.L., Wang, J., Wu, C., Lu, T.B., Chen, J.M., Cocrystallization of axitinib with carboxylic acids: preparation, crystal structures and dissolution behavior. *CrystEngComm.*, 23, 32, 5504–15, 2021.
75. Hayashizaki-Someya, Y., Kurosaki, E., Takasu, T., Mitori, H., Yamazaki, S., Koide, K., Takakura, S., Ipragliflozin, an SGLT2 inhibitor, exhibits a prophylactic effect on hepatic steatosis and fibrosis induced by choline-deficient l-amino acid-defined diet in rats. *Eur. J. Pharmacol.*, 754, 19–24, 2015.
76. Pekamwar, S.S. and Kulkarni, D.A., Development and evaluation of bicomponent cocrystals of aceclofenac for efficient drug delivery with enhanced solubility and improved dissolution. *Indian Drugs*, 58, 8, 54–60, 2021.
77. Sarkar, A. and Rohani, S., Cocrystals of acyclovir with promising physicochemical properties. *J. Pharm. Sci.*, 104, 1, 98–105, 2015.
78. Panzade, P., Shendarkar, G., Shaikh, S., Rathi, P.B., Pharmaceutical cocrystal of piroxicam: design, formulation and evaluation. *Adv. Pharm. Bull.*, 7, 3, 399, 2017.
79. Patel, R.D. and Raval, M.K., Formulation of diacerein cocrystal using β-resorcylic acid for improvement of physicomechanical and biopharmaceutical properties. *Org. Process Res. Dev.*, 25, 3, 384–94, 2020.
80. Latif, S., Abbas, N., Hussain, A., Arshad, M.S., Bukhari, N.I., Afzal, H., Riffat, S., Ahmad, Z., Development of paracetamol-caffeine cocrystals to improve compressional, formulation and *in vivo* performance. *Drug Dev. Ind. Pharm.*, 44, 7, 1099–108, 2018.
81. Basavoju, S., Boström, D., Velaga, S.P., Indomethacin–saccharin cocrystal: design, synthesis and preliminary pharmaceutical characterization. *Pharm. Res.*, 25, 530–41, 2008.
82. Butreddy, A., Sarabu, S., Bandari, S., Dumpa, N., Zhang, F., Repka, M.A., Polymer-assisted aripiprazole–adipic acid cocrystals produced by hot melt extrusion techniques. *Cryst. Growth Des.*, 20, 7, 4335–45, 2020.
83. Gautam, M.K., Besan, M., Pandit, D., Mandal, S., Chadha, R., Cocrystal of 5-fluorouracil: characterization and evaluation of biopharmaceutical parameters. *AAPS PharmSciTech.*, 20, 1–7, 2019.

84. Abidi, S.S., Azim, Y., Gupta, A.K., Pradeep, C.P., Mechanochemical synthesis and structural characterization of three novel cocrystals of dimethylglyoxime with N-heterocyclic aromatic compounds and acetamide. *J. Mol. Struct.*, 15, 1150, 103–11, 2017 Dec.
85. Chadha, R., Bhalla, Y., Nandan, A., Chadha, K., Karan, M., Chrysin cocrystals: Characterization and evaluation. *J. Pharm. Biomed. Anal.*, 134, 361–71, 2017.
86. Lin, Y., Yang, H., Yang, C., Wang. Preparation, J., characterization, and evaluation of dipfluzine–benzoic acid cocrystals with improved physicochemical properties. *Pharm. Res.*, 31, 566–78, 2014.
87. Smith, A.J., Kavuru, P., Wojtas, L., Zaworotko, M.J., Shytle, R.D., Cocrystals of quercetin with improved solubility and oral bioavailability. *Mol. Pharm.*, 8, 5, 1867–76, 2011.
88. Luo, C., Liang, W., Chen, X., Wang, J., Deng, Z., Zhang, H., Pharmaceutical cocrystals of naringenin with improved dissolution performance. *CrystEngComm.*, 20, 22, 3025–33, 2018.
89. Kharisma, R.M. and Sopyan, I., Dissolution rate repairing of simvastatin as a new approach in cocrystallization. *Der Pharmacia Lettre.*, 9, 6, 18–27, 2017.
90. Shimpi, M.R., Alhayali, A., Cavanagh, K.L., Rodríguez-Hornedo, N., Velaga, S.P., Tadalafil–malonic acid cocrystal: Physicochemical characterization, pH-solubility, and supersaturation studies. *Cryst. Growth Des.*, 18, 8, 4378–87, 2018.

5

A Brief on Analytical and Characterization Techniques for Evaluation of Pharmaceutical Cocrystals

Suraj R. Chaudhari[1], Shashikant B. Bagade[2], Sonali V. Chaudhari[3] and Atul A. Shirkhedkar[1*]

[1]*Department of Pharmaceutical Chemistry, R. C. Patel Institute of Pharmaceutical Education and Research, Shirpur, Maharashtra, India*
[2]*SVKM's NMIMS, School of Pharmacy & Technology Management, Shirpur, Maharashtra, India*
[3]*Department of Petrochemical Engineering, Dr. Babasaheb Ambedkar Technological University, Lonere, Maharashtra, India*

Abstract
Cocrystals are formed by energetically combining a molecular or ionic therapeutic agent with another substance known as a cocrystal precursor. This precursor remains in a solid state, which holds a significant function. Understanding the properties of cocrystals, including their thermal behavior and energy fluctuations, is really crucial in pharmaceutical and other fields. Research can enhance cocrystal quality; specifically, they are being extensively studied in industries like drug administration. This study is focused on enhancing drug delivery and bioavailability. Furthermore, when investigating solid-state transformations, for instance, polymorphic transitions and amorphization, it provides valuable insights into how cocrystals may evolve gradually over time, impacting their long-term performance and stability, especially during storage. It is essential to employ various sophisticated and advanced characterization techniques to optimize a drug formulation. This chapter offers an overview of vital analytical methods used to characterize cocrystals and to detail and explain strategic approaches employed for evaluating cocrystals. Also, physical parameters involved in the formulation's development phase may be monitored, investigated, and analyzed using these approaches. General knowledge is obtained from exploring these analytical and

**Corresponding author*: shirkhedkar@gmail.com

Inderbir Singh, Subrata Mallick and Ritu Rathi (eds.) Cocrystals in Pharmaceutical Sciences: Design to Applications, (121–152) © 2025 Scrivener Publishing LLC

characterization techniques, which are vital in the development and utilization of cocrystals across various scientific and industrial applications.

***Keywords*:** Active pharmaceutical ingredient, cocrystal, thermal studies, vibrational spectroscopy, x-ray diffraction

5.1 Introduction

In the pursuit of advancing pharmaceutical innovation, the primary focus of novel drug development research is to attain an active pharmaceutical ingredient (API) that embodies optimal physiochemical, pharmaceutical, and biological attributes [1]. It is significant for progress in drug research that APIs have greater attributes than just solubility, bioavailability, and stability. The similarities in particle size, hygroscopicity, and crystallinity are highly significant drug substance properties. When drug properties are improved, treatments work better, and the economy benefits, leading to enhanced patient outcomes, reduced healthcare costs, and increased productivity [2]. Researchers and pharmaceutical companies are keen on modifying APIs, often by incorporating cocrystals. This is done to change the quality of the drug so that it works better with the therapeutic solution and, most importantly, for the person undergoing treatment [1, 3].

Pharmaceutical cocrystals involve an API co-formulated using a cocrystallizing agent. These are solid products, particularly at room temperature (RT), and are usually bonded with noncovalent forces [2].

Due to the limited number of new drug discoveries in past years, pharmaceutical makers and researchers have focused on modifying existing active drug molecules into cocrystals [1–4]. A well-known challenge in developing pharmaceutical formulations is the issue of insoluble drug molecules. Studies have shown that more than 90% of drugs have been approved since 1995, and approximately 40% of newly developed chemical compounds encounter significant challenges due to their poor water solubility. This lack of solubility can lead to issues with drug absorption and bioavailability, ultimately reducing their effectiveness and making it difficult to achieve the desired therapeutic effects [5]. The bioavailability of API taken orally is greatly influenced by their solubility in water. As a result, these drugs may not get absorbed well in the body as they are poorly water-soluble [6, 7]. Therefore, from a pharmaceutical and pharmacological perspective, the main goal is to improve the characteristics of drugs, particularly their solubility [5].

However, cocrystal formation is still highly beneficial, offering significant improvements in solubility that can enhance drug performance and bioavailability, making it a valuable alternative when salt formation is unsuitable [1, 8].

This chapter discusses various analytical methods for evaluating solid-state drug forms, such as cocrystals. Also, it discusses the few analytical approaches implemented to characterize cocrystals. Many methods, such as spectroscopy, crystallography, thermal analysis, microscopy, and other physical techniques, are commonly used or combined to examine how cocrystals are structured and work in solid form [9, 10]. This chapter provides comprehensive information on various commonly utilized methods, encompassing conventional techniques and innovative approaches, to gain insight into cocrystals. We provide a concise overview of these methods, emphasizing their application in gaining insights into the structure and properties of cocrystals. These techniques can also be used to keep following up and exploring the physical characteristics of developing formulations.

A complete analysis of cocrystals has been understood by exploring analytical methodologies such as X-ray powder diffraction (XRPD), single-crystal X-ray diffraction (SCXRD), differential scanning calorimetry (DSC), thermogravimetric analysis (TGA), infrared spectroscopy (IR), nuclear magnetic resonance (NMR) spectroscopy, solid-state nuclear magnetic resonance (ssNMR), scanning electron microscopy (SEM), transmission electron microscopy (TEM), Raman spectroscopy, powder X-ray crystallography, and dynamic vapor sorption significantly contributes to the well-informed plan and application of cocrystals in miscellaneous scientific and industrial domains.

5.2 Analytical Techniques

5.2.1 X-Ray Powder Diffraction (XRPD)

It has been explored to investigate the structural characteristics of crystalline materials. It works by aiming X-rays at a finely ground powder sample and measuring its diffraction pattern. It delivers essential insights regarding the atomic layout inside the matrix. By analyzing the peaks in the pattern, researchers can determine how well the atoms are organized in the crystal structure. The XRPD helps determine the structures of powdered samples, learn about new crystals, and comprehend the transitions between materials' phases. It is utilized for component identification, crystal structure determination, quantitative analysis, stability and polymorphism, interaction research, and phase transitions in various domains, including geology, medicine, and materials science. Herein, the single crystal and powder methods need the diffraction condition met, as Bragg's law suggested [10–12].

$$n\lambda = 2d.\sin\theta \tag{5.1}$$

where "n" is the integer, "d" is the distance relating crystal planes, "θ" is the striking angle, and "λ" is the X-ray frequency wavelength.

Various literature reports consider it the "gold standard" method for identifying and confirming cocrystal forms. Yuta Otsuka and Satoru Goto investigated the effects of mechanochemical synthesis (dry and wet) on mixtures comprised of piroxicam and saccharin. XRPD analysis showed that grinding reduced the crystallinity of both compounds without causing peak shifts or new peaks, indicating no transition to an amorphous state after 4 hours. However, grinding an equimolar mixture resulted in peak broadening and significant intensity reduction, suggesting the amorphization of most particles. The study concludes that while individual grinding did not amorphized the samples, mixed grinding led to amorphization, requiring further thermal analysis to differentiate cocrystallization from individual amorphization [13].

The cocrystal of ketoprofen was made with succinic acid and saccharin conformers in 1:1:1 and 2:1:1 molar ratios and subsequently characterized. The diffraction peak intensity of the cocrystal is reduced compared to the reference standard (RS) owing to a change in crystal structure. In the diffractograms of RS-ketoprofen and the cocrystal compound of interest, an attenuation in strength was recorded at 6.3126°, 18.3094°, and 22.9618°, indicating a structural alteration [14]. In another study, Urszula Maciołek and Iwona Budziak-Wieczorek synthesized and characterized a cocrystal of (−)-epicatechin and barbituric acid. X-ray crystallography was explored to elucidate the structure and intermolecular relationship. Insights from the analysis of the crystal lattice packing and interactions suggested that the molecules were aligned into tapes through O–H···O contacts involving (−)-epicatechin and coformers [15].

5.2.2 Single-Crystal X-Ray Diffraction (SCXRD)

It is helpful to researchers from many different disciplines. Researchers can crystallize the overall crystal structure of cocrystals at the atomic level. This will help determine the specificity of the cocrystal's properties and functions. It will give a clear picture of the molecular structure and their placement pattern, essential for the formation of cocrystals in which a specific shape and function prevail. This technique lets us see where each atom is within the structure, making the atomic arrangement in crystals of cocrystal. Measuring bond length and angle is vital for scientists to understand molecules' stability and geometric arrangement. In addition, the lattice cell size is the fundamental reproducible element of the crystal, another essential detail that SCXRD provides. This provides valuable insights into the structure and geometry of the cocrystal.

Further understanding of the general structure and organization of the cocrystal can be obtained by examining the diffraction pattern, which can reveal the lattice symmetry of the crystal. Researchers can analyze the diffraction pattern and create an electron density map showing the distribution of electrons within the crystal. This map provides a more comprehensive view of the molecular structure. Furthermore, crystal XRD offers knowledge about the molecular scaffolding in the crystal structure, including details about vacancies and intermolecular interactions. This information helps us understand crystal packing. Finally, the method detects anomalies in atomic arrangement and isolates fluctuations in the thermal motion of atoms because it is sensitive to disorder and anisotropy [16–19].

A one-pot, three-component Betti reaction created a cocrystal of 8-hydroxyquinoline and acetone-(2,4-dinitrophenyl) hydrazone. The chemical was analyzed, revealing that noncovalent interactions are the primary forces driving the development of the cocrystal [20].

In another approach, a cocrystal of nateglinide was prepared to address its poor water solubility, significantly limiting its oral bioavailability. Although various techniques exist to enhance drug solubility, each method has limitations. According to SCXRD results, the newly developed crystal material is classified as associated with the P21 monoclinic crystal arrangement, formed from nateglinide and isonicotinamide in a stoichiometric ratio of 2:2 [21]. The slow solvent evaporation technique synthesized a novel energetic cocrystal of trinitrotoluene and pyrene in a 1:1 molar ratio. SCXRD analysis and noncovalent interaction calculations revealed that the donor primarily governs the structure of the trinitrotoluene and pyrene cocrystal–acceptor π-π interactions [22].

In a vital effort aiming to enhance X-ray absorption efficiency in pure luminescent materials, a highly fluorescent dye, 2,5-di(4-pyridyl)thiazolo[5,4-d]thiazole (Py_2TTz), was produced. Following that, cocrystals were produced with 1,4-diiodotetrafluorobenzene (I_2F_4B) and 1,3,5-trifluoro-2,4,6-triiodobenzene (I_3F_3B), yielding Py2TTz-I2F4 and Py2TTz-I3F3. Py2TTz-I2F4 stood out for its outstanding scintillation properties. The structural integrity of these compounds was determined using XRPD and SCXRD [23].

5.2.3 Differential Scanning Calorimetry (DSC)

Researchers extensively used the thermal analysis approach to examine the thermal aspects of components and measure heat flow around physical and chemical changes functionally related to temperature.

In DSC, temperature deviation (ΔT) between the sample and the reference material is evaluated, and the two materials must be subjected to the same temperature followed by a recorded heat flow to maintain a uniform temperature until a thermal event occurs in the sample. DSC thermograms reveal peaks or valleys corresponding to thermal events, and the resulting heat flux is plotted against temperature to create a curve to interpret the data. However, offering advantages in providing information about energy changes associated with phase transitions and high sensitivity for detecting subtle changes in heat flux, DSC has limitations of widespread contact effect. It proves invaluable in examining cocrystals, which are crystalline structures resulting from the combination of diverse molecules. When specifically applied to cocrystals, DSC enables exploring their thermal behavior by measuring heat flow during temperature changes. It determines the most critical points in the transformation, like melting points and various crystal formations. The analysis enables you to know about the structure and corresponding properties. Through DSC, pharmaceuticals address drug delivery problems by analyzing the nature of cocrystal formulations concerning transition temperatures and other thermal properties. Furthermore, DSC helps decode many solid-state transitions, such as polymorphism and amorphization, which usually affect the crystals' overall quantities and stability [24–28].

DSC has been applied to understand the crystalline behavior of poorly water-soluble drugs, primarily in multicomponent solid forms [29], with Garbacz and Wesolowski demonstrating the effectiveness of an extended method incorporating DSC-assisted FTIR and Raman spectroscopy as a tool reliable for the cocrystallization test, a set of 72 binary mixtures consisting of benzodiazepine derivatives, and nine cofactors were used as model materials. DSC supported the analysis by reflecting on the phase transformations that occur during cocrystallization [30]. Garbacz *et al.* confirmed the formation of cocrystals for two benzodiazepines: chlordiazepoxide with p-aminobenzoic acid and lorazepam with nicotinamide, employing DSC analysis. For chlordiazepoxide cocrystals, DSC curves consistently showed an endothermic peak followed by an exothermic peak, with slight variations based on the preparation method and solvent used. Cocrystals prepared by slurry evaporation had melting points at 219.6°C and 221.7°C, while those prepared by liquid-assisted grinding melted at 220.4°C and 223.1°C. The DSC method confirmed cocrystallization and indicated decomposition upon melting. For lorazepam cocrystals, DSC curves consistently showed endothermic peaks above 177°C, with higher melting points for those prepared by slurry evaporation. Cocrystals obtained by liquid-assisted grinding exhibited additional endothermic peaks, with an exothermic effect confirming cocrystallization and decomposition upon melting [31]. In a study by Machado Cruz *et al.*, the cocrystal approach was successfully employed to facilitate the solubility of the antifungal drug itraconazole by crystallizing it with terephthalic acid, marking the first successful conversion of itraconazole into a multicomponent solid phase. DSC analysis revealed that pure itraconazole melts at approximately 166°C, while terephthalic acid melts with sublimation at around 350°C. Thermal investigation of the binary itraconazole–terephthalic acid combination post-processing showed a distinct endothermic peak at 198°C, which was missing in the parent compounds, indicating the melting of the newly formed phase. Additionally, a broad peak at about 310°C for the itraconazole–terephthalic acid 1:1 sample and at around 320°C for the 1:2 systems was detected, suggesting the possible stoichiometry of the cocrystal seems to be 2:1 due to the excess terephthalic acid [32].

5.2.4 Thermogravimetric Analysis (TGA)

It is used to determine the weight variation of a sample concerning temperature and time during the process, all within a controlled environment. The sample is progressively subjected to a programmed temperature ramping process in the TGA, and the results recorded in the form of the weight changes at different temperatures are always made with the greatest attention to precision and accuracy. There are various stages of temperature change, for example, combustion (or oxidation) and melting (or volatilization). The TGA results describe the material's structure, thermal stability, and composition and provide information about the thermal behavior of a sample under study. TGA is the most widely used instrument for analyzing the thermal properties of materials in diverse fields, including chemistry, materials science, and pharmaceuticals. It serves as the backbone of thermo-analytical investigations, and its high-resolution capabilities reveal the complex behavior of a compound subjected to temperature variation. From the existing literature, we can point out three types of TGA.

First is operating TGA, where the sample undergoes a changeful temperature increase, usually linear over time. Another type is isothermal, or the static TGA, during which the sample is kept at a fixed acknowledged proportionality, and the ensuing weight change is observed. Besides that, quantitative TGA, qualitative TGA, a process of heating the sample to a constant weight at different temperature intervals during a continuous increase in temperature, is the other type of TGA [33–36]. TGA determines the thermal stability of the cocrystal by measuring mass variations in response to changes in temperature. This aids in determining the temperature range within which the cocrystal is stable. Thus, the (−)-epicatechin–barbituric acid cocrystal thermal properties were addressed using the TGA-DSC analysis related to the individual parts aimed at exploring stability and phase shifts. (−)-epicatechin and barbituric acid revealed one endothermic peak at 245.04°C and 252.64°C, correspondingly, uniform with the stated melting point. In the DSC analysis of the epicatechin and barbituric acid cocrystal, the endothermic peak at around 121.34°C is attributed to the water/solvent loss. The other at 214.95°C implies the melting point of the cocrystal, this is considerably dissimilar to the characteristics of (−)-epicatechin and the coformer. The variation in melting transitions among the cocrystal and each component highlights a distinct thermal character demonstrating the association of a new solid phase [15]. Two solid pharmaceuticals cocrystals, EFZA (efavirenz with zidovudine) and NEF (efavirenz with nevirapine), have been synthesized by Ngilirabanga *et al.* The purity of each cocrystal was validated using TGA. The TGA curves for both cocrystals had no mass loss before their respective melting temperature ranges (96°C to 102°C for EFZA and 123°C to 127°C for NEF), indicating a lack of solvent. The individual steps of weight loss observed in the curves indicate mass loss due to post-melt decomposition, a result visually confirmed by hot-phase microscopy. The total mass loss that occurred was about 70%, and 93% was recorded in cocrystals of EFZA and NEF, respectively, indicating significant drug degradation [37].

5.2.5 Spectroscopic Methods (Infra-red, Raman, and Nuclear Magnetic Resonance)

Pharmaceutically significant solid chemical substances like cocrystals can be examined using spectroscopic methods such as IR, Raman, and NMR. These approaches allow researchers to uncover the details about the structural characteristics and properties of the materials. This analytical approach proves invaluable in studying and analyzing various pharmaceutical compounds in their solid forms. Vibrational spectra, including mid-infrared and Raman spectra, are like unique fingerprints for molecules, revealing their identity through characteristic bond stretching and bending frequencies. The mid-IR range (4000 to 400 cm^{-1}) and Raman spectra are especially useful in determining the solid-state form of a substance, helping in phase identification, and studying hydrogen bonding in solid materials. This is essential for understanding the arrangement of molecules within cocrystals. When Raman spectroscopy looks at lower wavenumbers (400 to 50 cm^{-1}), it can study how molecules move together, which helps us understand different crystal shapes. Although mid-IR and Raman spectroscopy serve similar purposes, they complement each other as vibrations need specific conditions to be either mid-IR or Raman active, depending on changes in dipole moment or polarizability.

ssNMR is an approach that explains the structure of organic and pharmaceutical substances in a solid form, like crystals or complexes. It checks out tiny bits of powdered material without damaging them. Compared to infrared, Raman spectroscopy, and powder X-ray diffraction, ssNMR usually gives us more detailed information. These benefits are handy when studying pharmaceutical cocrystals. These substances are frequently developed using methods that are not appropriate for studying using X-ray techniques on individual crystals. This methodology aids in the acquisition of pertinent data such as local atomic environment, molecular interactions, and crystal packing arrangement.

Scientists used ssNMR to analyze organic cocrystals and complexes, investigating dipolar correlation experiments (1H-1H, 1H-13C, and 19F-13C) to study hydrogen bonding, molecular connections, and particle dispersion. Their work aimed to determine crystal lattice packing geometry by examining molecular force interactions, though prompt validation of these interactions remains a challenge [38–44]. The different solid-state analytical methods suitable for defining cocrystals (as given in [45] and explained by Johan Wouters and Luc Quéré), as shown in Figure 5.1, are discussed below. Many spectroscopic techniques such as Raman, NMR, FTIR, terahertz (THz) spectroscopy, and many advanced applications of these approaches are put into practice to study the structure of cocrystals and investigate the intermolecular forces between the components of cocrystals [46]. In the research conducted by Maciołek *et al.*, four novel cocrystals of quercetin with N-aromatic derivatives as coformers were fabricated and examined. The FTIR and FT-Raman spectra provided evidence for forming intermolecular O–H...N or N–H...O hydrogen bonds [47]. Multivariate curve resolution (MCR) was employed by Queiroz *et al.* to establish the phase purity of pharmaceutical cocrystals of hydrochlorothiazide with

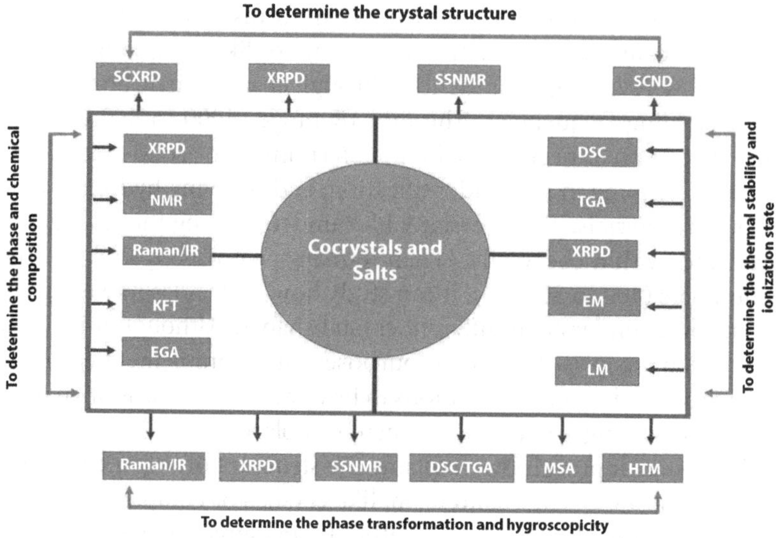

*Single crystal neutron diffraction (SCND), Evolved gas Analysis (EGA), Karl fisher titration (KFT), Moisture sorption analysis (MSA), Hot stage microscopy (HTM), single crystal X-ray diffraction (SCXRPD), powder X-ray diffraction (XRPD), solid-state NMR spectroscopy (SSNMR), Infrared Spectroscopy (IR), Differential Scanning Calorimetry (DSC), Thermogravimetric Analysis (TGA), EM: electron Microscopy, and Light Microscopy (LM)

Figure 5.1 Various analytical approaches explored for the characterization of salts and cocrystals.

nicotinamide from mid-infrared spectra. An *in silico* coformer screening, utilizing quantum chemistry simulations, was conducted to select one coformer from 10 potential candidates. This analysis predicted the thermodynamic propensity of coformers to form cocrystals with the hydrochlorothiazide [48]. The cocrystal of atorvastatin calcium was prepared and characterized using the liquid-assisted grinding method, with citric acid and nicotinamide as coformers, to enhance the model drug's solubility. FTIR spectroscopy confirms the structure of cocrystals and distinguishes the cocrystals from salts. FTIR spectrum of pure atorvastatin calcium was recorded at 3364 cm^{-1} (N-H Stretching), 1650 cm^{-1} (C=O stretching), and 1215 cm^{-1} (C-N stretching). In the proposed work, the FTIR spectrum of optimized atorvastatin calcium–citric acid cocrystals showed a shift in the O–H stretching frequency from 3275 cm^{-1} to 3212 cm^{-1}, indicating hydrogen bonding. At the same time, other significant peaks remained stable, suggesting drug stability despite decreased peak intensity. For atorvastatin calcium–nicotinamide cocrystals, the O–H frequency shifted to 2922 cm^{-1}, indicating hydrogen bonding. The FTIR spectrum retained all other drug absorption peaks, confirming its stability [49]. The oral bioavailability of diflunisal is significantly hindered by its low solubility and enhanced membrane permeability. The low solubility and high membrane permeability significantly impact its oral bioavailability as a nonsteroidal anti-inflammatory drug.

Conversely, pyrazinamide, an essential antituberculosis drug, is associated with several undesired effects. These inadequacies restrict usage. A novel solid-phase drug-drug cocrystal of diflunisal and pyrazinamide in a 1:1 molar ratio was formulated using the mechanical grinding methodology. The terahertz absorption features of diflunisal and pyrazinamide, with their combination, and drug-drug cocrystals developed from the liquid-assisted grinding process in the observed spectral interval of 0.2 to 1.6 THz. The terahertz experimental outcomes exhibited that the fundamental reagents successfully form the cocrystalline structure with the assistance of noncovalent interactions. The primary instigators for this formation are strong intermolecular hydrogen bonds. Raman spectroscopic results confirmed that the complexity of diflunisal and pyrazinamide drug cocrystals is mainly driven by hydrogen bonding interactions, which are essential for cocrystal formation [50].

5.3 Physicochemical Characterization of Cocrystals

5.3.1 Assessment of Solubility and Dissolution Characteristics

During drug development, dissolution is essential, and appropriate dissolution methods must be selected to check drug release. The main factor affecting drug absorption is the extent of solubility of the API. The new dosage forms are required to improve the solubility of drugs that do not dissolve easily. This is important because solubility and permeability affect the drug's bioavailability. In general, when administered orally, drugs are made more soluble to improve their bioavailability. The primary goal in exploring cocrystals is to increase the dissolution profile of poorly soluble compounds. Certainly, cocrystals of neutral molecules might expand the range of solid forms available for development, whereas cocrystals of free acids or free bases may enhance the dissolution profile [1]. Solubility and dissolution investigations characterize material behavior and predict *in vivo* performance. There are two significant ways to measure solubility. Thermodynamic equilibrium techniques offer insights into cocrystal solubility and the origins of solution behavior under varying conditions. Kinetic approaches like phase solubility studies provide details on the time-dependent aspects of these processes. Cocrystal solubility can be determined by identifying the transition or eutectic point at the equilibrium between the cocrystal, drug, or coformer solid phases and solution phase [51–53].

5.3.2 Cocrystal Stability

Drug stability is a crucial characteristic after solubility and can be modified with the help of appropriate cocrystals. The stability of cocrystals can be evaluated in terms of many factors, such as thermodynamic stability, kinetic stability, environmental stability, and mechanical stability. Several factors affect the stability of cocrystals [1]. First, the stability of crystal lattices is greatly affected by strong and special intermolecular interactions. Second, molecules' crystal structure or arrangement within the crystal lattice affects stability; immaculate, closely packed structures are more stable. Third, choosing a coformer is very important. It must form a stable network with the API while still being nontoxic and pharmaceutically acceptable. Fourth, polymorphism is vital because multiple crystal forms can lead to different stabilities between the polymorphs. Fifth, the solubility and behavior of cocrystals in solution affect their stability, especially in aqueous conditions, where more soluble cocrystals can separate faster. Finally, the manufacturing process involves conditions during cocrystal formation. Subsequent processing, such as drying and grinding, can affect stability. Solid-state stability of crystals of drugs is studied using direct and indirect methods, while solution stability can be evaluated using solubility equilibriums, kinetic dissolution, and internal dissolution rate testing. Hirshfeld surface analysis can also be useful in evaluating stability, especially in the case of cocrystals. The most common procedures for assessing the physical stability of pharmaceutical solids are DSC and TGA, detailed in sections 2.2 and 2.3. Thermal stress detection is frequently included, especially in the case of hydrates/solvents and thermally unstable materials. Physical stress stability studies aim to provide evidence of how different environmental conditions affect the physical quality of pharmaceutical compounds. Moisture absorption studies are a critical evaluation of crystals, especially in pharmaceutical formulations, as they reveal how they interact with moisture and how this interaction affects their stability, physical properties, and performance. Moisture absorption can cause variations in the physical form of the crystal, such as melting or wetting, and these changes can affect crystal stability and shelf life [54–57].

5.3.3 Discussion on Exploration of Analytical Techniques for Characterization of Cocrystals

Many drug cocrystals have been analyzed in the literature using different techniques to evaluate their applicability for routine use compared to conventional dosage forms. A concise overview of the documented reports is discussed in the subsequent paragraphs. The details regarding the characterization of pharmaceutical cocrystals by DSC, FTIR, and solubility are presented in Table 5.1.

Table 5.1 The details regarding the characterization of pharmaceutical cocrystals by DSC, FTIR, and solubility.

Sr. no.	Analytes	Conformers	DSC [°C]	FTIR	Solubility [mg/mL]	Ref.
1.	Gliclazide	Catechol	122	1687 cm^{-1} (red-shift of C=O stretching)	7113 mg/mL	[58]
		Resorcinol	117	1688.5 cm^{-1} (red-shift of C=O stretching)	4130 mg/mL	
		p-Toluenesulfonic acid	178	1741 cm^{-1} (blue-shift of C=O stretching)	3129 mg/mL	
		Piperazine	157	1595.5 cm^{-1} (red-shift of C=O stretching)	7710 mg/mL	
2.	Tolbutamide	Piperazine (I) Piperazine (II)	113 104	1588 cm^{-1} (C=O stretching) 1587 cm^{-1} (C=O stretching)	331896 mg/mL 9987 mg/mL	
3.	Glipizide	Piperazine	178 Wide peak at 60	1668 cm^{-1} (C=O stretching) 1579 cm^{-1} (C=O stretching)	-	
4.	Diacerein	Urea	137.5	1678.07 cm^{-1} (C=O stretching)	21.0 µg/mL	[59]
		Tartaric acid	174.4	1693.5 cm^{-1} (C=O l stretching)	13.125 µg/mL	

(*Continued*)

Table 5.1 The details regarding the characterization of pharmaceutical cocrystals by DSC, FTIR, and solubility. (*Continued*)

Sr. no.	Analytes	Conformers	DSC [°C]	FTIR	Solubility [mg/mL]	Ref.
5.	Flufenamic acid	2-chloro-4-nitrobenzoic acid	143.93	1670 cm^{-1} (C=O stretching)	266.0 (x 5.47)	[60]
		Ethenzamide	97.28	1668 cm^{-1} (blue-shift of C=O stretching)	253.0 (x 5.47)	
6.	Glipizide	Oxalic acids	173.26	1654.98 cm^{-1} (C=O stretching)	15.1 µg/mL	[61]
		Malonic acids	174.46	1653.05 cm^{-1} (C=O stretching)	30.38 µg/mL	
		Stearic acids	169.18	1651.12 cm^{-1} (C=O stretching)	40.5 µg/mL	
		Benzoic acids	176.13	1653.05 cm^{-1} (C=O stretching)	17.55 µg/mL	
7.	Ticagrelor	L-Tartaric acid	131.99	1741 cm^{-1} (C=O stretching)	6.8 µg/mL	[62]
8.	Quercetin	Succinic acid	251.3	1830 cm^{-1} (C=O stretching) 3424 cm^{-1} (-OH)	-	[63]
9.	Etraverine	Tartaric acid	77.89 208.82	1741 cm^{-1} (C=O stretching)	3.6 fold increase	[65]
10.	Simvastatin	Nicotinamide	105.7	NA	30 µg/mL	[66]
11.	Aripiprazole	Orcinol	184-185	NA	NA	[67]
12.	Phenylacetic acid	Acridine	452.7	1698 cm^{-1} (C=O band)	-	[68]
		Caffeine	359.9	1700, 1678 and 1644 cm^{-1} (C=O band)	-	

(*Continued*)

Table 5.1 The details regarding the characterization of pharmaceutical cocrystals by DSC, FTIR, and solubility. (*Continued*)

Sr. no.	Analytes	Conformers	DSC [°C]	FTIR	Solubility [mg/mL]	Ref.
		Isonicotinamide	363.1	3369 and 3158 cm^{-1} (H-bonded NH_2 stretching modes)	-	
		Nicotinamide	344.5	3369 and 3158 cm^{-1} (H-bonded NH_2 stretching modes)	-	
13.	Gliclazide	Sebacic acid	153.62	1714 (C=O band stretching) 3297 (-OH- carboxylic group)	5.12 mg/mL	[69]
		α-hydroxyacetic acid	148.86	1716 cm^{-1} (C=O band stretching) 3285 cm^{-1} (-OH- alcohol)	6.35 mg/mL	
14.	Tramadol HCl	Celecoxib	164	3481.6 cm^{-1} (-OH streatching)	-	[71]
15.	Nevirapine	Maleic Acid,	-	1640.58 cm^{-1} (C=O band stretching) 1694.10 cm^{-1}(C=O band stretching)	-	[72]
		Salicylic Acid	-	-	-	
		Glutaric Acid	-	1638.15 cm^{-1} and 1719.23 cm^{-1} (C=O band stretching)	-	
		Rac-Tartaric Acid	-	1637.25 cm^{-1} and 1708.50 cm^{-1} C=O band stretching)	-	

(*Continued*)

Table 5.1 The details regarding the characterization of pharmaceutical cocrystals by DSC, FTIR, and solubility. (*Continued*)

Sr. no.	Analytes	Conformers	DSC [°C]	FTIR	Solubility [mg/mL]	Ref.
		Saccharin	-	-	-	
16.	Febuxostat	Isonicotinamide	166.8	3299 cm^{-1} (N-H stretch) 2968 cm^{-1} (O-H stretch) 2221 cm^{-1} (C≡N stretch) 1694 cm^{-1} (carboxylic C=O stretch)	13.1 mg/mL	[73]
		Arginine	199.3	3368 cm^{-1} (N-H stretch) 2963 cm^{-1} (O-H stretch) 2229 cm^{-1} (C≡N stretch) 1644 cm^{-1} (carboxylic C=O stretch)	570.8 mg/mL	
17.	Carbamazepine	Nicotinamide	-	-	1.95	[74]
		Saccharin	-	-	0.99	
		Succinic Acid	-	-	0.55	
		fumaric acid	-	-	1.24	
18.	Ivabradine hydrochloride	(S)-mandelic acid	160.2	3500 cm^{-1} (secondary alcohol) 2500 cm^{-1} (N–H+ functional group)	-	[77]
		(*R*)-mandelic acid	131.5	1450 cm^{-1} (C-H deformation vibrations)	-	

(*Continued*)

Table 5.1 The details regarding the characterization of pharmaceutical cocrystals by DSC, FTIR, and solubility. (*Continued*)

Sr. no.	Analytes	Conformers	DSC [°C]	FTIR	Solubility [mg/mL]	Ref.
19.	Ticagrelor	Nicotinamide	122.4	1704 cm^{-1}(C=O band stretching) 1094 cm^{-1} (hydrogen bond interaction N–H…O)	pH 2 acidic medium significant solubliity	[79]
		Nicotinamide hydrate	120.5			
20.	Sulfaguanidine	thiobarbutaric acid	170	3340 cm^{-1} (NH2) stretching 1176 cm^{-1} sulfonyl group 3191 cm^{-1} NH-O interaction	-	[80]
		1,10-phenanthroline	217	3340 cm^{-1} (NH2) stretching 1231 cm^{-1} sulfonyl group 1420 cm^{-1} (C=N) 1620 cm^{-1} (C=C)	-	
21.	Zonisamide	179caffeine	179	-	0.7418 mg/mL	[81]
		5-methyl-2-pyridone	148.2		-	
		L-proline	157.7		0.2758 mg/mL	
		catechol	107		0.7924 mg/mL	
22.	Ketoconazole	oxalic acid	186	1686 cm^{-1} (C=O)	22.34 mg/mL	[82]
		fumaric acid	174	1730.80 cm^{-1} (C=O)	43.92 mg/mL	
		nicotinamide	136	1626 cm^{-1} (C=O)	33.42 mg/mL	

(*Continued*)

Table 5.1 The details regarding the characterization of pharmaceutical cocrystals by DSC, FTIR, and solubility. (*Continued*)

Sr. no.	Analytes	Conformers	DSC [°C]	FTIR	Solubility [mg/mL]	Ref.
23.	Dapsone	caffeine	176.6	3372 and 3356 cm^{-1} (Aromatic Amino group) 1283, 1136, and 1099 cm^{-1} (sulfonyl group)	506.5 μg/mL	[83]
24.	Indole-3-acetic acid	2-aminopyrimidine	122-126	3318 cm^{-1} (NH) stretching 3173 cm^{-1} (O-H) stretching 1687 cm^{-1} (C=O) stretching	-	[84]
		2-amino-4,6-dimethylpyrimidine	151-153	3319 cm^{-1} (NH) stretching 3142 cm^{-1} (O-H) stretching 1664 cm^{-1} (C=O) stretching	--	
		2,4,6-triaminopyrimidine	249-251	1631 cm^{-1} (carboxylate group) 1334 cm^{-1} (proton transfer x-ray crystal structure)	-	
25.	Indole-3-butyric acid	2-amino-4,6-dimethylpyrimidine	151-153	3340 cm^{-1} (NH) stretching 3164 cm^{-1} (O-H) stretching 1671 cm^{-1} (C=O) stretching	-	
26.	Riluzole	ferulic acid	159.4	3279 cm^{-1} (N-H) stretching 1606.5 cm^{-1} (N-H) bending	-	[85]

(*Continued*)

Table 5.1 The details regarding the characterization of pharmaceutical cocrystals by DSC, FTIR, and solubility. (*Continued*)

Sr. no.	Analytes	Conformers	DSC [°C]	FTIR	Solubility [mg/mL]	Ref.
		syringic acid	142.4		-	
		fumaric acid,	163.1		-	
		cinnamic acid	145.5		-	
		2,4- dihydroxy benzoic acid	157.3		-	
		proline	206.9		-	
		vanillic acid	139.7		-	
27.	Tadalafil	Malonic Acid	192.7		-	[86]
28.	Zoledronic Acid	DL-tartaric acid Cocrystal-I Cocrystal-II Cocrystal-III	 130.40 121.14 118.83	1739 cm^{-1} (C=O) ketone 3574.10 cm^{-1} (-OH)	 6.74 mg/mL 8.03 mg/mL 29.06 mg/mL	[87]
		nicotinamide	131.76	1681.93 cm^{-1} (C=O) amide group 3574.10 cm^{-1} (-OH)	97.19 mg/mL	
29.	Salicylic acid	3-Nitrobenzamide	114.5 115.6	-	-	[88]
30.	Valsartan	succinic acid	93.63	2922.24 cm^{-1} (C-H) stretching 1552.28 cm^{-1} (C-N) stratching 1733.94 cm^{-1} (C=O) stretching 1472.80 cm^{-1} (C=C) stretching	472.58 ± 0.25 μg/mL	[89]
		Fumaric acid	94.84		383.11 ± 0.23 μg/mL	
		Oxalic acid	95.05		355.35 ± 0.24 μg/mL	

(*Continued*)

Table 5.1 The details regarding the characterization of pharmaceutical cocrystals by DSC, FTIR, and solubility. (*Continued*)

Sr. no.	Analytes	Conformers	DSC [°C]	FTIR	Solubility [mg/mL]	Ref.
31.	Meloxicam	salicylic acid	207	1653 cm^{-1} (C=O) stretching absent 3290 cm^{-1} (N-H) stretching absent SO_2 change from 1263 cm^{-1} to 1277 cm^{-1}	Similar solubility like drug	[90]
		fumaric acid	243	1662 cm^{-1} (C=O) stretching 3290 cm^{-1} (N-H) stretching	higher solubility	
		malic acid	211 226	3290 cm^{-1} (O-H) absent from drug 3442 cm^{-1} (O-H) absent from conformers	Similar solubility like drug	

*Differential Scanning Calorimetry: DSC, Thermal Gravimetric Analysis: TGA, Fourier Transform Infra-Red Spectroscopy: FTIR, Powder X-Ray Diffraction: Single Crystal X-ray Diffraction: PXRD Single Crystal X-ray Diffraction: SCXRD, hydrochloric acid: HCl.

The authors, Ali Samie *et al.*, made several new salts with three essential medicine chemical components for an antidiabetic medicament (gliclazide, glipizide, and tolbutamide). The salts were based on CAC (catechol), REC (resorcinol), piperazine (PPZ), and p-toluenesulfonic acid (PTSA). The findings established that PPZ's salt generated had remarkable solubility [58]. Thus, in the work of Thenge *et al.*, the design and characterization of the diacerein cocrystals with molecules such as urea and tartaric acid were carried out, and the conformer form was established. The expert team noticed that their compound had a phenomenon of rapid degradation, which was more than the typical compound [59]. Ionic cocrystals (ICCs) were prepared by Shemchuk *et al.* using grinding and mixing in a solution. These crystals included L-histidine, DL-histidine, and calcium salts such as calcium chloride, bromide, and iodide. X-ray diffraction techniques using either powder or single-crystal analysis were used to characterize the structure. For data collection and structure solution utilizing SCXRD, ICCs produced by solvent evaporation worked well; in all other cases, PXRD structure solution was successfully used [64]. A novel cocrystal involving aripiprazole (ARI) with orcinol as a coformer was synthesized, characterized, and compared to other ARI cocrystals using dihydroxy- and trihydroxy-benzene coformers (catechol, resorcinol, and phloroglucinol) were made by Cho *et al.* The exact structure of the ARI-ORC cocrystal was figured out using a method called SCXRD. Surprisingly, the cocrystal powder exhibited no detectable degradation after exposure to a temperature of 80°C and 98% (relative humidity) for 1 week. It matches with its relatively high melting point [67]. In the present study, conducted by researchers Pan *et al.*, two polymorphs (α and β forms) of azelnidipine (AZD) were developed.

Additionally, 12 complexes of AZD with oxalic acid became synthesized with AZD with oxalic acid molar ratios of 2:1, 1:1, and 1:2. The confirmation of the formation of 2 cocrystals and 10 co-amorphous structures was achieved through XRD, TMDSC, and DSC-TG. Further insights into intermolecular connections were explored using NMR and FTIR. In different media, the solubility and the dissolution rate in 0.1 mol·$^{L-1}$ HCl of the complexes were also investigated as part of the study [70]. Almansa *et al.* prepared a new kind of cocrystal by mixing rac-tramadol hydrochloride and celecoxib in equal amounts. This cocrystal, showcasing four distinct mechanisms of analgesic action, maintains a 1:1 molecular ratio corresponding to a 1:1.27 weight ratio, ensuring that a dose (100 mg) comprised of rac-tramadol hydrochloride (44 mg) and celecoxib (56 mg), aligning with typical clinical doses. The developed ionic cocrystal was thoroughly characterized, revealing a well-structured DSC profile and a supramolecular 3D structure. The three components were associated with hydrogen bonding, with chloride ions forming a crucial intermolecular interaction with proximal drug molecules [71] and Mohammad Khairool Azizul and colleagues investigated the metastable zone width and nucleation order of carbamazepine–saccharin cocrystals through slow cooling crystallization. Their study employed the Kashchiev–Borissova–Hammond–Roberts method to determine kinetic parameters, including induction time, critical nucleus radius, and the interfacial energy of the cocrystals [75].

Andrea Perrin *et al.* investigated various cocrystals formed from bis(2-pyrrolidone) and ε-caprolactam in combination with boric acid and 1,4-phenylenediboronic acid. The substantial polarity of the lactam carbonyl group demonstrated a marked affinity for the hydrogen bond donor properties of boronic acid. These cocrystals exhibited cyclic boronic acid tape motifs, with an uncommon donor-donor pattern defined by the robust perpendicular interaction with lactam attributes. Notably, all cocrystals can serve as contributors to boric or boronic acid cross-linkers in the process of guar gelation [76].

The researchers tested how helpful multivariate statistical techniques were in detecting cocrystal formation, using indomethacin with saccharin and furosemide with p-aminobenzoic acid as examples. These samples were made at different ratios of the API and the coformer (1:1, 2:1, and 1:2). The researchers used data from DSC curves, FTIR, and Raman spectroscopy to analyze cocrystals (CA and PCA). They found that comparing cocrystals with physical mixtures gave a clearer picture of how cocrystallization works compared to only looking at the drug molecule with the coformer or the drug molecule with physical mixtures alone. They also observed that the results of these analyses could differ depending on the instrument used, and the calculation method employed [78]. Aitipamula *et al.* developed four cocrystals of zonisamide, an antiepileptic drug. They used different conformers (L-proline, catechol, 5-methyl-2-pyridone, and caffeine), and cocrystal screening was conducted using solvent-evaporation and solid-state grinding techniques. Initial characterization with DSC and TGA showed no weight loss, indicating the solid surface's absence of solvates/hydrates or solvent/moisture absorption.

Further analysis with PXRD confirmed the presence of unique peaks, confirming cocrystal formation. The powder obtained from grinding was used for subsequent experiments. The drug-drug cocrystal with caffeine exhibited promise for developing a sustained-release formulation of zonisamide. It demonstrated stability under various conditions and presented a lower dissolution rate and solubility than the original zonisamide [81]. L.H. do Amaral *et al.* developed the cocrystal of dapsone with caffeine [83]. Here, the cocrystals were prepared using large-scale preparation techniques, which were explored for oral inhaled powder form with postulated cocrystals. A (1:1) M cocrystal ratio could be obtained by slow evaporation from acetone and ethyl acetate. A range of mixtures, including acetone, ethyl acetate, and ethanol, possessed cocrystals using the said methods. The samples prepared were characterized utilizing XPRD, DSC, TGA, and FTIR. The median particle sizes (D50) obtained through laser diffraction analysis for spray-dried powders made with acetone, ethanol, and ethyl acetate were 5.4 ± 0.7, 5.2 ± 0.1, and 5.1 ± 0.0 μm, correspondingly, making them suitable for use with dry powder inhalers for pulmonary delivery.

The investigation carried out by Zhou *et al.* revealed a rare type of cocrystal involving salicylic acid and 3-nitrobenzamide (SA-3NBZ which is formed by thermal means). Calculations based on the theory were carried out to explain the examined structure. The DSC and hot-stage microscopy were utilized to identify polymorphic cocrystals of salicylic acid and 3-nitrobenzamide in 1:1:2 molar ratios observed by SCXRD and their confirmation. The ability to stabilize the energy of the polymorphic cocrystals of the salicylic acid was studied experimentally using the density functional theory with the dispersion correction. A weak stabilization effect was observed in both 1:1 and 2:2 cocrystals, resulting in an energy decline (-0.01 and -0.05 eV per molecule), correspondingly, compared to the initial materials.

Furthermore, analysis of the noncovalent interactions index also suggested that these weak interactions were influenced by intralayer hydrogen bonds and van der Waals forces [88]. Similarly, three new ethenzamide cocrystals with glutaric, maleic, and malonic acids were synthesized by A. Kozak *et al.* The main goal was to analyze changes in the ethenzamide molecule geometry and interactions induced by different acids in its crystalline state and correlate these with the physicochemical properties of pure ethenzamide [91]. Wang *et al.* showed an advanced approach in their research. They used both Raman and terahertz time-domain spectroscopic approaches to analyze the structure and vibrational modes of a pharmaceutical cocrystal made of acetazolamide and 4-hydroxybenzoic acid. Both spectra provided unique fingerprinting information for the cocrystal and its parent materials. The experiment observations showed distinct peaks (absorption) at 0.82, 1.16, 1.28, and 1.64 THz, which were different from the prime substances, within the frequency range of 0.2 to 1.8 THz. These differences were also evident in Raman vibrational spectra. Using density functional theory, three theoretical cocrystal forms were simulated, and the form with C(N)NH···HOOC interactions demonstrated a higher consistency with investigational consequences. This study highlights the effectiveness of proposed methodologies in elucidating intermolecular hydrogen bonding interactions crucial for pharmaceutical cocrystallization [92].

5.4 Conclusion

Understanding how drug molecules pack together (their structure) is critical to unlocking valuable information about their behavior. This includes forms like crystals (polymorphs, solvates, and cocrystals) or formless solids (amorphous). To achieve a complete picture, scientists need to combine various analytical tools. Existing methods effectively analyze cocrystals, revealing detailed information about their structure, stability, and how they behave as drugs. The future holds promise for even deeper insights. The development of *in situ* and real-time characterization techniques is crucial. Current methods typically analyze samples post-formation. Techniques that monitor the cocrystal formation and transformation in real time will offer deeper insights into the mechanisms and enable better process optimization.

Data science and machine learning can also analyze vast datasets to predict successful cocrystal formation and accelerate discovery. Furthermore, researchers are exploring new types of cocrystal formers to create cocrystals with even more desirable properties. However, there are still knowledge gaps to address. We need a clearer understanding of how a cocrystal's structure directly affects its performance, such as how easily it dissolves. Predicting a cocrystal's long-term stability is also crucial for its use as a drug. Finally, scaling up cocrystal production from research labs to commercial manufacturing requires further development of efficient methods. Hence, scaling up production for commercial use of cocrystal remains challenging. Developing scalable and cost-effective processes is essential for broader pharmaceutical applications. By tackling these challenges, scientists can unlock the full potential of cocrystals and deliver improved pharmaceutical products.

Acknowledgment

Our sincere thanks to the principal of RCPIPER, Shirpur, India, for support in providing library resources.

References

1. Pindelska, E., Sokal, A., Kolodziejski, W., Pharmaceutical cocrystals, salts and polymorphs: Advanced characterization techniques. *Adv. Drug Deliv. Rev.*, 117, 111–146, 2017.
2. Vioglio, P.C., Chierotti, M.R., Gobetto, R., Pharmaceutical aspects of salt and cocrystal forms of APIs and characterization challenges. *Adv. Drug Deliv. Rev.*, 117, 86–110, 2017.
3. Rathi, R. and Singh. Multicomponent crystal compromising dasatinib, I., and selected co-crystals formers: a patent evaluation of EP2861589B1. *Pharm. Pat. Anal.*, 11, 1, 15–21, 2021.
4. Yadav, A.V., Shete, A.S., Dabke, A.P., Kulkarni, P.V., Sakhare, S.S., Cocrystals: a novel approach to modify physicochemical properties of active pharmaceutical ingredients. *Indian J. Pharm. Sci.*, 71, 4, 359, 2009.
5. Lin, S.Y., Simultaneous screening, and detection of pharmaceutical cocrystals by the one-step DSC–FTIR microspectroscopic technique. *Drug Discov. Today*, 22, 4, 718–728, 2017.
6. Rathi, R., Kaur, S., Singh, I., A Review on Co-crystals of Herbal Bioactives for Solubility Enhancement: Preparation Methods and Characterization Techniques. *Cryst. Growth. Des.*, 22, 3, 2023–42, 2022.

7. Kalepu, S. and Nekkanti, V., Insoluble drug delivery strategies: review of recent advances and business prospects. *Acta Pharm. Sin. B.*, 5, 5, 442–453, 2015.
8. Serajuddin, A.T., Salt formation to improve drug solubility. *Adv. Drug Deliv. Rev.*, 59, 7, 603–616, 2007.
9. Rathi, R. and Sanshita, I.S., Hot Melt Extrusion Technique for Developing Pharmaceutical Co-crystals: A Review. *Drug Deliv. Lett.*, 14, 2, 97–108, 2024.
10. Izutsu, K.I., Koide, T., Takata, N., Ikeda, Y., Ono, M., Inoue, M., Fukami, T., Yonemochi, E., Characterization and quality control of pharmaceutical cocrystals. *Chem. Pharm. Bull.*, 64, 10, 1421–1430, 2016.
11. Das, R., Ali, E., Abd Hamid, S.B., Current applications of x-ray powder diffraction-a review. *Rev. Adv. Mater. Sci.*, 38, 2, 95, 2014.
12. Bunaciu, A.A., UdriŞTioiu, E.G., Aboul-Enein, H.Y., X-ray diffraction: instrumentation and applications. *Crit. Rev. Anal. Chem.*, 45, 4, 289–299, 2015.
13. Otsuka, Y. and Goto, S., Dry and wet mechanochemical synthesis of piroxicam and saccharin cocrystals and evaluation by powder X-ray diffraction, thermal analysis and mid-and near-infrared spectroscopy. *J. Pharm. Sci.*, *111*, 1, 88–94, 2022.
14. Imanto, T.E.G.U.H., Wikantyasning, E.R., Nurwaini, S.E.T.Y.O., Amalia, M.O.N.I.C.A., Sambudi, N.S., Harun, N.Y., Preparation and Solid-State Characterization of Ketoprofen-Succinic Acid-Saccharin cocrystal with Improved Solubility. *Int. J. Appl. Pharm.*, *16*, 1, 275–279, 2024.
15. Budziak-Wieczorek, I. and Maciołek, U., Synthesis and Characterization of a (–)-Epicatechin and Barbituric Acid Cocrystal: Single-Crystal X-ray Diffraction and Vibrational Spectroscopic Studies. *ACS Omega*, 6, 12, 8199–8209, 2021.
16. Boffa Ballaran, T., Kurnosov, A., Trots, D., Single-crystal X-ray diffraction at extreme conditions: a review. *High Pressure Res.*, 33, 3, 453–465, 2013.
17. Rathi, R., Kushwaha, R., Goyal, A., Singh, I., Oxaliplatin-flavone pharmaceutical co-crystal-CN111205332A: patent spotlight. *Pharm. Pat. Anal.*, 11, 5, 147–154, 2022.
18. Gründer, Y. and Lucas, C.A., Surface X-ray diffraction studies of single crystal electrocatalysts. *Nano Energy*, 29, 378–393, 2016.
19. Pauchet, M., Gervais, C., Courvoisier, L., Coquerel, G., Successful application of the derived crystal packing (DCP) model in resolving the crystal structure of a metastable polymorph of (±) modafinil. *Cryst. Growth Des.*, 4, 6, 1143–1151, 2004.
20. Zerrouki, K., Bouchene, R., Tüzün, B., Retailleau, P., Novel supramolecular cocrystal of 8-hydroxyquinoline with acetone-(2, 4-dinitrophenyl) hydrazone: One pot synthesis, structural characterization, Hirshfeld surface and energy framework analysis, computational investigation and molecular docking study. *J. Mol. Struct.*, *1300*, 137290, 2024.

21. Gao, P., Yang, Z., Zhang, Y., Xu, H., Li, M., Zhu, Y.J., Zhou, X.T., Ding, D.Z., Li, L., Improving the equilibrium solubility of Nateglinide by synthesizing a novel Nateglinide cocrystal with a single-crystal structure. *J. Mol. Struct.*, *1309*, 138030, 2024.
22. Şen, N., Aslan, N., Yüksel, B., Teciman, I., Characterization and properties of a new insensitive explosive cocrystal composed of trinitrotoluene and pyrene. *Struct. Chem.*, *35*, 2, 553–567, 2024.
23. Chen, Y.H., Zhang, G.Z., Chen, F.H., Zhang, S.Q., Fang, X., Chen, H.M., Lin, M.J., Halogen-bonded charge-transfer cocrystal scintillators for high-resolution X-ray imaging. *Chem. Sci.*, 15, 7659–7666, 2024.
24. Abd-Elghany, M. and Klapötke, T.M., A review on differential scanning calorimetry technique and its importance in the field of energetic materials. *Phys. Sci. Rev.*, 3, 4, 1 - 14, 2018.
25. Yu, L., Huang, J., Jones, K.J., Measuring free-energy difference between crystal polymorphs through eutectic melting. *J. Phys. Chem. B*, 109, 42, 19915–19922, 2005.
26. Spink, C.H., Differential scanning calorimetry. *Methods Cell Biol.*, 84, 115–141, 2008.
27. Cooper, A. and Johnson, C.M., Differential scanning calorimetry, in: *Microscopy, optical spectroscopy, and macroscopic techniques*, vol. 22, C. Jones, B. Mulloy, A.H. Thomas, (Eds.), pp. 125–136, Humana Press, Humana Totowa, NJ, 1994.
28. Freire, E., Differential scanning calorimetry, in: *Protein stability and folding: Theory and practice*, vol. 40, B.A. Shirley (Ed.), pp. 191–218, Humana Press, Humana Totowa, NJ, 1995.
29. Patel, R.D. and Raval, M.K., Differential scanning calorimetry: A screening tool for the development of diacerein eutectics. *Results Chem.*, 4, 100315, 2022.
30. Garbacz, P. and Wesolowski, M., Benzodiazepines cocrystals screening using FTIR and Raman spectroscopy supported by differential scanning calorimetry. *Spectrochim Acta A Mol. Biomol. Spectrosc.*, *234*, 118242, 2020.
31. Garbacz, P., Paukszta, D., Sikorski, A., Wesolowski, M., Structural characterization of cocrystals of chlordiazepoxide with p-aminobenzoic acid and lorazepam with nicotinamide by dsc, x-ray diffraction, ftir and raman spectroscopy. *Pharmaceutics*, *12*, 7, 648, 2020.
32. Machado Cruz, R., Boleslavská, T., Beránek, J., Tieger, E., Twamley, B., Santos-Martinez, M.J., Dammer, O., Tajber, L., Identification and pharmaceutical characterization of a new itraconazole terephthalic acid cocrystal. *Pharmaceutics*, *12*, 8, 741, 2020.
33. Coats, A.W. and Redfern, J.P., Thermogravimetric analysis. A review. *Analyst*, 88, 1053, 906–924, 1963.
34. Mansa, R. and Zou, S., Thermogravimetric analysis of microplastics: A mini review. *Environ. Adv.*, 5, 100117, 2021.

35. Loganathan, S., Valapa, R.B., Mishra, R.K., Pugazhenthi, G., Thomas, S., Thermogravimetric analysis for characterization of nanomaterials, in: *Thermal and rheological measurement techniques for nanomaterials characterization*, S. Thomas, R. Thomas, A.K. Zachariah, R.K. Mishra (Eds.), pp. 67–108, Elsevier, Amsterdam, Netherlands, 2017.
36. Yin, S.X., Scaringe, R.P., Malley, M.F., Gougoutas, J.Z., *In-situ* variable temperature powder X-ray diffraction and thermal analysis-applications in the pharmaceutical industry. *Am. Pharm. Rev.*, 8, 56–58, 2005.
37. Ngilirabanga, J.B., Rosa, P.P., Aucamp, M., Kippie, Y., Samsodien, H., Dual-drug cocrystal synthesis for synergistic *in vitro* effect of three key first-line antiretroviral drugs. *J. Drug Deliv. Sci. Technol.*, 60, 101958, 2020.
38. Erxleben, A., Application of vibrational spectroscopy to study solid-state transformations of pharmaceuticals. *Curr. Pharm. Des.*, 22, 32, 4883–4911, 2016.
39. Du, Y. and Xue, J., Investigation of polymorphism and cocrystallization of active pharmaceutical ingredients using vibrational spectroscopic techniques. *Curr. Pharm. Des.*, 22, 32, 4917–4928, 2016.
40. Chavan, R.B., Bhargavi, N., Lodagekar, A., Shastri, N.R., Near infra-red spectroscopy: a tool for solid state characterization. *Drug Discov. Today*, 22, 12, 1835–1843, 2017.
41. Iwata, K., Karashima, M., Ikeda, Y., Isotope-edited infrared spectroscopy for efficient discrimination between pharmaceutical salts and cocrystals. *Mol. Pharmaceutics*, 14, 7, 2350–2358, 2017.
42. Parmar, V.K. and Shah, S.A., hydrochloride salt cocrystals: preparation, characterization, and physicochemical studies. *Pharm. Dev. Technol.*, 18, 2, 443–453, 2013.
43. Vogt, F.G., Clawson, J.S., Strohmeier, M., Edwards, A.J., Pham, T.N., Watson, S.A., Solid-state NMR analysis of organic cocrystals and complexes. *Cryst. Growth Des.*, 9, 2, 921–937, 2009.
44. Khan, F.M., Ahmad, M., Idrees, H.A., Simvastatin-nicotinamide cocrystals: formation, pharmaceutical characterization and *in vivo* profile. *Drug Des. Devel. Ther.*, 4303–4313, 2020.
45. Wouters, J. and Quéré, L. (Eds.), *Pharmaceutical salts and cocrystals*, Royal Society of Chemistry, 2012.
46. Ali, A., Kuznetsov, A., Ibrahim, M., Abbas, A., Akram, N., Maqbool, T., Chemistry and modern techniques of characterization of Cocrystals, in: *Drug Formulation Design*, R. Shukla, A. Kuznetsov, A.A. XXXA. (Eds.), IntechOpen, 2022.
47. Maciołek, U., Mendyk, E., Kuśmierz, M., Kozioł, A.E., Binary Co-Crystals of Quercetin: Synthesis, Structure, and Spectroscopic Characterization. *ChemPlusChem.*, *88*, 7, e202300166, 2023.
48. Queiroz, A.L.P., Rodrigues, M., Zeglinski, J., Crean, A.M., Sarraguça, M.C., Vucen, S., Determination of cocrystal phase purity by mid infrared spectroscopy and multiple curve resolution. *Int. J. Pharm.*, *595*, 120246, 2021.

49. Naqvi, A., Ahmad, M., Minhas, M.U., Khan, K.U., Batool, F., Rizwan, A., Preparation and evaluation of pharmaceutical cocrystals for solubility enhancement of atorvastatin calcium. *Polym. Bull.*, *77*, 6191–6211, 2020.
50. Wu, X., Wang, Y., Xue, J., Liu, J., Qin, J., Hong, Z., Du, Y., Solid phase drug-drug pharmaceutical cocrystal formed between pyrazinamide and diflunisal: Structural characterization based on terahertz/Raman spectroscopy combining with DFT calculation. *Spectrochim Acta A Mol. Biomol. Spectrosc.*, *234*, 118265, 2020.
51. Tomaszewska, I., Karki, S., Shur, J., Price, R., Fotaki, N., Pharmaceutical characterisation and evaluation of cocrystals: importance of *in vitro* dissolution conditions and type of coformer. *Int. J. Pharm.*, *453*, 2, 380–388, 2013.
52. Thakuria, R., Delori, A., Jones, W., Lipert, M.P., Roy, L., Rodríguez-Hornedo, N., Pharmaceutical cocrystals and poorly soluble drugs. *Int. J. Pharm.*, *453*, 1, 101–125, 2013.
53. Good, D.J. and Rodríguez-Hornedo, N., Solubility advantage of pharmaceutical cocrystals. *Cryst. Growth Des.*, *9*, 5, 2252–2264, 2009.
54. Baertschi, Alsante, Reed, *Pharmaceutical stress testing: predicting drug degradation*, vol. 210, S.W., K.M. , R.A. (Eds.), Informa Healthcare, London, 2011.
55. Loftsson, T. (Ed.), *Drug stability for pharmaceutical scientists*, Elsevier, Kidlington, Oxford, UK, 2014.
56. Chemburkar, S.R., Bauer, J., Deming, K., Spiwek, H., Patel, K., Morris, J., Henry, R., Spanton, S., Dziki, W., Porter, W., Quick, J., Dealing with the impact of ritonavir polymorphs on the late stages of bulk drug process development. *Org. Process Res. Dev.*, 4, 5, 413–417, 2000.
57. Bolla, G., Mittapalli, S., Nangia, A., Celecoxib cocrystal polymorphs with cyclic amides: synthons of a sulfonamide drug with carboxamide coformers. *Cryst. Eng. Comm.*, 16, 1, 24–27, 2014.
58. Samie, A., Desiraju, G.R., Banik, M., Salts and cocrystals of the antidiabetic drugs gliclazide, tolbutamide, and glipizide: solubility enhancements through drug–coformer interactions. *Cryst. Growth Des.*, 17, 5, 2406–2417, 2017.
59. Thenge, R.R., Patond, V.B., Ajmire, P.V., Barde, L.N., Mahajan, N.M., Tekade, N.P., Preparation and characterization of cocrystals of diacerein. *Indones. J. Pharm.*, 28, 1, 34, 2017.
60. Nechipadappu, S.K., Tekuri, V., Trivedi, D.R., Pharmaceutical cocrystal of flufenamic acid: synthesis and characterization of two novel drug-drug cocrystal. *J. Pharm. Sci.*, 106, 5, 1384–1390, 2017.
61. Pandey, N.K., Sehal, H.R., Garg, V., Gaur, T., Kumar, B., Singh, S.K., Gulati, M., Gowthamarajan, K., Bawa, P., Rajesh, S.Y., Sharma, P., Stable cocrystals of glipizide with enhanced dissolution profiles: preparation and characterization. *AAPS PharmSciTech.*, 18, 2454–2465, 2017.
62. Shane, N.L.J., Chamle, A.H., Vasantharaju, A., Pai, G., Sathyanarayana, M.B., Fabrication and solid state characterization of ticagrelor cocrystals with

improved solubility and dissolution. *Indian J. Pharm. Qual. Assur.*, 20, 14.54, 13–98, 2017.
63. Setyawan, D., Oktavia, I.P., Farizka, R., Retno, S.A.R.I., Physicochemical characterization and *in vitro* dissolution test of quercetin-succinic acid cocrystals prepared using solvent evaporation. *Turk. J. Pharm. Sci.*, 14, 3, 280, 2017.
64. Shemchuk, O., Degli Esposti, L., Grepioni, F., Braga, D., Ionic cocrystals of enantiopure and racemic histidine with calcium halides. *Cryst. Eng. Comm.*, 19, 42, 6267–6273, 2017.
65. Aithal, K., Pai, A., Pai, G., Muddukrishna, B.S., Preparation, solid state characterization of Etravirine cocrystals with improved solubility. *Lat. Am. J. Pharm.*, 36, 5, 972–979, 2017.
66. Sopyan, I., Fudholi, A., Muchtaridi, M., Sari, I.P., Simvastatin-nicotinamide cocrystal: design, preparation and preliminary characterization. *Trop. J. Pharm. Res.*, 16, 2, 297–303, 2017.
67. Cho, M.Y., Kim, P., Kim, G.Y., Lee, J.Y., Song, K.H., Lee, M.J., Yoon, W., Yun, H., Choi, G.J., Preparation and characterization of aripiprazole cocrystals with coformers of multihydroxybenzene compounds. *Cryst. Growth Des.*, 17, 12, 6641–6652, 2017.
68. Noa, F.M.A. and Jacobs, A., Phenylacetic acid cocrystals with acridine, caffeine, isonicotinamide and nicotinamide: Crystal structures, thermal analysis, FTIR spectroscopy and Hirshfeld surface analysis. *J. Mol. Struct.*, 1139, 60–66, 2017.
69. Chadha, R., Rani, D., Goyal, P., Supramolecular cocrystals of gliclazide: synthesis, characterization and evaluation. *Pharm. Res.*, 34, 552–563, 2017.
70. Pan, Y., Pang, W., Lv, J., Wang, J., Yang, C., Guo, W., Solid state characterization of azelnidipine–oxalic acid cocrystal and co-amorphous complexes: The effect of different azelnidipine polymorphs. *J. Pharm. Biomed. Anal.*, 138, 302–315, 2017.
71. Almansa, C., Merce, R., Tesson, N., Farran, J., Tomas, J., Plata-Salaman, C.R., Cocrystal of Tramadol hydrochloride–Celecoxib (CTC): a novel API–API cocrystal for the treatment of pain. *Cryst. Growth Des.*, 17, 4, 1884–1892, 2017.
72. Samsodien, H., Bapoo, R., Doms, T.I., Harneker, Z., Louw, A.S., Scheepers, I.C., Sonday, A.B., Geldenhuys, B., FTIR, dissolution and anti-viral activity of nevirapine cocrystals. *Pharm. Anal. Acta.*, 8, 9, 1-10, 2017.
73. Kang, Y., Gu, J., Hu, X., Syntheses, structure characterization and dissolution of two novel cocrystals of febuxostat. *J. Mol. Struct.*, 1130, 480–486, 2017.
74. Ramle, N.A., Rahim, S.A., Anuar, N., El-Hadad, O., *Solubility of carbamazepine cocrystals in ethanolic solution. AIP Conference Proceedings of AIP Publishing*, pp. 040001-1- 040001-7, August 2017.
75. Mohammad, K.A., Abd Rahim, S., Abu Bakar, M.R., Kinetics and nucleation mechanism of carbamazepine–saccharin cocrystals in ethanol solution. *J. Therm. Anal. Calorim.*, 130, 1663–1669, 2017.

76. Perrin, A., Goodwin, M.J., Musa, O.M., Yufit, D.S., Steed, J.W., Boric acid cocrystals in guar gelation. *Cryst. Eng. Comm.*, 19, 47, 7125–7131, 2017.
77. Sládková, V., Dammer, O., Sedmak, G., Skořepová, E., Kratochvíl, B., Ivabradine hydrochloride (S)-mandelic acid cocrystal: *in situ* preparation during formulation. *Crystals*, 7, 1, 13, 2017.
78. Garbacz, P., Wesolowski, M., DSC, F.T.I.R., and Raman Spectroscopy coupled with multivariate analysis in a study of cocrystals of pharmaceutical interest. *Molecules*, 23, 9, 2136, 2018.
79. Inam, M., Wu, J., Shen, J., Phan, C.U., Tang, G., Hu, X., Preparation and characterization of novel pharmaceutical cocrystals: Ticagrelor with nicotinamide. *Crystals*, 8, 9, 336, 2018.
80. Abidi, S.S.A., Azim, Y., Khan, S.N., Khan, A.U., Sulfaguanidine cocrystals: Synthesis, structural characterization and their antibacterial and hemolytic analysis. *J. Pharm. Biomed. Anal.*, 149, 351–357, 2018.
81. Aitipamula, S., Cadden, J., Chow, P.S., Cocrystals of zonisamide: physicochemical characterization and sustained release solid forms. *Cryst. Eng. Comm.*, 20, 21, 2923–2931, 2018.
82. Kamble, R.N.,., Bothiraja, C., Mehta, P.P., Varghese, V., Synthesis, solid state characterization and antifungal activity of ketoconazole cocrystals. *J. Pharm. Investig.*, 48, 541–549, 2018.
83. do Amaral, L.H., do Carmo, F.A., Amaro, M.I., de Sousa, V.P., da Silva, L.C.R.P., de Almeida, G.S., Rodrigues, C.R., Healy, A.M., Cabral, L.M., Development and characterization of dapsone cocrystal prepared by scalable production methods. *AAPS Pharm. Sci. Tech.*, 19, 2687–2699, 2018.
84. Abidi, S.S.A., Azim, Y., Gupta, A.K., Pradeep, C.P., Cocrystals of indole-3-acetic acid and indole-3-butyric acid: Synthesis, structural characterization and Hirshfeld surface analysis. *J. Mol. Struct.*, 1166, 202–213, 2018.
85. Yadav, B., Balasubramanian, S., Chavan, R.B., Thipparaboina, R., Naidu, V.G.M., Shastri, N.R., Hepatoprotective cocrystals and salts of riluzole: prediction, synthesis, solid state characterization, and evaluation. *Cryst. Growth Des.*, 18, 2, 1047–1061, 2018.
86. Shimpi, M.R., Alhayali, A., Cavanagh, K.L., Rodríguez-Hornedo, N., Velaga, S.P., Tadalafil-malonic acid cocrystal: Physicochemical characterization, pH-solubility, and supersaturation studies. *Cryst. Growth Des.*, 18, 8, 4378–4387, 2018.
87. Varma, A., Laxmi, P., Pai, A., Pai, G., Vasantharaju, S.G., Sathyanarayana, M.B., Designing of stable cocrystals of zoledronic acid using suitable coformers. *Chem. Pharm. Bull.*, 67, 8, 816–823, 2019.
88. Zhou, Z., Calatayud, M., Contreras-García, J., Li, L., Tong, H.H., Zheng, Y., X-Ray Diffraction and Theoretical Calculation–Supported Formation of Polymorphic Cocrystals Discovered Through Thermal Methods: A Case Study. *J. Pharm. Sci.*, 108, 10, 3340–3347, 2019.
89. Thenge, R., Patel, R., Kayande, N., Mahajan, N., Cocrystals of valsartan: preparation and characterization. *Int. J. Pharm. Biol. Sci.*, 9, 274–283, 2019.

90. Fernandes, R.P., do Nascimento, A.L.C.S., Carvalho, A.C.S., Teixeira, J.A., Ionashiro, M., Caires, F.J., Mechanochemical synthesis, characterization, and thermal behavior of meloxicam cocrystals with salicylic acid, fumaric acid, and malic acid. *J. Therm. Anal. Calorim.*, 138, 765–777, 2019.
91. Kozak, A., Marek, P.H., Pindelska, E., Structural characterization and pharmaceutical properties of three novel cocrystals of ethenzamide with aliphatic dicarboxylic acids. *J. Pharm. Sci.*, 108, 4, 1476–1485, 2019.
92. Wang, Y., Xue, J., Qin, J., Liu, J., Du, Y., Structure and spectroscopic characterization of pharmaceutical cocrystal formation between acetazolamide and 4-hydroxybenzoic acid. *Spectrochim. Acta Part A Mol. Biomol. Spectrosc.*, 219, 419–426, 2019.

6

Advancements and Applications of Herbal Cocrystals

Rajat Goyal[1], Pankaj Popli[2], Kashish Wilson[1], Hitesh Chopra[3*], Sonia Dhiman[2] and Sumeet Gupta[1]

[1]*MM College of Pharmacy, Maharishi Markandeshwar (Deemed to be University), Mullana-Ambala, Haryana, India*
[2]*Chitkara College of Pharmacy, Chitkara University, Punjab, India*
[3]*Department of Biosciences, Saveetha School of Engineering, Saveetha Institute of Medical and Technical Sciences, Chennai, Tamil Nadu, India*

Abstract
Cocrystals have drawn the attention of the pharmaceutical sector through the introduction of regulatory guidelines by the US Food and Drug Administration (FDA) because of their growing patent portfolios. The drug's physicochemical properties depend on the qualities of the active pharmaceutical ingredient (API) that it possesses, as well as its formulation type. The existing and newly discovered APIs present significant challenges in drug development. These challenges are mostly caused by APIs' poor water solubility which inevitably leads to poor bioavailability of the corresponding drugs in the patient. As a result, the preformulation of APIs mainly requires a brief knowledge of their solid-state features and physicochemical properties. Cocrystallization defies a promising approach for transforming API physical and pharmacological properties, *viz*, solubility, rate of dissolution, stability, hygroscopicity, tablet ability, mechanical characteristics, bioavailability, and efficient therapeutic potential. Herbal bioactives are promising prospective for cocrystallization to improve solubility, taste masking, and bioavailability. This chapter provides an extensive overview of the literature on the development of cocrystals from various herbal bioactives, like curcumin, quercetin, berberine, resveratrol, naringenin, and various other herbal chemical substances. Various techniques are currently explored for producing medicinal cocrystals. Instrumental and physicochemical characterization methods have been also included. In conclusion, cocrystallization of herbal bioactives is a possible way to

**Corresponding author*: chopraontheride@gmail.com

Inderbir Singh, Subrata Mallick and Ritu Rathi (eds.) Cocrystals in Pharmaceutical Sciences: Design to Applications, (153–188) © 2025 Scrivener Publishing LLC

improve the physical characteristics of the aforementioned substances. However, toxicological, stability, regulatory, and scale-up difficulties need to be sufficiently addressed before commercial viability can be anticipated.

Keywords: Herbal bioactives, cocrystals, physicochemical characterization, bioavailability, solubility

6.1 Introduction

6.1.1 Definition and Overview

Crystal engineering finds diverse applications, including the utilization of complementary molecules to create binary compounds [1]. In the pharmaceutical realm, molecules possess inherent recognition sites that selectively bind to biomolecules, offering opportunities to construct supramolecular self-assembling complexes with tailored physiochemical attributes [2]. A notable example is pharmaceutical cocrystals, which are crystalline molecular complexes comprising two or more neutral molecules. These cocrystals represent a cutting-edge approach in supramolecular chemistry, offering enhancements in various active pharmaceutical ingredient (APIs) characteristics such as dissolution rate, mechanical characteristics, and stability, as shown in Figure 6.1 [3]. Compared to traditional solid-state modification techniques like salt formation, cocrystal formation presents several potential advantages. For instance, virtually all types of drug molecules have the potential to form cocrystals, obviating the need for covalent modification of APIs [4]. Moreover, herbal medicines have a rich history of use dating back to ancient times, particularly prominent in regions like India where traditional systems such as Siddha and Ayurveda rely heavily on herbal preparations [5].

In modern pharmaceutical practice, herbal drugs hold a significant position due to their well-known effects and minimal side effects. Additionally, herbal drugs offer a promising avenue for cocrystal formation compared to synthetic counterparts [6]. Despite their broad pharmacological potential across various diseases, herbal drugs often exhibit limited effectiveness within the human biological system due to factors such as poor absorption, inability to penetrate lipid membranes, large molecular size, and reduced bioavailability [7]. As a result of these restrictions, some extracts are not employed in medicinal applications. Cocrystals have evolved as a different approach for enhancing the kinetic and dynamic characteristics of herbal medicine molecules within biological systems to solve these problems [8, 9].

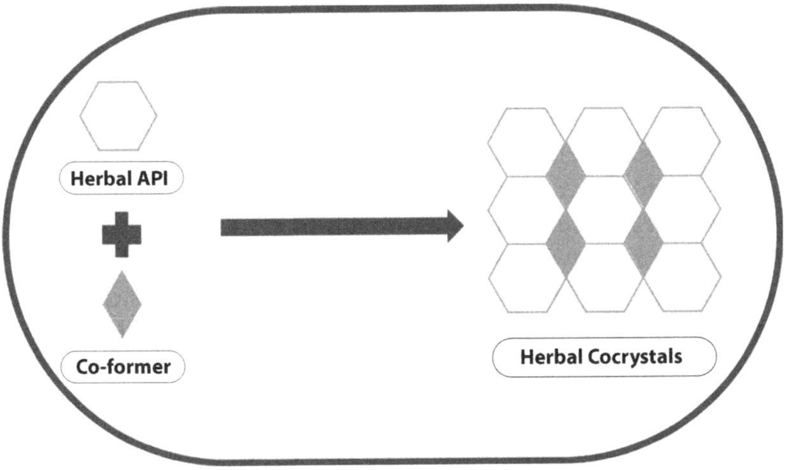

Figure 6.1 Herbal API and coformer forming the cocrystal formation.

In designing dosage forms, cocrystals are becoming a more desirable option than polymorphs, salts, solvates, and crystal habit manipulation because they offer the advantage of altering the bulk material and physical and chemical properties of the medicinal product without sacrificing its inherent action [10]. Utilizing cocrystals to convert herbal medicine into a solid form offers multiple advantages, one of which is the potential to achieve a high dissolution rate that is equivalent to the amorphous form while keeping the long-term chemical and physical stability related to crystalline forms [11]. Furthermore, cocrystallization can enhance functionality and processability downstream. Pharmaceutical businesses may profit from prospects for intellectual property protection, including the opportunity to prolong the life cycles of current APIs through this method [12].

6.1.2 Fundamentals of Cocrystal Formation

APIs can currently be modified physicochemically using pharmaceutical cocrystallization, a relatively recent technique that provides several benefits over salt forms. Initially, cocrystal formation's value potential increases by its mere existence that it can be utilized for a variety of APIs, such as basic, acidic, and nonionizable compounds. Secondly, when compared to salt forms, the range of potential "counter molecules" for cocrystal production is broader due to their abundance, particularly nontoxic compounds [13].

Analogous to salt selection, which relies on pKa considerations for acid-base pairs, cocrystals offer a method to concurrently approach many functional areas within a pharmaceutical compound [14]. Unlike salts, which typically target an individual basic or acidic functional group, cocrystals can accommodate tertiary and quaternary combinations. Notably, cocrystals enable tailored adjustments to the solubility and dissolution-pH dependence of APIs, even for nonionizable molecules [15].

Differentiating between solvates and cocrystals independently depends on each component's physical condition. Solvates, formed when a single component is liquid at room temperature, are often unstable due to solvent loss during storage [16]. While solvates may enhance drug dissolution rates, their instability, and potential toxicological concerns limit their regulatory acceptance. In contrast, cocrystals, typically a product of deliberate design, offer greater stability as the cocrystallizing agents are solid at room temperature. Although polymorphic cocrystals are rare, they present a more controlled alternative to solvates. Cocrystals, distinct from solid dispersions, provide a crystalline single-phase multicomponent system [17]. They are distinguished from solid dispersions by the interactions between components. Solid dispersions, utilized to enhance drug dissolution rates and absorption, involve the active ingredients' dispersion within an inert carrier. However, solid dispersions may inadvertently form during attempts for cocrystal preparation from solution [18].

Regarding polymorphism, while proving the absence of polymorphism is challenging, preliminary evidence suggests that polymorphic substances may serve as suitable candidates for cocrystallization [19]. For instance, carbamazepine, with multiple well-characterized polymorphs, has demonstrated diverse cocrystal formations, illustrating the potential to mitigate polymorphism practically through cocrystals [20]. The key advantages of cocrystallization lie in its versatility across different types of drug molecules and the abundance of potential counter-molecules for synthesis. This approach offers opportunities for the pharmaceutical industry to address intellectual property concerns by revitalizing old APIs, thus extending their life cycles [21]. To summarize, cocrystallization in the pharmaceutical sector provides an effective means for modifying the characteristics of therapeutic solids. In contrast to conventional methods like salt production and solvates, it offers several advantages, like flexibility, stability, and the ability to minimize polymorphism [22]. Furthermore, cocrystallization is an effective strategy for pharmaceutical development because of its broad spectrum of applications and intellectual property prospects [23].

6.1.3 Selection of Coformer

A pharmaceutical cocrystal comprises two neutral molecules, out of which one serves as an API while the other as a coformer. The USFDA has cataloged an extensive list of thousands of compounds suitable to be employed as coformers for medicinal cocrystals [24]. Preferably, the cocrystal should be listed on the USFDA "Everything added to food in the United States" (EAFUS) list, which includes over 3000 compounds recognized as acceptable for use as food additives or have been approved as generally regarded as safe (GRAS). The coformer utilized as a non-API constituent must be safer for consumption and devoid of any adverse reactions [25]. Cocrystals comprise multiple components that are bound together by supramolecular synthons. The primary challenge in synthesizing cocrystals lies in selecting coformers which compatible with specific APIs, necessitating the presence of functional groups in both the API and coformer capable of forming supramolecular hetero- or homosynthons [26]. Researchers employ diverse knowledge-based methodologies to screen coformers, including the following:

- *In silico* screening involves using computational tools to predict interactions between coformers and API. Coformers selected for screening must have hydrogen donor or acceptor groups to form hydrogen bonds with the API. Molecular docking simulations, using software like Autodock4, Glide, or GOLD, determine the most stable conformation and orientation with the lowest energy interaction. The binding free energy is calculated to identify coformers that can successfully form cocrystals with the API [27, 28].
- Hydrogen bonding is crucial in cocrystal formation, driving molecular arrangements through noncovalent interactions like π–π stacking and van der Waals forces. Effective hydrogen bond formation follows guidelines that prioritize six-membered ring interactions. Using the Cambridge structure database, hydrogen bond propensity values (0 to 1) predict bond formation. This approach quantitatively evaluates interactions, aiding in the efficient screening and selection of drug-coformer systems for cocrystal development [29, 30].
- The Cambridge structure database is essential for analyzing intramolecular interactions in crystals. It aids in identifying packing motifs and functional groups, supporting

drug discovery by generating coformer collections for specific APIs, streamlining research, and reducing resource use [31, 32].

- The pKa approach predicts herbal cocrystals or salt formation by assessing proton transfer between an acid and base. It calculates ΔpKa by subtracting the acid pKa from the base. A ΔpKa over 2 to 3 suggests salt formation, while a negative value indicates cocrystals. The value between 0.3 creates a continuum, not clearly distinguishing between cocrystals and salts [29, 33].
- The Hansen solubility parameter assesses molecular compatibility based on solubility it calculates Δδt, the difference in solubility parameters between an API and a coformer, to predict cocrystal formation. Values below 7MPO suggest higher cocrystal likelihood, though practical results often deviate from predictions [34, 35].
- COSMO-RS (conductor-like screening model for real solvents) accurately predicts the thermodynamic properties of liquids. Using the COSMOtherm program, it evaluated conformer feasibility in the supercooled liquid phase for APIs. COSMO-RS, based on statistical physics, quantifies interaction energies, including polar and hydrogen bonds. It calculates excess enthalpy for stoichiometric mixtures, distinguishing hydrogen bonding from overall enthalpic contributions [36, 37].
- In 1995, Desiraju introduced "supramolecular synthon" to describe the organization of crystalline structures in supramolecules formed through specific intramolecular interactions. Crystal engineering leverages supramolecular chemistry principles, focusing on noncovalent interactions to create new entities. Originally introduced by Corey in organic chemistry, synthons are structural units in supramolecules [38, 39].

6.1.4 Methods of Cocrystal Synthesis

These are various methods are utilized for cocrystal production including the solid-state method and green-production method as demonstrated in Figure 6.2.

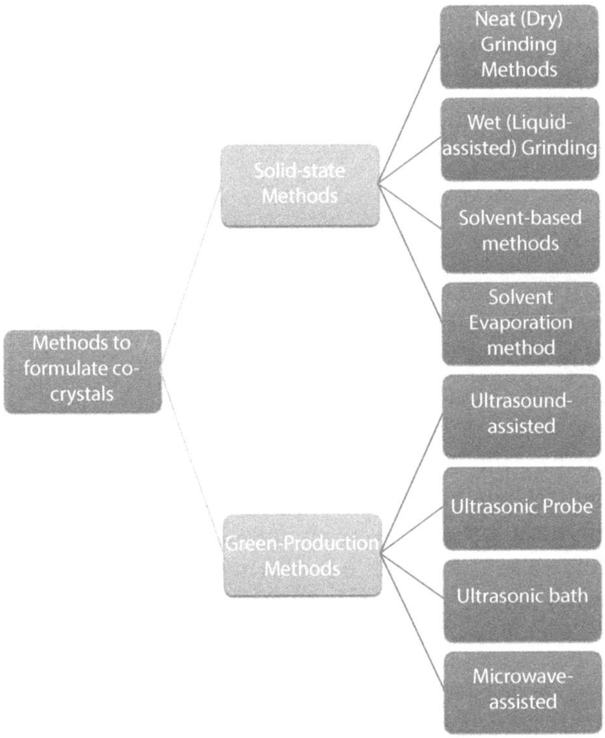

Figure 6.2 Various methods of production of cocrystals.

6.1.4.1 *Solid-State Methods*

Neat (Dry) Grinding: In this method, pressure is applied on a solid combination of coformers and APIs to facilitate cocrystal formation. A mortar and pestle can be utilized for manual mixing, while a vibratory or ball mill can be utilized for mechanical mixing [40]. The amount of time spent on grinding, which typically takes 30 to 60 min, is controlled to maximize the quality of the final product. This technique's disadvantage lies in the comparatively poor effectiveness of cocrystal formation, yet it ensures stability simply by doing the requirement for solvents [41]. A portion of the product could turn amorphous after dry mixing, adding contaminants that require further removal.

Wet (Liquid-Assisted) Grinding: In this approach, solid powders are initially milled with a small amount of liquid added. Various solvents such as water, ethanol, or toluene can be used, requiring just a couple of droplets or microliters. Also known as solvent-drop grinding, this approach enhances cocrystal formation by acting as a solid powder binder rather than a dissolution medium [42]. Its advantages include simplicity, effectiveness, environmental friendliness, cost-effectiveness, and higher cocrystal purity compared to other methods. Additionally, it improves the yield percentage compared to dry grinding [43].

6.1.4.2 *Solvent-Based Methods*

These procedures rely on the inclusion of a significant solvent amount in which APIs and coformers dissolve. The selection of solvent is critical, with adjustments in solvent type and temperature aimed at achieving optimal cocrystal characteristics [44]. Compound crystallization from solution develops in solvent-dependent cocrystallization when the solvent becomes supersaturated by the cocrystal. The rate-limiting step for solvent-based methods occurs when the rate of cocrystallization is greater than the degree of solvent unsaturation. Various techniques fall under solvent-based methods, including solvent evaporation, slurring, active cocrystallization, and the antisolvent method, with solvent evaporation being the most common.

6.1.4.3 *Solvent Evaporation Method*

This prevalent approach involves mixing equimolar amounts of API and coformer through solvent or solvent mixture. The solvent promotes bond formation and molecular interaction, with subsequent evaporation revealing a developed cocrystal. Careful monitoring of the evaporation rate is crucial; rapid evaporation typically yields numerous small cocrystals, whereas slow evaporation produces fewer, larger cocrystals [45]. Solvent evaporation can occur spontaneously at room temperature or through accelerated techniques involving increased temperature and/or decreased pressure, like rotary evaporators as well as filtration by vacuum [46].

6.1.4.4 *Miscellaneous Methods*

Spray-drying represents a widely acknowledged approach employed for converting emulsions, liquids, and suspensions into dry powder [47]. Employing advanced equipment enables precise regulation of atmospheric conditions and the application of hot steam air to rapidly evaporate the solvent. This approach significantly outpaces previous solution cocrystallization procedures in the context of speed and purity of cocrystal manufacturing. Electrospray technology entails the utilization of an electric field to generate a fine spray of charged droplets. This approach involves the application of electrical impulses to disintegrate the liquid into minute droplets [48]. The fluid is expelled through a narrow nozzle subjected to elevated electric potential, thereby leading to dispersion into a thin mist of highly charged droplets. Following the completion of the drying process, the resultant particles are collected employing an electrostatic powder collector. Acoustic resonant mixing (RAM) involves the utilization of sound waves to homogenously blend the API and coformer formed in the presence of a solvent, facilitating the formation of a cocrystal without the requirement of a grinding medium. Acoustic power is imparted to a moist powdered combination, which ensures a thorough amalgamation of all the components [49].

Nano-cocrystallization in the pharmaceutical holds promise for improving the dispensability of poorly soluble drugs, a longstanding challenge for the pharmaceutical industry. The research in this domain centers on the manufacturing of nanococrystals which possess significant commercial worth. Scalable production of nanocrystals can be achieved through methods like media grinding, precipitation, and high-pressure homogenization [50]. Nanocrystals exhibit notable effectiveness for therapy when administered intravenously, orally, or via the pulmonary route. Furthermore, they offer targeted delivery capabilities [51]. The production process typically employs a top-down flow process using a ball grinding technique to induce shear stresses, generating uniform nanosized particles. Precipitation, a bottom-up growth process, is also commonly employed. Freeze-drying, which is also referred to as lyophilization, is a widely adopted preservation method for various materials, including pharmaceuticals. The process involves subjecting the material solution to a regulated and reduced surrounding pressure, inducing sublimation of frozen water within the materials [52]. This approach also holds the potential for creation of the solid-state cocrystal structures.

6.1.5 Characterization Techniques

Cocrystal characterization techniques include structural methods like single-crystal X-ray crystallography, powder X-ray diffraction, and infrared spectroscopy, as well as physical property analyses like melting point determination, differential scanning calorimetry (DSC), and thermogravimetric analysis (TGA) as described in Figure 6.3 [53]. It has been demonstrated that near-infrared (NIR) spectroscopy may be used for cocrystal screening, with Raman spectroscopy acting as a comparison technique. Whether in single crystal or powder form, X-ray diffraction detects X-ray

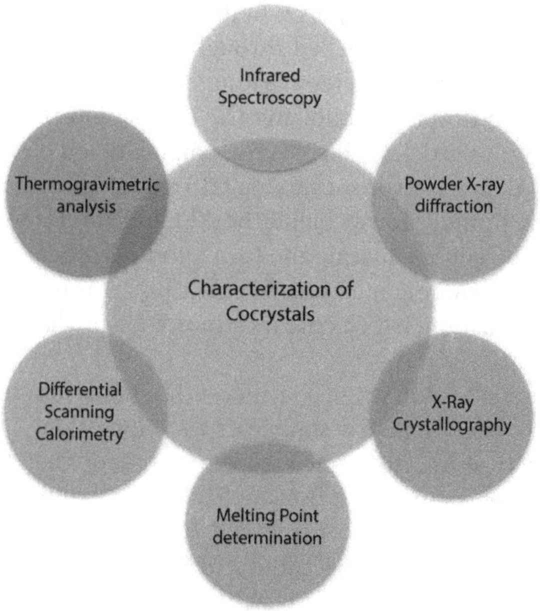

Figure 6.3 Characterization method of cocrystals.

pattern changes relative to the original material pattern, which gives a clear indicator of new cocrystal formation [54].

Modifications in the material's vibrational mode brought on by light scattering can reveal the existence of cocrystals, which makes Raman spectroscopy an invaluable diagnostic tool [55]. FTIR spectroscopy represents functional groups as peaks within certain ranges, providing a distinct "fingerprint" for every chemical [56]. Further research to confirm cocrystal formation may be necessary if FTIR detects any new bands or changes in previously identified peaks that may indicate the establishment of hydrogen bonds. When choosing appropriate coformers from a range of chemicals for cocrystal building, FTIR is very useful.

The thermal analysis method depends on variations in the enthalpy and melting temperatures of cocrystals, as well as on the physical mixes of these cocrystal constituents (active medicinal substances and coformers) [57]. TGA and DSC are the two methods used in this analysis. Cocrystal formation in DSC is indicated by the existence of an endothermic peak at a temperature that is lower or intermediate than the melting temperatures of the raw components [58]. DSC measures phase changes like melting, polymorphism, and the loss of volatile components in cocrystals. It tracks power changes with temperature [59].

On the other hand, TGA measures mass loss with temperature and is effective in identifying the precise temperature at which the cocrystal begins to decompose and lose its volatile component. The amount of mass loss in TGA offers confirmed information about the stoichiometric ratio of the starting materials in the cocrystal [60]. Unique cocrystals, such as a stable salt-cocrystal of tiotropium fumarate with fumaric acid, were characterized, revealing their distinct structures and properties [61]. Investigations into the thermodynamic stability of cocrystals and their component forms provided valuable insights into cocrystal formation processes [62].

6.2 Benefits and Applications of Herbal Cocrystals

6.2.1 Enhanced Solubility and Bioavailability

The Biopharmaceutics Classification System (BCS) categorizes drugs into four classes based on their solubility and permeability, revealing that approximately 40% of marketed drugs have poor solubility [63]. Researchers have concentrated on enhancing the solubility of such drugs using various methods, including cocrystal formulation. Shimpi *et al.* developed a cocrystal of tadalafil, a phosphodiesterase inhibitor, with malonic acid. This cocrystal exhibited a significantly higher aqueous solubility across a wide pH range compared to the drug alone. Additionally, a eutectic mixture of tadalafil with a coformer demonstrated a notable increase in solubility compared to pure tadalafil [64]. Similarly, Chen *et al.* successfully prepared a cocrystal of the blood coagulation inhibitor, apixaban, with oxalic acid. They discovered that this cocrystal significantly raised the intrinsic dissolving rate and solubility in buffer solution, which improved the pharmacokinetic characteristics in Beagle dogs [65]. Additionally, McNamara *et al.* combined glutaric acid with a class II low water-soluble sodium channel blocker API, a derivative of pyrimidine-4-carboxamide. Dogs' *in vivo* plasma concentration studies showed that this cocrystal greatly increased solubility and intrinsic dissolution rate [66].

Furthermore, ezetimibe, an antihyperlipidemic drug, was investigated by Sugandha and colleagues as a cocrystal with methylparaben. When compared to free ezetimibe, cocrystals made using various techniques showed more solubility, which improved dissolution and bioavailability [67]. To overcome the restricted solubility, dissolution, and bioavailability of piperine, a secondary metabolite generated from plants, Zaini and associates synthesized it as a cocrystal with succinic acid. This new cocrystal was shown to have a notable rise in solubility along with enhancements in dissolution and bioavailability. It is believed that the water-soluble coformer succinic acid enabled the development of channels between piperine molecules [68].

6.2.2 Improved Stability and Shelf Life

The use of cocrystals in the formulation of several medications has been shown to significantly increase stability. For instance, flufenamic acid, a nonsteroidal anti-inflammatory medication, was combined with nitrobenzoic acid and ethenzamide to generate cocrystals. Weak halogen bonds were used in this formulation to stabilize the flufenamic acid for more than 6 months under ambient settings. Yang *et al.* produced berberine, a naturally occurring isoquinoline alkaloid extract with a variety of pharmacological properties, as a cocrystal with fumaric acid. Even at high temperatures and humidity, this cocrystal demonstrated improved solubility and stability. The dynamic vapor sorption (DVS) analysis and stress testing were part of the stability evaluation [69].

6.3 Challenges and Limitations of Herbal Cocrystals, Solubility Issues, and Incompatibilities

Herbal cocrystals, despite their broad applicability, face challenges in commercial-scale production, limiting their market presence. Their development requires precision, as creating a commercially viable, stable form is guaranteed. Key obstacles include determining accurate drug dosage and selecting suitable coformers from many potential options, necessitating effective screening tools. Current methods like hydrogen bond propensity and pKa rules aid in coformer identification, but more efficient screening techniques are needed. Additionally, issues with polymorphism, solubility, and scaling up remain significant. Nonetheless, herbal cocrystals hold promise for enhancing drug properties and offer a cost-effective approach to new pharmaceutical developments [12, 70].

6.3.1 Scale-Up and Manufacturing Challenges

The FDA defines a batch as a specific quantity of drug produced with consistent quality, while a "lot" refers to part of a batch with similar consistency. The pharmaceutical industry, traditionally using batch production can benefit from adopting, energy efficiency, and improved quality control.

Continuous processes enable real-time monitoring and produce more consistent outcomes, though they involve complex technology and higher transition costs compared to batch methods [49, 71].

6.3.2 Safety Concerns

The development of herbal cocrystals represents an emerging area of research. However, this necessitates caution owing to its limited knowledge surrounding its safety profile. Here are numerous incompatibilities linked to safety that must be kept in mind, which include the following: inadequate scientific research has been conducted to determine the safety of the majority of herbal cocrystals [72]. Also, choosing a cocrystal possibility is the safety of the coformer, which is another crucial consideration as this can result in added regulatory requirements. The safety of both the inactive along inactive components in a pharmaceutical product is a necessity for obtaining market approval [50]. The coformer employed for the selected cocrystal for the development of the product must be nontoxic and safer to use when administered in the required amount for therapeutic doses of the medicine [73, 74]. Before formulating a cocrystal for any medication, it is crucial to carefully evaluate the safety of the coformer and determine the appropriate quantity to be used in the cocrystal product. Thereby, employing excipients that are already recorded in the inactive ingredient database (IID) as coformers could be used as it would necessitate a less thorough regulatory evaluation as their safety in a specific quantity for certain methods of administration and dosage type may have previously been verified [29, 75].

6.3.3 Quality Control

Quality control is an essential aspect of pharmaceutical cocrystal production. These cocrystals are manufactured employing unit procedures commonly used for producing APIs and their formulations [76]. Crystallization from solutions is expected to yield high-quality solids, allowing for thorough characterization of the cocrystals, as outlined in the recommendations provided by the European Medicines Agency. However, cocrystals produced utilizing conventional formulation methods, including grinding and spray-drying, may have smaller dimensions, lower purity, and contamination, making it challenging to fully analyze them [77].

To ensure consistent production meeting quality standards, a tailored control strategy aligned with the properties of the product and manufacturing processes is necessary. Integrating this approach with effective manufacturing protocols can provide a consistent production of pharmaceutical cocrystals and formulations that meet the desired quality standards [78]. Control measures may vary depending on whether cocrystals are employed for a specific formation or distributed as raw materials for many generic formulations. Designating cocrystals as APIs and providing active substance master files can facilitate their distribution for formulation manufacture. Clarifying terminology about the quality of medicinal cocrystals would facilitate their utilization.

6.3.4 Regulatory Considerations

Crystal engineering of pharmaceutical cocrystals offers a promising approach to improving solubility. Regulatory agencies like the FDA and EMA have issued guidelines for herbal cocrystals as intermediates, similar to API-excipient complexes; the EMA aligns this documentation with salts. Regulatory approval requires demonstrating stability, dissolution, and safety of coformers. Both agencies emphasize that cocrystals enhance drug development but require strict compliance with manufacturing and characterization standards [12, 44, 79].

6.4 Reported Herbal Cocrystals

The description of herbal cocrystals and conformers with their methodology of formation and characteristic features is mentioned in Table 6.1.

6.5 Future Perspectives and Trends

6.5.1 Advances in Cocrystal Technology

New developments in the field of pharmaceutical materials science have expanded our understanding of the relationship between the functional capabilities of APIs and their molecular and crystal structures. The intentional creation of medicinal materials has become possible as a result of this increased understanding. The influence of APIs' solid-state and physicochemical features on the effectiveness of pharmaceutical formulations is becoming more well-acknowledged [102]. Cocrystals, consisting of several coformers and APIs, have shown promise in customizing the characteristics of solid dosage forms [103].

Table 6.1 Description of herbal cocrystals and conformers with their methodology of formation and characteristic features.

Sr. no.	Herbal cocrystals (API)	Coformer	Method	Inference	References
1.	Berberine chloride	Citric acid	Manual grinding using an agate mortar	• Enhanced thermal stability, along with resistance toward changes in the humidity, was noted. • Furthermore, enhancement in tablet ability, solubility, and dissolving rates.	[80]
2.	Berberine chloride	Fumaric acid	Recrystallization and slurry method	• A notable improvement in the speed at which the substance dissolves was observed. • Enhanced durability when exposed to high levels of humidity and temperature.	[81]
3.	Berberine chloride	Myricetin and dihydromyricetin	Slurry method	• Improvement in water solubility was observed with the use of myricetin, resulting in synergistic anticancer actions for the HT-29 cell line in both cocrystals.	[67]
4.	Chrysin	Berberine	Ball mill	• Almost 1.7-times increase in bioavailability was observed in rats in comparison to that of the pure form.	[82]

(Continued)

Table 6.1 Description of herbal cocrystals and conformers with their methodology of formation and characteristic features. (*Continued*)

Sr. no.	Herbal cocrystals (API)	Coformer	Method	Inference	References
5.	Apigenin	4,4'-bipyridine	Solvent-drop grinding	• Improvement in the solubility profile, followed by increased efficacy and lower toxicity, were seen, thereby resulting in a significantly enhanced anti-inflammatory impact.	[83]
6.	Baicalein	Isoniazid, isonicotinamide, caffeine, theophylline, and theophylline trihydrate	Reaction crystallization and liquid diffusion	• The greatest improvement in the rate of dissolution was observed with caffeine. • 4.1-factor increase in the values of the area under the curve and better bioavailability.	[84]
7.	Chrysin	Cytosine and thiamine hydrochloride	Solvent-drop grinding	• Achieved a three- to fourfold increase in the dissolution and solubility rate. • Along with the improvement in the *in vivo* and *in vitro* characteristics such as antihemolytic, antioxidant, and anti-inflammatory properties.	[85]

(*Continued*)

Table 6.1 Description of herbal cocrystals and conformers with their methodology of formation and characteristic features. (*Continued*)

Sr. no.	Herbal cocrystals (API)	Coformer	Method	Inference	References
8.	Curcumin	Resorcinol, catechol, hydroxyquinol, hydroquinone, and pyrogallol	Rapid solvent evaporation	• Compound stability was strengthened when exposed to high humidity conditions. • The compound's tablet-forming ability improved compared to its pure form. • The rate of dissolution increased by a factor of 7 when it was in a cocrystal form with hydroxyquinol.	[86]
9.	Genistein	4,4'-bipyridine	Solution evaporation	• Enhanced antibacterial efficacy, along with increased solubility when compared to the pure form was observed.	[63]
10.	Hesperetin	Picolinic acid, nicotinamide, and caffeine	Liquid-assisted grinding	• Achieved an increase of 4 to 5 factors in dissolution profile. • Thereby, resulting in increased safety against inflammation because of enhanced antioxidant and antihemolytic characteristics.	[64]

(*Continued*)

Table 6.1 Description of herbal cocrystals and conformers with their methodology of formation and characteristic features. (*Continued*)

Sr. no.	Herbal cocrystals (API)	Coformer	Method	Inference	References
11.	Isoniazid	Quercetin	Ball milling	• The dissolution rate was enhanced by 51.67 times, resulting in significant improvements. • The bioavailability was boosted by a factor of 28.91, leading to increased effectiveness. • Additionally, enhancement in the hepatoprotective activity of quercetin was also seen.	[65]
12.	Curcumin	Ascorbic acid	Solvent evaporation	• Significant rise in aqueous solubility of the substance, by an increase of 576 times concluded. • Therefore, better ability of the substance to dissolve also led to improvement in biological effects.	[66]
13.	Emodin	Berberine chloride dihydrate	Stirring	• Enhancements in the maximum concentration (Cmax), along with area under the curve (AUC) values when comparing it to pure emodin.	[67]

(*Continued*)

Table 6.1 Description of herbal cocrystals and conformers with their methodology of formation and characteristic features. (*Continued*)

Sr. no.	Herbal cocrystals (API)	Coformer	Method	Inference	References
14.	Berberine	3-methyl cinnamic acid	Solvent evaporation	• Observation of increased solubility in several polar solvents was noted. • A 7.7-fold increase in absorption was found.	[68]
15.	Luteolin	Caffeine and isoniazid	Liquid-assisted grinding method	• Increased solubility by a factor of 3 was reported when isoniazid was combined. • Both coformers showed a spike in AUC by 2.7- and 1.4-fold, respectively.	[69]
16.	Quercetin	Nicotinamide	Gas antisolvent process	• Superior dissolution performance at pH levels of 1.2 and 6.8 compared to a pure drug, achieving at least twice times that of pure drug release.	[87]
17.	Quercetin	Pentoxifylline, betaine, theophylline, and L-pyrroline	Solvent-drop grinding	• Enhanced the dissolution rate of quercetin compared to raw quercetin • In Sprague-Dawley rats, the oral bioavailability of quercetin increased by 64, 27, 10, and 7 times by quercetin designed with pentoxifylline, betaine, theophylline, and L-pyrroline	[88]

(*Continued*)

Table 6.1 Description of herbal cocrystals and conformers with their methodology of formation and characteristic features. (*Continued*)

Sr. no.	Herbal cocrystals (API)	Coformer	Method	Inference	References
18.	Quercetin and apixaban	Acetonitrile and ethyl acetate	Solution crystallization	• Enhanced the dissolution rates due to the presence of solvents • Remained stable at high temperatures and illumination conditions	[89]
19.	Curcumin and resveratrol	CO_2 and acetone	Cocrystallization with Supercritical solvent	• Antioxidant activity and dissolution rate were enhanced • Solubility of curcumin and resveratrol ameliorated by 1.5 and 2 times • Behavioral tests in mice model indicated an increase in antinociceptive and anti-inflammatory potency	[90]
20.	Curcumin	N-acetylcysteine	Antisolvent gas technique	• Demonstrated greater solubility in acidic and neutral pH environments compared to pure drug • Inhibit neutrophil chemotaxis showed in *in vitro* testing. • *In vivo* tests revealed enhanced antinociceptive and anti-inflammatory effects compared to pure curcumin.	[91]

(*Continued*)

Table 6.1 Description of herbal cocrystals and conformers with their methodology of formation and characteristic features. (*Continued*)

Sr. no.	Herbal cocrystals (API)	Coformer	Method	Inference	References
21.	Curcumin	4,4-bipyridine	-	• The powder dissolution experiment demonstrated a notable spring effect and achieved approximately seven times the apparent solubility of raw curcumin.	[92]
22.	Curcumin	Pyrogallol	Liquid-assisted grinding	• The wound was treated with cocrystals closed by day 12 after surgery. • Notably, by 15 days post-injury, showed a minimal scar range. • Additionally, histopathological assessments, including inflammation, granulation tissue formation, re-epithelialization, and collagen deposition, highlighted the significantly superior wound-healing capability.	[93]

(*Continued*)

Table 6.1 Description of herbal cocrystals and conformers with their methodology of formation and characteristic features. (*Continued*)

Sr. no.	Herbal cocrystals (API)	Coformer	Method	Inference	References
23.	Curcumin	L-pyrroline and piperine	Solvent evaporation technique	• After 8 hours, the cumulative drug release was 20% for pure curcumin, whereas cocrystals achieved a remarkable 71%. • In a permeability study at pH 6.8, curcumin showed a permeability of less than 0.05 mg/ml, while cocrystals demonstrated a fivefold increase, reaching 0.2545 mg/ml.	[94]
24.	Curcumin	Isonicotinamide	Solvent evaporation technique	• The solubility of cocrystals in water was 73.1 μg/ml, marking a 14-fold increase compared to the pure drug. • The cocrystals demonstrated 130% greater potency in inhibiting edema in mice compared to the pure drug.	[95]
25.	Curcumin	L-carnitine	Liquid-assisted grinding	• Demonstrated faster dissolution rates compared to pure drugs. • Pharmacokinetics studies showed that $AUC_{0\text{-}t}$ was 6.3 times higher and a C_{max} that was 10.7 times higher than the pure drug.	[96]

(*Continued*)

Table 6.1 Description of herbal cocrystals and conformers with their methodology of formation and characteristic features. (*Continued*)

Sr. no.	Herbal cocrystals (API)	Coformer	Method	Inference	References
26.	Hesperidin	L-arginine	-	• Significantly improved solubility and dissolution rate compared to pure drug. • Enhanced the *in vitro* anti-inflammatory and antioxidant activity of hesperidin compared to pure drugs.	[97]
27.	Hesperetin	Piperine	-	• Solubility experiment conducted on the powder cocrystals in simulated gastrointestinal fluid demonstrated that they significantly enhanced the dissolution behavior of the drug. • The bioavailability of the drug in cocrystals was found to be six times greater than that of the unprocessed drug.	[98]

(*Continued*)

Table 6.1 Description of herbal cocrystals and conformers with their methodology of formation and characteristic features. (*Continued*)

Sr. no.	Herbal cocrystals (API)	Coformer	Method	Inference	References
28.	Resveratrol	Piperazine	Mechanochemical synthesis	• Relatively short reaction time (30 to 60 min) for completion of cocrystallization on a large scale using a planetary ball mill and a plant reactor. • The herbal cocrystals enhanced the aqueous solubility of resveratrol by 6% to 16% at pH 6.8	[99]
29.	Resveratrol	Piperazine	Ultrasound and microwave-assisted technique	• Samples were obtained with high purity with a yield of 52.9% to 68.1%. • The solubility and *in vitro* antibacterial activity of the herbal cocrystals demonstrated promising results.	[100]
30.	Ferulic acid	Nicotinamide	Solvent evaporation technique	• The formulation method proved to be reproducible and yielded approximately 99%. • The herbal cocrystals exhibited a 70% improvement in ferulic acid solubility. • Demonstrated superior dissolution performance compared to the physical mixture at pH 6.8.	[101]

The main driving force for the investigation of pharmaceutical cocrystals is the possibility of altering their physical characteristics, namely, their rate of dissolution and hygroscopicity/physical stability. This section explains how these particular qualities of solid dosage forms can be improved by using cocrystals. Combinatorial chemistry has led to an increase in the predominance of APIs with low water solubility, which the BCS classifies as class II medicines [104]. Improving clinical efficacy requires tackling the problem of delayed dissolution in biological fluids [105]. Conventional methods like micronization and salt production frequently fail to significantly increase bioavailability [106]. Medicinal cocrystals, on the other hand, provide a different set of benefits and function as a substitute for altering the solid-state characteristics of APIs [107]. As an illustration, cocrystals of itraconazole with different carboxylic acids have shown better rates of dissolution and solubility than the free base. You may customize these cocrystal dissolving profiles by choosing cocrystal formers with the appropriate solubility ranges. However, more research is needed to establish a definitive correlation between cocrystal dissolution rates and cocrystal former solubility [108]. Furthermore, cocrystals offer a potential solution to the challenge of moisture-induced phase transformations, a critical consideration for pharmaceutical processing and storage. Cocrystallization with suitable coformers can inhibit unwanted solid-phase transformations, thereby enhancing the physical stability of moisture-sensitive pharmaceutical materials [109]. While studies on pharmaceutical cocrystals have yielded promising results, there is a need for comprehensive research encompassing the entire development cycle, from cocrystal formation to the final product [110]. Regulatory considerations regarding the eligibility of cocrystals for abbreviated new drug approval mechanisms also warrant attention, as they could significantly impact the adoption of cocrystal technology in the pharmaceutical industry [111].

6.5.2 Potential for New Herbal Therapeutics

The majority of pharmaceutical actives are typically formulated into solid dosage forms due to their favorable patient compliance, ease of administration, and cost-effectiveness [112]. However, during the development of pharmaceutical dosage forms, the bioavailability of a drug is often hindered by its low solubility in solid dosage forms [113]. For pharmaceutical actives with limited solubility and subsequent bioavailability, a significant challenge lies in enhancing drug solubility without compromising stability and other physicochemical characteristics [114].

Crystal engineering is a valuable technique for tailoring the properties of a solid dosage form at the molecular level. Cocrystallization, a method within crystal engineering, has been widely employed in the field of pharmaceutical science [115]. Pharmaceutical cocrystals are particularly intriguing as they offer the advantage of producing various solid-state forms from drugs that lack ionizable functional groups necessary for salt formation. It is well-established that the formation of cocrystals with suitable coformers can significantly improve the solubility, dissolution kinetics, thermal properties, mechanical attributes, physicochemical stability, and bioavailability of a drug. Consequently, the concept of cocrystals has gained rapid acceptance and recognition in both academic research and industrial applications [116].

6.6 Conclusion

Herbal cocrystals represent an innovative convergence of traditional herbal therapy with that of pharmaceutical sciences, therefore offering great potential to revolutionize drug delivery modalities and therapeutic efficacy. By amalgamating the therapeutic attributes of herbal compounds with the crystalline lattice structures of cocrystals, these novel formulations show improvement in absorption, durability, and dissolvability of active herbal constituents. Leveraging concepts from supramolecular chemistry are adept at optimizing drug formulations, mitigating challenges like limited aqueous solubility and low bioavailability often observed in herbal therapeutics. Moreover, the utilization of cocrystals might alleviate any adverse reactions and foster patient compliance through accurate dosing and controlled release kinetics. In addition to their use in pharmaceuticals, herbal cocrystals offer the potential for developing novel nutraceuticals and functional foods, catering to the escalating need for natural and sustainable healthcare alternatives. As ongoing research progresses, herbal cocrystals stand positioned to transform conventional herbal medicine by heralding in a new era of personalized treatments that offer improvement in effectiveness and safety profiles [1].

References

1. Desiraju, G.R., Crystal engineering: from molecule to crystal. *J. Am. Chem. Soc.*, 135, 27, 9952–67, 2013.
2. Wang, H., Mills, J., Sun, B., Cui, H., Therapeutic supramolecular polymers: Designs and applications. *Prog. Polym. Sci.*, 148, 101769, 2023.
3. Kukreti, N., Rani, R., Varshney, V.K., Chitme, H.R., Important medicinal plants recommended in management of rheumatoid arthritis. *Bangladesh Pharma. J.*, 25, 2, 125–36, 2022.
4. Tiwari, R.V., Patil, H., Repka, M.A., Contribution of hot-melt extrusion technology to advance drug delivery in the 21st century. *Expert Opin. Drug Deliv.*, 13, 3, 451–64, 2016.
5. Saggar, S., Mir, P.A., Kumar, N., Chawla, A., Uppal, J., Kaur, A., Traditional and herbal medicines: opportunities and challenges. *Pharmacogn. Res.*, 14, 2, 107–114, 2022.
6. Sharma, M., Rathi, R., Kaur, S., Singh, I., Abd Kadir, E., Chahardehi, A.M., Lim, V., Antiinflammatory activity of herbal bioactive-based formulations for topical administration, in: *Recent Developments in Anti-Inflammatory Therapy*, pp. 245–277, Academic Press, Cambridge, United States, 2023.
7. Bonifacio, B.V., da Silva, P.B., Ramos, MAS., Negri, K.M.S., Bauab, T.M., Chorilli, M., Nanotechnology-based drug delivery systems and herbal medicines: a review. *Int. J. Nanomed.*, 9, 1–15, 2014.
8. Liu, L., Wang, J.-R., Mei, X., Enhancing the stability of active pharmaceutical ingredients by the cocrystal strategy. *Cryst. Eng. Comm.*, 24, 11, 2002–22, 2022.
9. Nechipadappu, S.K., Tekuri, V., Trivedi, D.R., Pharmaceutical co-crystal of flufenamic Acid: Synthesis and characterization of two novel drug-drug co-crystal. *J. Pharm. Sci.*, 106, 1384–1390, 2017.
10. Nagpal, K., Pawar, P., Rathi, R., Gaur, N., Singh, I., Anthelmintic Potential of Herbal Drugs, in: *Herbal Drugs for the Management of Infectious Diseases*, pp. 341–57, Academic Press, Wiley online library, New Jersey, USA, 2022.
11. Chavan, R.B., Thipparaboina, R., Yadav, B., Shastri, N.R., Continuous manufacturing of co-crystals: challenges and prospects. *Drug Deliv. Transl. Res.*, 8, 1726–39, 2018.
12. Karagianni, A., Malamatari, M., Kachrimanis, K., Pharmaceutical cocrystals: New solid phase modification approaches for the formulation of APIs. *Pharmaceutics*, 10, 1, 18, 2018.
13. Berry, D.J. and Steed, J.W., Pharmaceutical cocrystals, salts and multicomponent systems; intermolecular interactions and property based design. *Adv. Drug Deliv. Rev.*, 117, 3–24, 2017.
14. Elder, D.P., Patterson, J.E., Holm, R., The solid-state continuum: a perspective on the interrelationships between different solid-state forms in drug substance and drug product. *J. Pharm. Pharmacol.*, 67, 6, 757–72, 2015.

15. Chettri, A., Subba, A., Singh, G.P., Bag, P.P., Pharmaceutical co-crystals: A green way to enhance drug stability and solubility for improved therapeutic efficacy. *J. Pharm. Pharmacol.*, 76, 1, 1–12, 2024.
16. Araya-Sibaja, A.M., Fandaruff, C., Wilhelm, K., Vega-Baudrit, J.R., Guillén-Girón, T., Navarro-Hoyos, M., Crystal engineering to design of solids: from single to multicomponent organic materials. *Mini-Rev. Org. Chem.*, 17, 5, 518–38, 2020.
17. Nair, A.R., Lakshman, Y.D., Anand, V.S.K., Sree, K.N., Bhat, K., Dengale, S.J., Overview of extensively employed polymeric carriers in solid dispersion technology. *AAPS Pharm. Sci. Tech.*, 21, 1–20, 2020.
18. Lu, J. and Rohani., S., Polymorphism and crystallization of active pharmaceutical ingredients (APIs). *Curr. Med. Chem.*, 16, 7, 884–905, 2009.
19. Hickey, M.B., Peterson, M.L., Scoppettuolo, L.A., Morrisette, S.L., Vetter, A., Guzmán, H., *et al.*, Performance comparison of a co-crystal of carbamazepine with marketed product. *Eur. J. Pharm. Biopharm.*, 67, 1, 112–9, 2007.
20. Singh, M., Shen, K., Ye, W., Gao, Y., Lv, A., Liu, K., *et al.*, Achieving High-Temperature Phosphorescence by Organic Cocrystal Engineering. *Angew. Chem.*, 136, e202319694, 2024.
21. Thayyil, A.R., Juturu, T., Nayak, S., Kamath, S., Pharmaceutical co-crystallization: Regulatory aspects, design, characterization, and applications. *Adv. Pharm. Bull.*, 10, 2, 203, 2020.
23. Ngilirabanga, J.B. and Samsodien, H., Pharmaceutical co-crystal: An alternative strategy for enhanced physicochemical properties and drug synergy. *Nano Select*, 2, 3, 512–26, 2021.
24. Buddhadev, S.S. and Garala, K.C., Pharmaceutical cocrystals—a review. *Proc. MDPI*, 62, 1, 14, 2021.
25. Angevine, D., *Crystal Engineering of Next Generation Harm Reduction Nicotine Salts, Co-Crystals & Polymorphs Thereof*, State University of New York, Buffalo, America, 2023, Phd thesis.
26. Tharik, A.A. and Meyyanathan, S.N., Recent patents of pharmaceutical co-crystals: product development on anti-cancer drugs and beyond. *Recent Pat. Anticancer Drug Discov.*, 18, 3, 246–67, 2023.
27. Kumar, A. and Nanda, A., *In-silico* methods of cocrystal screening: A review on tools for rational design of pharmaceutical cocrystals. *J. Drug Deliv. Sci. Technol.*, 63, 102527, 2021.
28. Chang, Y., Hawkins, B.A., Du, J.J., Groundwater, P.W., Hibbs, D.E., Lai, F., A guide to *in silico* drug design. *Pharmaceutics*, 15, 1, 49, 2023.
29. Khan, S., Zahoor, M., Rahman, M.U., Gul., Z., Cocrystals; basic concepts, properties and formation strategies. *Z. Phys. Chem.*, 237, 3, 273–332, 2023.
30. Galek, P.T., Fábián, L., Motherwell, W.S., Allen, F.H., Feeder, N., Knowledge-based model of hydrogen-bonding propensity in organic crystals. *Acta Crystallogr. B Struct. Sci.*, 63, 5, 768–82, 2007.
31. Taylor, R. and Wood, P.A., A million crystal structures: The whole is greater than the sum of its parts. *Chem. Rev.*, 119, 16, 9427–77, 2019.

32. Aitipamula, S., Zaiqun, Y., Chow, P.S., Tan., R.B., *Pharmaceutical Co-crystals—Molecular Design and Process Development*, Royal Society of Chemistry, London, UK, 2018.
33. Cruz-Cabeza, A.J., Lusi, M., Wheatcroft, H.P., Bond, A.D., The role of solvation in proton transfer reactions: implications for predicting salt/co-crystal formation using the Δp K a rule. *Faraday Discuss.*, 235, 446–66, 2022.
34. Abbott, S. and Hansen, C.M., Hansen solubility parameters in practice. *Hansen-Solubility*, 1, 1–30, 2008.
35. Ancheria, R.K., Jain, S., Kumar, D., Soni, S.L., Sharma, M., An overview of pharmaceutical co-crystal. *Asian J. Pharm. Res. Develop.*, 7, 2, 39–46, 2019.
36. Sulaimon, A.A. and Tajuddin, M.Z., Application of COSMO-RS for pre-screening ionic liquids as thermodynamic gas hydrate inhibitor. *Fluid Ph. Equilib.*, 450, 194–9, 2017.
37. Sarathi, P. and Padhi, S., Insight of the various *in silico* screening techniques developed for assortment of cocrystal formers and their thermodynamic characterization. *Drug Dev. Ind. Pharm.*, 47, 10, 1523–34, 2021.
38. Aitipamula, S., Antonijevic, I., Baruah, J.B., Berryman, O.B., Boldyreva, E.V., Decato, D.A., Fourmigue, M., Groeneman, R.H., Gryl, M., Hutchins, K.M., Johmoto, K., *Multi-component crystals: synthesis, concepts, function*, Walter de Gruyter GmbH & Co KG, Germany, 2017.
39. Manin, A.N., Drozd, K.V., Churakov, A.V., Perlovich, G.L., Hydrogen bond donor/acceptor ratios of the coformers: do they really matter for the prediction of molecular packing in cocrystals? The case of benzamide derivatives with dicarboxylic acids. *Cryst. Growth Des.*, 18, 9, 5254–69, 2018.
40. Aljohani, M., *Enhancing the bioavailability of active pharmaceutical ingredients through cocrystallization and coamorphization*, NUI Galway, 2019.
41. Rathi, R. and Singh., I., Multicomponent crystal compromising dasatinib and selected co-crystals formers: a patent evaluation of EP2861589B1. *Pharm. Pat. Anal.*, 11, 1, 15–21, 2021.
42. Berner, M., Zarko, V.E., Talawar, M.B., Nanoparticles of energetic materials: synthesis and properties. *Combust. Explos. Shock Waves*, 49, 625–47, 2013.
43. Hossain Mithu, M.S., Economidou, S., Trivedi, V., Bhatt, S., Douroumis, D., Advanced methodologies for pharmaceutical salt synthesis. *Cryst. Growth Des.*, 21, 2, 1358–74, 2021.
44. Healy, A.M., Worku, Z.A., Kumar, D., Madi., A.M., Pharmaceutical solvates, hydrates and amorphous forms: A special emphasis on cocrystals. *Adv. Drug Deliv. Rev.*, 117, 25–46, 2017.
45. Anand, R., Kumar, A., Nanda., A., Pharmaceutical Co-Crystals-Design, Development and Applications. *Drug Deliv. Lett.*, 10, 3, 169–84, 2020.
46. Mendoza-Muñoz, N., Alcalá-Alcala, S., Quintanar-Guerrero, D., Preparation of polymer nanoparticles by the emulsification-solvent evaporation method: from vanderhoff's pioneer approach to recent adaptations, in: *Polymer Nanoparticles for Nanomedicines: A Guide for their Design, Preparation and Development*, pp. 87–121, 2016.

47. Tasharrofi, N., Nourozi, M., Marzban, A., How liposomes pave the way for ocular drug delivery after topical administration. *J. Drug Deliv. Sci. Technol.*, 67, 103045, 2022.
48. Gang, W., Jie, W.J., Ping, Z.L., Ming, D.S., Ying, L.J., Lei, W., Fang, Y., Liposomal quercetin: evaluating drug delivery *in vitro* and biodistribution *in vivo*. *Expert Opin. Drug Deliv.*, 9, 6, 599–613, 2012.
49. De Crozals, G., Bonnet, R., Farre, C., Chaix, C., Nanoparticles with multiple properties for biomedical applications: A strategic guide. *Nano Today*, 11, 4, 435–63, 2016.
50. Liu, R., Luo, C., Pang, Z., Zhang, J., Ruan, S., Wu, M., Wang, L., Sun, T., Li, N., Han, L., Shi, J., Advances of nanoparticles as drug delivery systems for disease diagnosis and treatment. *Chin. Chem. Lett.*, 34, 2, 107518, 2023.
51. Rathi, R. and Singh, I., Hot Melt Extrusion Technique for Developing Pharmaceutical Co-crystals: A Review. *Drug Deliv. Lett.*, 14, 2, 97–108, 2024.
52. Hazarika, H., Krishnatreyya, H., Chattopadhyay, P., Saha, A., Pathak, Y.V., Zaman, M.K., Nanoemulsion Delivery of Herbal Products: Prospects and Challenges. *Nano Med. Nano Saf. Recent Trend. Clin. Evid.*, 1, 267–88, 2020.
53. Aneef Mohammed, Y., *Solubility Enhancement of Co-Crystal Based Solid Dosage Form*, JKK Nattraja College of Pharmacy, Komarapalayam, 2012.
54. Erxleben, A., Application of vibrational spectroscopy to study solid-state transformations of pharmaceuticals. *Curr. Pharm. Des.*, 22, 32, 4883–911, 2016.
55. Banedar, P.N., *A raman spectroscopic study of solid dispersions and co-crystals during the pharmaceutical hot melt extrusion process*, University of Bradford, 2015.
56. Dutta, A., Fourier transform infrared spectroscopy. *Spectrosc. Methods Nanomater. Character.*, 2, 73–93, 2017.
57. Patel, D.J. and Puranik, P.K., Pharmaceutical Co-crystal: An Emerging Technique to enhance Physicochemical properties of drugs. *Int. J. Chem. Tech. Res.*, 13, 283–90, 2020.
58. Liu, H., Tong, H.H., Zhou, Z., Feasibility of thermal methods on screening, characterization and physicochemical evaluation of pharmaceutical cocrystals. *J. Therm. Anal. Calorim.*, 147, 23, 12947–63, 2022.
59. Jagia, M., Kale, D.P., Bansal, A.K., Patel., S., Novel co-crystals and eutectics of febuxostat: Characterization, mechanism of formation, and improved dissolution. *AAPS Pharm. Sci. Tech.*, 23, 1–17, 2022.
60. Reutzel-Edens., S.M., Analytical techniques and strategies for salt/co-crystal characterization. *Pharm. Salts Co-cryst.*, 16, 212–46, 2011.
61. Nanjwade, V.K., Manvi, F., Nanjwade, B.K., Maste, M.M., New trends in the co-crystallization of active pharmaceutical ingredients. *J. Appl. Pharm. Sci.*, 1, 01–5, 2011.
62. Gagnière, E., Mangin, D., Puel, F., Rivoire, A., Monnier, O., Garcia, E., *et al.*, Formation of co-crystals: kinetic and thermodynamic aspects. *J. Cryst. Growth*, 311, 9, 2689–95, 2009.

63. Zhang, Y.N., Yin, H.M., Zhang, Y., Zhang, D.J., Su, X., Kuang, H.X., Preparation of a 1: 1 cocrystal of genistein with 4, 4′-bipyridine. *J. Cryst. Growth*, 458, 103–9, 2017.
64. Chadha, K., Karan, M., Bhalla, Y., Chadha, R., Khullar, S., Mandal, S., Vasisht, K., Cocrystals of hesperetin: structural, pharmacokinetic, and pharmacodynamic evaluation. *Cryst. Growth Des.*, 17, 5, 2386–405, 2017.
65. Liu, F., Wang, L.Y., Yu, M.C., Li, Y.T., Wu, Z.Y., Yan, C.W., A new cocrystal of isoniazid-quercetin with hepatoprotective effect: The design, structure, and *in vitro/in vivo* performance evaluation. *Eur. J. Pharm. Sci.*, 144, 105216, 2020.
66. Pantwalawalkar, J., More, H., Bhange, D., Patil, U., Jadhav, N., Novel curcumin ascorbic acid cocrystal for improved solubility. *J. Drug Deliv. Sci. Technol.*, 61, 102233, 2021.
67. Li, P., Ramaiah, T., Zhang, M., Zhang, Y., Huang, Y., Lou, B., Two cocrystals of berberine chloride with myricetin and dihydromyricetin: crystal structures, characterization, and antitumor activities. *Cryst. Growth Des.*, 20, 1, 157–166, 2019.
68. He, Y., Chen, S., Li, M., Gao, Y., Feng, H., Umar, Q., Feng, Y., Novel co-crystal of 3-methylcinnamic acid with berberine (1: 1): synthesis, characterization, and intestinal absorption property. *Drug Dev. Ind. Pharm.*, 49, 10, 617–627, 2023.
69. Luo, Y., Chen, S., Zhou, J., Chen, J., Tian, L., Gao, W., Zhou, Z., Luteolin cocrystals: Characterization, evaluation of solubility, oral bioavailability and theoretical calculation. *J. Drug Deliv. Sci. Technol.*, 50, 248–254, 2019.
70. Rathi, R., Kaur, S., Singh, I., A review on co-crystals of Herbal Bioactives for solubility enhancement: preparation methods and characterization techniques. *Cryst. Growth Des.*, 22, 3, 2023–42, 2022.
71. Guo, C., Zhang, Q., Zhu, B., Zhang, Z., Bao, J., Ding, Q., Mei, X., Pharmaceutical cocrystals of nicorandil with enhanced chemical stability and sustained release. *Cryst. Growth Des.*, 20, 10, 6995–7005, 2020.
72. Dutt, B., Choudhary, M., Budhwar, V., A comparative study of selected drug delivery systems: Key emphasis on cocrystallization. *Drug Deliv. Lett.*, 11, 2, 136–55, 2021.
73. Veisi, H., Karmakar, B., Tamoradi, T., Hemmati, S., Hekmati, M., Hamelian, M., Biosynthesis of CuO nanoparticles using aqueous extract of herbal tea (Stachys Lavandulifolia) flowers and evaluation of its catalytic activity. *Sci. Rep.*, 11, 1, 1983, 2021.
74. Azizkhani, M., Jafari Kiasari, F., Tooryan, F., Shahavi, M.H., Preparation and evaluation of food-grade nanoemulsion of tarragon (*Artemisia dracunculus* L.) essential oil: antioxidant and antibacterial properties. *J. Food Sci. Technol.*, 58, 1341–1348, 2021.
75. Firooziyan, S., Amani, A., Osanloo, M., Moosa-Kazemi, S.H., Basseri, H.R., Hajipirloo, H.M., Sedaghat, M.M., Preparation of nanoemulsion of Cinnamomum zeylanicum oil and evaluation of its larvicidal activity against

a main malaria vector Anopheles stephensi. *J. Environ. Health Sci. Eng.*, 19, 1025–1034, 2021.

76. Shimpi, M.R., Alhayali, A., Cavanagh, K.L., Rodríguez-Hornedo, N., Velaga, S.P., Tadalafil–malonic acid cocrystal: Physicochemical characterization, pH-solubility, and supersaturation studies. *Cryst. Growth Des.*, 18, 8, 4378–87, 2018.
77. Chen, Y., Li, L., Yao, J., Ma, Y.Y., Chen, J.M., Lu, T.B., Improving the solubility and bioavailability of apixaban via apixaban–oxalic acid cocrystal. *Cryst. Growth Des.*, 16, 5, 2923–30, 2016.
78. McNamara, D.P., Childs, S.L., Giordano, J., Iarriccio, A., Cassidy, J., Shet MS, M.S., *et al.*, Use of a glutaric acid cocrystal to improve oral bioavailability of a low solubility API. *Pharm. Res.*, 23, 1888–97, 2006.
79. Alamgir, A.N.M. and Alamgir, A.N.M., Herbal drugs: their collection, preservation, and preparation; evaluation, quality control, and standardization of herbal drugs. *Ther. Use Med. Plants Their Extracts Vol. 1 Pharmacogn.*, 1, 453–495, 2017.
80. Lu, Q., Dun, J., Chen, J.M., Liu, S., Sun, C.C., Improving solid-state properties of berberine chloride through forming a salt cocrystal with citric acid. *Int. J. Pharm.*, 554, 14–20, 2019.
81. Yang, D., Cao, J., Jiao, L., Yang, S., Zhang, L., Lu, Y., Du, G., Solubility and stability advantages of a new cocrystal of berberine chloride with fumaric acid. *ACS Omega*, 5, 14, 8283–8292, 2020.
82. Sa, R., Zhang, Y., Deng, Y., Huang, Y., Zhang, M., Lou, B., Novel salt cocrystal of chrysin with berberine: preparation, characterization, and oral bioavailability. *Cryst. Growth Des.*, 18, 8, 4724–4730, 2018.
83. Ma, X.Q., Zhuang, C., Wang, B.C., Huang, Y.F., Chen, Q., Lin, N., Cocrystal of apigenin with higher solubility, enhanced oral bioavailability, and anti-inflammatory effect. *Cryst. Growth Des.*, 19, 10, 5531–5537, 2019.
84. Zhu, B., Zhang, Q., Wang, J.R., Mei, X., Cocrystals of baicalein with higher solubility and enhanced bioavailability. *Cryst. Growth Des.*, 17, 4, 1893–1901, 2017.
85. Chadha, R., Bhalla, Y., Nandan, A., Chadha, K., Karan, M., Chrysin cocrystals: Characterization and evaluation. *J. Pharm. Biomed. Anal.*, 134, 361–371, 2017.
86. Wong, S.N., Hu, S., Ng, W.W., Xu, X., Lai, K.L., Lee, W.Y.T., Chow, S.F., Cocrystallization of curcumin with benzenediols and benzenetriols via rapid solvent removal. *Cryst. Growth Des.*, 18, 9, 5534–5546, 2018.
87. Dias, J.L., Rebelatto, E.A., Hotza, D., Bortoluzzi, A.J., Lanza, M., Ferreira, S.R., Production of quercetin-nicotinamide cocrystals by gas antisolvent (GAS) process. *J. Supercrit. Fluids*, 188, 105670, 2022.
88. Haskins, M.M., Kavanagh, O.N., Sanii, R., Khorasani, S., Chen, J.M., Zhang, Z.Y., Dai, X.L., Ren, B.Y., Lu, T.B., Zaworotko, M.J., Tuning the pharmacokinetic performance of quercetin by cocrystallization. *Cryst. Growth Des.*, 23, 8, 6059–66, 2023.

89. Zhang, L., Kong, D., Wang, H., Jiao, L., Zhao, X., Song, J., Yang, D., Yang, H., Yang, S., Du, G., Lu, Y., Cocrystal of apixaban–quercetin: Improving solubility and bioavailability of drug combination of two poorly soluble drugs. *Molecules*, 26, 9, 2677, 2021.
90. Dal Magro, C., dos Santos, A.E., Ribas, M.M., Aguiar, G.P., Volfe, C.R., Lopes, M.L., Siebel, A.M., Müller, L.G., Bortoluzzi, A.J., Lanza, M., Oliveira, J.V., Production of curcumin-resveratrol cocrystal using cocrystallization with supercritical solvent. *J. Supercrit. Fluids*, 171, 105190, 2021.
91. Paulazzi, A.R., Alves, B.O., Zilli, G.A., Dos Santos, A.E., Petry, F., Soares, K.D., Danielli, L.J., Pedroso, J., Apel, M.A., Aguiar, G.P., Siebel, A.M., Curcumin and n-acetylcysteine cocrystal produced with supercritical solvent: characterization, solubility, and preclinical evaluation of antinociceptive and anti-inflammatory activities. *Inflammopharmacology*, 30, 1, 327–41, 2022.
92. Shen, R., Zhang, J., Wang, X., Cocrystal of curcumin with 4, 4′-bipyridine toward improved dissolution: Design, structure analysis, and solid-state characterization. *J. Mol. Struct.*, 1284, 135348, 2023.
93. Samie, A., Alavian, H., Vafaei-Pour, Z., Mohammadpour, A.H., Jafarian, A.H., Danesh, N.M., Abnous, K., Taghdisi, S.M., Accelerated Wound Healing with a Diminutive Scar through Cocrystal Engineered Curcumin. *Mol. Pharm.*, 20, 10, 5090–107, 2023.
94. Singh, M., Takawale, S., Patil, R., Chaudhari, S., Pathan, A., Sangshetti, J., Moin, A.T., Zubair, T., Uddin., M.H., Development and Characterization of Co-crystals Assisted with *In-silico* Screening for Solubility and Permeability Enhancement of Curcumin. *BioRxiv.*, 1, 2023–09, 2023.
95. Wicaksono, Y., Barikah, K.Z., Della Yudatama, A., Maulia, H., Setyawan, D., Increasing the Solubility and Anti-Inflammatory Activity of Curcumin by Cocrystallization. *Sci. Technol. Indones.*, 8, 3, 501–8, 2023.
96. Wang, H., Zheng, C., Tian, F., Xiao, Z., Sun, Z., Lu, L., Dai, W., Zhang, Q., Mei, X., Improving the dissolution rate and bioavailability of curcumin via co-crystallization. *Pharmaceuticals*, 17, 4, 489, 2024.
97. Elshaer, M., Osman, S.K., Mohammed, A.M., Zayed, G., Co-crystallization of Hesperidin with different co-formers to enhance solubility, antioxidant and anti-inflammatory activities. *Pharm. Dev. Technol.*, 29, 1–2, 2024.
98. Liu, Y., Yang, F., Zhao, X., Wang, S., Yang, Q., Zhang, X., Crystal structure, solubility, and pharmacokinetic study on a hesperetin cocrystal with piperine as coformer. *Pharmaceutics*, 14, 1, 94, 2022.
99. Mitran, R.A., Ioniţă, S., Lincu, D., Soare, E.M., Atkinson, I., Rusu, A., Pandele-Cuşu, J., Iordache, C., Pongratz, I., Pop, M.M., Fruth, V., Mechanochemical Synthesis of Resveratrol–Piperazine Cocrystals. *Mater.*, 17, 13, 3145, 2024.
100. Ioniţă, S., Pătraşcu, M., Soare, E.M., Lincu, D., Atkinson, I., Rusu, A., Pop, M.M., Iordache, C., Uşurelu, C.D., Baltac, A.S., Mitran, R.A., Rapid Synthesis and Evaluation of Resveratrol-Piperazine Cocrystals by Ultrasound and Microwave Methods. *Pharm. Res.*, 7, 1–1, 2024.

101. Júnior, J.V., Dos Santos, J.A., Lins, T.B., de Araújo Batista, R.S., de Lima Neto, S.A., de Santana Oliveira, A., Nogueira, F.H., Gomes, A.P., de Sousa, D.P., de Souza, F.S., Aragão, C.F., A new ferulic acid–nicotinamide cocrystal with improved solubility and dissolution performance. *J. Pharm. Sci.*, 109, 3, 1330–7, 2020.
102. Lieberman, H. and Vemuri., N.M., Chemical and Physicochemical Approaches to solve formulation problems. *Pract. Med. Chem.*, 1, 767–91, 2015.
103. Bidhuri, N. and Padhi, S., Review on comprehensive description of development and assessment of co-crystal drug delivery system. *Clin. Trials*, 13, 16, 2023.
104. Sathisaran, I. and Dalvi, S.V., Engineering cocrystals of poorly water-soluble drugs to enhance dissolution in aqueous medium. *Pharmaceutics*, 10, 3, 108, 2018.
105. Wen, H., Jung, H., Li, X., Drug delivery approaches in addressing clinical pharmacology-related issues: opportunities and challenges. *AAPS J.*, 17, 1327–40, 2015.
106. Al-Kassas, R., Bansal, M., Shaw, J., Nanosizing techniques for improving bioavailability of drugs. *J. Control. Release*, 260, 202–12, 2017.
107. Gadade, D.D. and Pekamwar, S.S., Pharmaceutical cocrystals: regulatory and strategic aspects, design and development. *Adv. Pharm. Bull.*, 6, 4, 479, 2016.
108. Shete, A., Murthy, S., Korpale, S., Yadav, A., Sajane, S., Sakhare *et al.*, S., Cocrystals of itraconazole with amino acids: Screening, synthesis, solid state characterization, *in vitro* drug release and antifungal activity. *J. Drug Deliv. Sci. Technol.*, 28, 46–55, 2015.
109. Thakur, T.S. and Thakuria, R., Crystalline multicomponent solids: An alternative for addressing the hygroscopicity issue in pharmaceutical materials. *Cryst. Growth Des.*, 20, 9, 6245–65, 2020.
110. Garg, U. and Azim., Y., Challenges and opportunities of pharmaceutical cocrystals: a focused review on non-steroidal anti-inflammatory drugs. *RSC Med. Chem.*, 12, 5, 705–21, 2021.
111. McKellar, S.C., *Crystal engineering approaches to solid-state pharmaceutical systems*, University of Strathclyde, UK, 2012, PhD thesis.
112. Arshad, M.S., Zafar, S., Yousef, B., Alyassin, Y., Ali, R., AlAsiri, A., *et al.*, A review of emerging technologies enabling improved solid oral dosage form manufacturing and processing. *Adv. Drug Deliv. Rev.*, 178, 113840, 2021.
113. Allen Jr., L.V., Dosage form design and development. *Clin. Ther.*, 30, 11, 2102–11, 2008.
114. Zhang, X., Xing, H., Zhao, Y., Ma., Z., Pharmaceutical dispersion techniques for dissolution and bioavailability enhancement of poorly water-soluble drugs. *Pharmaceutics*, 10, 3, 74, 2018.

115. Rathi, R., Guggal, Y., Twinkle, Sandhu, V., Singh, I., Exploring the Patent Landscape and Regulatory Prospective on Pharmaceutical Cocrystals. *J. Pharm. Technol. Res. Manag.*, 11, 49–60, 2023.
116. Ghadi R, G.A., Ghumre, S., Waghmare, N., Kadam, V.J., Co-crystals: emerging approach in pharmaceutical design. *Am. J. Pharm. Res.*, 4, 7, 3881–3893, 2014.

7

Drug-Drug Cocrystals: Advancements and Applications

Ritu Rathi[1], Sushant Sharma[1], Siddhartha Gautham Pal[1], Rohil Panwar[2], Taruna[3], Rakesh Pahwa[2] and Inderbir Singh[1*]

[1]*Chitkara College of Pharmacy, Chitkara University, Punjab, India*
[2]*Institute of Pharmaceutical Sciences, Kurukshetra University, Kurukshetra, Haryana, India*
[3]*Pt. C.L.S. Government College, Karnal, Haryana, India*

Abstract

Drug-drug cocrystals (DDCs) are a combination of two or more drugs arranged into a single crystalline structure. DDCs provide enormous benefits such as improved solubility, stability, and bioavailability without altering the chemical integrity of the individual components. This method provides an avenue for reformulating existing drugs but also facilitates the creation of synergistic therapeutic combinations that may enhance efficacy and reduce adverse effects. There are numerous methods for the preparation of DDCs including solvent-based, solvent-free approaches which are further classified in this chapter. During the preparation of DDCs, some of the parameters are most important to consider, such as critical process parameters (CPP), and quality target product profile (QTPP), which define the desired properties of the final product. It is crucial to consider the ability of DDC to withstand different storage conditions. Various evaluation tests for multiple drug cocrystals (MDC) are reviewed in this book chapter. Even with their potential benefits, the development of DDCs has several challenges. Some of the challenges, such as the selection of appropriate drugs and their combinations, require careful consideration of factors such as solubility profiles, molecular interactions, and compatibility between dosage forms. Additionally, there are a limited number of marketed drug substances that restrict the available combinations for DDC formation. To overcome these challenges, various strategies were explored, including selecting appropriate techniques for preparing DDCs and investigating different stoichiometric ratios in cocrystal formation. Here, we review the

**Corresponding author*: inderbir.singh@chitkara.edu.in

Inderbir Singh, Subrata Mallick and Ritu Rathi (eds.) Cocrystals in Pharmaceutical Sciences: Design to Applications, (189–230) © 2025 Scrivener Publishing LLC

DDCs that were reported in the last decade along with their patents and marketed products.

Keywords: Drug-drug cocrystal, multiple drug crystals, patents, marketed

7.1 Introduction

Monotherapies (i.e., targeting a specific receptor) are no longer effective in treating many complicated disorders, including HIV/AIDS, cancer, diabetes, cardiovascular disease, and infectious diseases. As an outcome, the combination of more than one therapeutic agent in unit dosage form has gained popularity [1]. The successful development of bilayer tablet products addresses the challenges of drug incompatibility and enables both rapid and slow drug release from a single tablet. However, this innovation introduces new challenges in formulation design, manufacturing process control, and product performance that are not seen in traditional monolithic tablets [2]. Therefore, it is necessary to develop new technologies to overcome these problems. Hence, the need for drug-drug cocrystals (DDCs) seems to be important. According to the FDA, cocrystals are defined as "dissociable multi-component solid crystalline supramolecular complexes composed of two or more components within the same crystal lattice where the components are in a neutral state and interact via nonionic interactions." Generally, cocrystals consist of an API and a coformer; when both the components of cocrystals are API (active pharmaceutical ingredient), it is called DDCs, also known as multiple drug cocrystals (MDCs) [3, 4]. DDCs can address the problems associated with traditional or conventional combination drug therapies. The fixed-dose combinations (FDC) of the multi-targeted drugs are observed to de-escalate the associated adverse outcomes, dosage frequency with good efficiency, and medicinal effects. In addition to this, DDCs decrease the occurrence of drug resistance which culminates in more pronounced and simplified disease management [5]. The first ever cocrystal was reported by Friedrich Wöhler in 1844, consisting of quinone and hydroquinone. Many DDCs were then discovered in the late 1800s and early 1900s reported in the book Organische Molekulverbindungen, published by Paul Pfeiffer in 1922. Throughout the 1900s, more cocrystals originated, while some were found by accident, others were found through screening methods. By understanding how intermolecular interactions impact crystal packing, cocrystals with specific physical and chemical features can be engineered to alter the physicochemical properties of drugs [6]. Therefore, the concept of DDCs has generated massive interest in the pharmaceutical field.

DDCs help to meet the criteria of innovation, as well as practicality for pharmaceutical development, which are necessary for patent eligibility. Multidrug cocrystals offer advantages such as increased bioavailability [7], enhanced solubility and dissolution of component drugs [8–10], potential stabilization through intermolecular interactions [11], synergistic or additive therapeutic effects [12], and opportunities for the life cycle management of existing pharmaceutical products [13]. Along with physiochemical properties, DDCs can enhance biological activities as well by adding an appropriate, therapeutically useful component without requiring any chemical alteration [14]. Another study comprising meloxicam and aspirin reported a 12-time decrease in the Tmax, which resulted in bioavailability enhancement fourfold [15]. Palash *et al.* reported DDC of curcumin with resorcinol and pyrogallol that showed improved drug solubility [16]. DDCs have also gained a new form of intellectual property protection that nanoparticles, microspheres, liposomes, and niosomes currently lack. This IP protection provides a regulatory and competitive advantage for DDC products over other novel drug delivery technologies in the market. DDCs can overcome the challenges of fixed-dose combinations and could overcome the additional indications by drug-drug interactions with modified physicochemical properties of each drug [17]. DDCs have many benefits and have outstanding development in the field of pharmaceutical sciences but they still have major challenges due to lack of commercial viability and economic factors and are difficult to evaluate [18]. There are limited drug-drug combinations available until now, which leads to challenges in the development of MDCs. One of the major factors is limited solubility; they possess less solubility than the parent drug, and their therapeutic effect is not reversible. However, DDCs cover the majority of challenges that are not present in traditional drugs. DDCs are explored in various therapeutic areas such as DDCs of nonsteroidal anti-inflammatory drugs (NSAIDs), DDCs of anti-tubercular drugs, and DDCs of diuretic drugs [19–21].

Despite having so many advantages, DDCs face their own development challenges. One of the challenges is its synthesizing and designing. DDCs mainly prepared to provide a synergistic effect because it is made up of the combination of two APIs but the combination of two drugs will not always provide a guarantee that DDCs will be effective and provide a better therapeutic effect than monolithic therapy. Therefore, it will not ensure that our combination is pharmaceutically acceptable. For example, a scientist named Tomita synthesized DDC in 1977 consisting of theophylline and phenobarbital. A combination of phenobarbital (a nervous system depressant) and theophylline (antiasthmatic) should provide a synergistic effect but there is no supporting evidence for synergism or any practical application in therapy. Furthermore, phenobarbital is a cytochrome P450 (CYP) inducer and theophylline undergoes first-pass metabolism, which can lead to decreased plasma concentration of theophylline [22]. Moreover, cocrystals consist of fixed stoichiometric ratios but the dose ratio obtained from DDCs may not be in accordance with the clinical applications. For example, among scientists, Desiraju *et al.* developed a DDC consisting of zidovudine and lamivudine as both of the drug substances are antiviral agents, so it may provide some benefit. However, the mass ratio of DDC consisting of lamivudine and zidovudine (1:1.2) was dissimilar from their oral dosage form having a mass ratio of 1:2 which could result in its difficulty in clinical application [23]. Another challenge is the quantification of multiple drugs, especially in the case of nutraceuticals having poor solution state stability. Furthermore, challenges are represented in the figure. Primarily, we can use three aspects to overcome these disadvantages. Firstly, it is important to select cocrystallization techniques that align with the properties, nature, and morphology of drugs involved. Therefore, choosing the most suitable method will result in more effective screening procedures and may increase the success rate of DDC preparation. Secondly, one of the problems associated with DDCs is different stoichiometric ratios which may be overcome if more emphasis is given to exploring DDCs with different stoichiometric ratios, which could be applied to triumph over the dosage problem. Thirdly, we can think of expanding the DDCs to drug- drug-drug-nutraceutical cocrystals, which may be beneficial and can ease the task of development.

At last, we need to consider several strategic methods such as investigating varied stoichiometric ratios, cocrystallization techniques, and also considering the combination of drug-drug nutraceutical cocrystals. These actions are critical for getting at most from DDCs, finally leading to improved patient care, treatment, safety, and effectiveness [24].

7.1.1 Preparation Techniques

Various methods are used for preparing DDCs, including solvent-based and solvent-free approaches with miscellaneous methods, as listed in Table 7.1. Solvent-based methods utilize high solvent consumption, with the choice of solvent impacting the cocrystallization outcome [25]. On the other hand, solvent-free (solid-state) methods offer the potential to eliminate the need for solvents in cocrystal synthesis. Solvent-based methods include solvent evaporation, vapor diffusion, slurry transformation

Table 7.1 Various methods of preparation techniques for drug-drug cocrystals.

	Methods of cocrystallization	Description
Solvent-based techniques	Solvent evaporation	API and coformer are dissolved in a common solvent. Then, the solution slowly evaporates over time. As the solvent evaporates, the drug and coformer molecules come together and form a cocrystal.
	Slurry conversion	In this method, API and coformer are mixed in a liquid solvent for the formation of slurry. Then, the slurry is stirred, and the solvent is allowed to evaporate to get the cocrystals.
	Vapor diffusion	API and coformer are dissolved in a common solvent. Then, the solution is placed in a closed container. After that, a vapor of a different solvent is added to the container. As the vapor diffuses into the solution, the formation of cocrystals takes place.
	High-pressure homogenization (HPH)	With high-pressure drug and coformer mixed. Then, the mixture is passed through a narrow nozzle under controlled pressure, which results in the nanosized cocrystal formation.
	High shear granulation (HSG)	The drug and coformer are mixed with a binder and placed in high shear force within the granulator. The speed of the impeller and chopper are maintained constant during granulation. The high-shear mixing results in the formation of granules of cocrystals.

(Continued)

Table 7.1 Various methods of preparation techniques for drug-drug cocrystals. (*Continued*)

	Methods of cocrystallization	Description
	Ultrasound-assisted	The drug and the coformer are dissolved in a solvent. Then, the solution is exposed to ultrasound waves, which results in crystal growth by increasing mass transfer and disrupting agglomerates. The formation of cocrystals takes place.
Solvent-free techniques	Neat grinding	With the application of manual or mechanical pressure, the API and coformer are combined with their dry solid form.
	Liquid-assisted grinding(LAG)	The API and coformer are combined in their solid forms. A catalytic quantity of liquid is added to the mixture by applying pressure through manual or mechanical means to promote cocrystal preparation.
	Hot-melt extrusion (HME)	Cocrystallization is formed by pressure and heat to melt API and coformer. Cocrystals developed without the use of solvents make it an eco-friendlier option.
	Matrix-assisted cocrystallization (MAC)	Equimolar amounts of the API (drug) and the **coformer** are mixed with a **matrix material** in the solid state. Then, the mixture is fed into a hot-melt extruder that is responsible for forming a cocrystal.
	Polymer-assisted grinding (POLAG)	API and the coformer are blended with a polymer in their solid state. The grinding process promotes the cocrystal formation.

(*Continued*)

Table 7.1 Various methods of preparation techniques for drug-drug cocrystals. (*Continued*)

	Methods of cocrystallization	**Description**
Miscellaneous techniques	Microwave-assisted	In this technique with or without the use of solvent, the drug with the coformer is taken in equimolar ratios. Then, it is placed in microwave irradiation in a microwave reactor.
	Freeze drying (lyophilization)	In the freeze-drying process, the drug and the coformer are dissolved in a solvent. Then, the solution is frozen to form a solid mass. When the frozen solid is placed in the vacuum, it undergoes a sublimation process, and the cocrystals collect as a dry powder.
	Electrospray	The drug and the coformer are dissolved in a solvent. Then, the solution passed through a nozzle with the influence of an electric field, and fine droplets formed. The solvent evaporation method results in the formation of cocrystals.
	Laser irradiation	In the laser irradiation process, the drug and the coformer are used in powder form. With the help of a high-power CO_2 laser, this process results in the formation of cocrystals.
	Resonant acoustic mixing (RAM)	It involves subjecting a mixture of the drug and the coformer to resonant acoustic waves. These pressure waves stimulate efficient mixing and cocrystal formation.

crystallization, high-shear granulation, high-pressure homogenization, and ultrasound-assisted cocrystallization [26]. In contrast, solid-based approaches consist of methods like pure grinding crystallization, liquid-assisted grinding, hot-melt extrusion, matrix-assisted cocrystallization, and polymer-assisted grinding. Additionally, microwave-assisted cocrystallization, freeze drying, electrospraying, laser irradiation, and resonant acoustic mixing form the miscellaneous techniques used in DDC production [27]. The different miscellaneous methods, i.e., the newer methods are used in large-scale manufacturing of DDC as compared with the conventional methods. Conventional methods are used for small-scale development of DDC. The different strategies for preparing cocrystals are given below [28, 29].

There are some critical process parameters in the preparation of DDCs. The process parameters and input variables validate the Critical Quality Attributes (CQAs) that define the desired properties of the final product [30]. It is crucial to consider the ability of DDC to withstand different storage conditions. Cocrystals tend to break down when exposed to higher temperatures and high humidity levels during storage, primarily due to the partial dissolution of the acids used as coformers. The speed and arrangement of the screw are crucial factors, especially in processes such as hot-melt extrusion [31, 32]. The quality target product profile (QTPP) outlines essential quality characteristics for DDC, encompassing factors such as solubility, stability, particle size, appearance, permeability, dissolution rate, as well as the pharmacokinetic and therapeutic behavior of cocrystals [33]. Critical Quality Attributes (CQAs) represent specific subsets of the QTPP that must adhere to defined ranges, limits, or distributions to ensure the desired quality traits of cocrystal. These attributes encompass the physical, chemical, biological, and microbiological properties of the final product [34]. Newer techniques such as hot-melt extrusion and screw melt extrusion have started swapping traditional solvent-based methods [35]. These new processes offer easier scalability, reduced development time, increased energy transfer rates, minimized plant space requirements, reduced waste production and energy usage, improved specificity, and consistent development conditions to prevent the buildup of harmful substances, all contributing to a more cost-effective manufacturing process [14].

7.1.2 Characterization of DDCS

Currently, there are various evaluation tests available for MDC; they are based on knowledge-based approaches. Some of the methods include synthon engineering, molecular complementarity, hydrogen bonding propensity, Hansen solubility parameter, pKa-based lattice energy calculations,

Table 7.2 Various instrumental techniques for characterization of DDCs.

Instrument	Working principle
Single-crystal XRD	Constructive interference is supported by the interaction between the monochromatic X-rays and the crystalline sample.
Powder XRD	Because the wavelength of X-rays and the interatomic spacing in crystals are comparable, X-rays are diffracted by cocrystals.
Raman spectroscopy, FTIR	Provide details regarding the stretching and vibration of molecules, atoms, or lattices.
Hirshfeld surface analysis	Investigation of the interactions between molecules within a crystal structure.
Terahertz spectroscopy (THz)	It uses electromagnetic fields that spin at a frequency ranging from several hundred gigahertz to several terahertz in order to examine the vibrations occurring both within and between molecules in molecular solids.
Solid-state nuclear magnetic resonance	Examining the structural characteristics based on the nucleus's spin, which produces the magnetic field.
Hot-stage microscopy	Examining materials at different times and temperatures utilizing both thermal analysis and microscopy.
Differential scanning calorimetry	The measurement of enthalpy changes caused by changes in a substance's physical and chemical properties over time or at different temperatures.
Thermogravimetric analysis	It is predicated on the composition and thermal stability of the substance being measured.
Scanning electron microscopy	SEM scans the surface of the sample using a focused electron beam.

and phase diagrams based on solubility and melting. These screening methods are being widely used for DDC systems [36]. Prashant *et al.* used a synthon-based retrosynthetic strategy to develop an MDC of lamivudine and zidovudine [23].

The characterization techniques for DDCs are similar to those used for common cocrystals. The analytical methods include microscopy, single-crystal x-ray diffraction, powder x-ray diffraction, solid-state nuclear magnetic resonance, infrared spectroscopy, Raman spectroscopy, thermogravimetric analysis, differential scanning calorimetry, dynamic vapor sorption, differential scanning calorimetry, thermogravimetric analysis, among others [37]. The analysis of compositions containing cocrystals is essential to evaluate their interactions with excipients and instabilities due to changes in temperature, pH, and dissociation [38, 39]. The various instrumentations and physicochemicals for characterization of DDCs are listed below in Tables 7.2 and 7.3.

Table 7.3 Various physicochemical techniques for characterization of DDCs.

Name	Description
Hansen solubility study (HSS)	This is an essential tool for predicting how well coformers and medications will work together throughout the crystallization process.
Hygroscopicity	It might change a drug's physiochemical characteristics, including its solubility, rate of dissolution, mechanical attributes, stability, and bioavailability.
Melting point	Drug cocrystallization may change the drugs' melting points and improve their thermal stability.
Stability	The cocrystals should be exposed to different stability conditions, including temperatures and humidity.
Photostability	For evaluating the photodegradation of the cocrystals and ensuring suitable storage, handling, and transportation conditions.
Tabletability	Evaluated for over-countering the scale-up issues, improved mechanical properties, and reducing tablet defects.

7.2 Reported Drug-Drug Cocrystals for Non-Steroidal Anti-Inflammatory Drugs (NSAIDs)

NSAIDs have been utilized for alleviating pain, reducing fever, and managing various inflammatory conditions [40]. The Biopharmaceutical Classification System (BCS) categorizes NSAIDs as class II, distinguished by having low solubility in aqueous solutions [41]. For increasing the solubility of drugs, cocrystals emerge as an innovative approach. In this context, Bhatia *et al.* synthesized cocrystals of ketoprofen-p-aminobenzoic acid (KP-PABA) through the solvent evaporation method employing a central composite experimental design. The cocrystals were evaluated for their analgesic and anti-inflammatory properties. The study found that KP-PABA cocrystals had a solubility of 60.11 µg/ml and a drug release rate of 93.45%. The optimized formulation had a lower IC_{50} value of 248.79 µg/ml compared to a pure drug (524.40 µg/ml), suggesting greater protection against protein denaturation compared to ketoprofen alone. *In vivo* studies revealed that cocrystals significantly reduced paw edema volume in Wistar rats, indicating a promising drug delivery system for anti-inflammatory activity [14]. Likewise, Charoenchaitrakool *et al.* fabricated cocrystals of mefenamic acid, nicotinamide, and paracetamol utilizing the gas anti-solvent technique. Box-Behnken design was implemented to assess the effect of process factors on the mefenamic acid dissolving rate and determine the ideal environment for preparing cocrystals with the extreme mefenamic acid dissolution rate. Findings indicated that the quickest dissolving time of cocrystals was 4.19 min while utilizing a precipitation temperature of 25°C. Prepared cocrystals had a dissolving rate 16.40 times higher than unprocessed mefenamic acid. Therefore, outcomes showed that the cocrystal can accelerate mefenamic acid's dissolution rate [42]. Similarly, Ma *et al.* synthesized meloxicam (MLX)-salicylic acid (SA) cocrystals using SA as the coformer. The dissolving curve demonstrated that the MLX-SA cocrystals increased the dissolution rate and solubility of MLX. Moreover, the gel was developed, which improved the transdermal penetration without influencing the *in vitro* release of MLX. The MLX-SA cocrystals incorporated gel successfully treated rheumatoid arthritis in rats [43]. In another study, Aramini *et al.* employed a slurry approach for producing the ketoprofen lysin salt (KLS) and gabapentin (GABA) cocrystals. KLS-GABA cocrystals enhanced KLS solubility and permeability in the gastrointestinal tract, as indicated by phytochemical, biodistribution, and pharmacokinetic studies. *In vivo,* the study revealed that KLS-GABA cocrystal improved systemic absorption of KLS as compared to KLS alone or combined with GABA. The prepared cocrystals produced the synergistic effect of KLS and GABA. In the inflammatory and neuropathic pain rat model, the KLS-GABA cocrystal alleviated pain and spinal neuroinflammation. The findings highlighted that KLS-GABA cocrystals were an effective strategy for chronic pain treatment [44]. Chen *et al.* utilized a melt crystallization technique to prepare celecoxib-carbamazepine cocrystals. Single-crystal X-ray diffraction (XRD) was used to examine cocrystal structure and growth dynamics with the aim of optimizing circumstances for the hot-melt extrusion method. The synthesized cocrystals altered the intrinsic dissolving rate of the celecoxib drug and showed a synchronized release pattern. This study unveiled that melt crystallization was a feasible technique for generating DDCs on a large scale [28]. Nechhaipadappu and associates synthesized cocrystals of flufenamic acid (FFA) containing 2-chloro-4-nitrobenzoid acid (CNB) and ethenzamide (ETZ) as coformers. The cocrystals were prepared by utilizing solvent evaporation and solvent-drop–assisted grinding techniques. In the FFA-CNB cocrystal, FFA's solubility doubled, while there was no rise in the FFA-ETZ cocrystals. At 75% relative humidity, the cocrystals remained stable for six months at room temperature without exhibiting any hygroscopic behavior [45].

Table 7.4 Salient outcomes of NSAID-based cocrystals.

Drug	Coformers	Method of preparation	Salient outcomes	Ref.
Diclofenac	Ciprofloxacin	Mechanochemical	20.8 times higher solubility of diclofenac without affecting the antibacterial efficacy of ciprofloxacin, reduced dose and side effects of parent drugs	[47]
Etodolac	Glutaric acid	Cooling crystallization	Increased two to three times the solubility of etodolac than pure drug	[48]

(*Continued*)

Table 7.4 Salient outcomes of NSAID-based cocrystals. (*Continued*)

Drug	Coformers	Method of preparation	Salient outcomes	Ref.
Etodolac	Isonicotinamide	Mechanical grinding	Enhanced dissolution rate, solubility, and permeability	[49]
Naproxen	Caprolactam	Liquid-assisted grinding	11.9 times increased solubility of naproxen and caprolactam	[50]
Niflumic acid	Caffeine	Mechanochemical	Improved the solubility and stability of the drug, reduced dose, and sustained drug release	[51]
Naproxen	L-alanine	Liquid-assisted grinding	Improved dissolution rate and oral bioavailability	[52]
Piroxicam	Febuxostat	---	Piroxicam solubility was improved 2.8-fold compared to pure drug	[53]
Ibuprofen	Nicotinamide	Slow evaporation	Enhanced solubility and analgesic activity	[54]
Niflumic acid	Caprolactam	Liquid-assisted grinding	Increased 1.4 times solubility and 2 times dissolution rate	[55]

(*Continued*)

Table 7.4 Salient outcomes of NSAID-based cocrystals. (*Continued*)

Drug	Coformers	Method of preparation	Salient outcomes	Ref.
Tramadol hydrochloride	Celecoxib	Wet grinding	Celecoxib released faster, while tramadol released more slowly from cocrystals	[56]
Naproxen	Nicotinamide	---	High dissolution rate	[57]
Tramadol	Celecoxib	---	Improved pharmacokinetic profile and analgesic activity	[58]

Devi and colleagues developed KP cocrystals with fumaric acid as a coformer. The cocrystals were prepared using a solvent-assisted grinding method. The formulation substantially improved solubility and dissolution rate, with IC_{50} values of 327.33 μg/ml and 556.11 μg/ml, respectively. Compared to the pure drugs, the cocrystal formulation demonstrated greater resistance to protein denaturation. *In vivo* experiments demonstrated that the cocrystal formulation produced a faster pharmacological effect in Wistar rats and albino mice than the standard drugs [46]. A list of recent DDCs of NSAIDs is represented in Table 7.4.

Therefore, cocrystals present a novel method to improve therapeutic efficacy and overcome the delivery challenges associated with NSAIDs.

7.3 Reported Drug-Drug Cocrystals for Anticancer Drugs

Cancer represents a significant global cause of mortality, and for the management of this, many anticancer drugs are used, but these have low solubility and bioavailability, which reduces their therapeutic efficiency. The current research focuses on enhancing oral absorption of these low-solubility drugs [59, 60]. To address these challenges, cocrystals emerge as a novel therapeutic strategy. Duan *et al.* prepared cocrystals of palbociclib with orcinol or resorcinol. The fabricated cocrystals were studied using a variety of methods, including XRD and thermogravimetric analysis. The cocrystals of palbociclib exhibited improved dissolution rates under pH conditions of 1.0 and 6.8, and increased solubility two to three times while maintaining high stability. The palbociclib-resorcinol cocrystals elevated relative bioavailability in rats by 1.24 times compared to palbociclib alone. Additionally, it displayed a higher IC_{50} value for healthy cells, indicating a better biosafety profile. So, cocrystallization might be an effective technique for enhancing palbociclib's physicochemical characteristics and pharmacokinetics profile [61]. Liu *et al.* developed a DDC hydrate by mixing dihydromyricetin (slightly water-soluble) with pentoxifylline (very water-soluble) in a 1:1 ratio. The cocrystals were produced through hydrogen bonding interactions between hydroxyl groups of dihydromyricetin and acceptor groups of pentoxifylline. The dihydromyricetin-pentoxifylline cocrystals decreased the hydrophilicity and solubility gap between these drugs and released simultaneously and continuously. At 100 μM dosage, the cocrystals had a synergistic anticancer effect on HepG2 cells. Therefore, the study identified cocrystallization as an appropriate strategy for the prolonged release of drugs with varied water solubility [62]. Yin *et al.* fabricated cocrystals of lobaplatin using flavonoids like fisetin, quercetin (QCT), naringenin myricetin, and luteolin as coformers. The anticancer activity of all prepared cocrystals was assessed through the CCK-8 assay. The outcomes unveiled that the cocrystals of lobaplatin-myricetin monohydrate, lobaplatin-luteolin monohydrate, and lobaplatin-fisetin monohydrate had higher efficacy than lobaplatin alone. These advancements may provide an appropriate solution for the overwhelming limitations of the parent drugs [63]. In another study, Yu *et al.* prepared 5-fluorouracil (5-FU)–L-proline zwitterionic cocrystals. Additionally, the prepared cocrystals were incorporated into carriers composed of poly(ethylene glycol)-poly(caprolactone) to form cocrystal micelles. Research findings demonstrated a substantial improvement in 4.6 times the solubility and 3.89 times the permeability of the pure 5-FU drug. Moreover, the drug loading and entrapment efficiency of 5-FU cocrystal micelles showed enhancement of 2.39 times and 1.74 times, respectively, in comparison to the FU micelles without cocrystals. Both the cocrystals and their micelles demonstrated outstanding antineoplastic activity with the cocrystal micelles exhibiting even more pronounced improvements. The findings suggest the potential utility of cocrystal-based formulations in cancer therapy [64]. Hao *et al.* synthesized cocrystals of gallic acid and 5-FU to improve the antitumor and synergistic action. The dissolution rate and solubility of the 5-FU–gallic acid cocrystals were less compared to pure 5-FU, but membrane permeability was improved. It significantly enhanced oral absorption, with an $AUC_{0\text{-}t}$

Table 7.5 Anticancer drug–based cocrystals for the management of various cancers.

Drug	Coformers	Preparation method	*In vivo* model	Key findings	Ref.
Panobinostat	Dabrafenib	Mechanochemical	$BRAF^{V600E}$	Improved dissolution rate and decreased survival rate of cancer cells.	[69]
5-FU	Trimethoprim	---	SW480, MDA-MB-231, and H1299	Increased aqueous solubility, effectively inhibited breast cancer cells and low toxicity.	[70]
Phenylahistin	Microtubule inhibitor	---	BALB/c mice	Improved pharmacokinetic profile, reduced toxicity	[71]
Lenvatinib	Salicylic acid and sulfamerazine	Slow solvent evaporation	Hep G2	Lenvatinib-salicylic acid CC showed 120 times higher solubility than lenvatinib alone. Exhibited 2.25 times more aqueous solubility compared to lenvatinib alone	[72]
Nandrolone	Salicylic acid	Grinding	HeLa cell	Enhanced anticancer activity	[73]
Curcumin	Ascorbic acid	Solvent evaporation	---	Increased dissolution rate and aqueous solubility	[74]

(*Continued*)

Table 7.5 Anticancer drug–based cocrystals for the management of various cancers. (*Continued*)

Drug	Coformers	Preparation method	*In vivo* model	Key findings	Ref.
Luteolin monohydrate	Caffeine and isoniazid	Liquid-assisted grinding and rapid solvent removal	Male Sprague-Dawley rats	Improved solubility and bioavailability	[75]
Niclosamide	Nicotinamide	Spray drying	A549	Enhanced bioavailability and reduced dose	[76]
Flucytosine	Sinapic acid and syringic acid	---	DU145	Upregulated the apoptotic pathway, flucytosine repurposed for anticancer activities	[77]
Curcumin	Trimesic acid	---	MDA-MB-231	Increased dissolution rate and bioavailability of curcumin	[78]
Berberine chloride	Myricetin and dihydromyricetin	Slurry	HT-29	Showed a synergistic effect	[79]
Palbociclib	Adipic acid	Slurry	---	Palbociclib solubility was improved 56-fold compared to pure drug	[80]
Gefitinib	Thiosemicarbazide	Recrystallization	A549 and H1299	Improved physiochemical and biopharmaceutical performance of gefitinib.	[81]

of 2.78 times that of 5-FU alone. Moreover, the cocrystals demonstrated greater cytotoxic activity against 4T1 mouse breast cancer cells [65]. In another study, a slurry liquid-assisted grinding and solvent evaporation approach was utilized to fabricate the cocrystals of rucaparib (Ruc)-theophylline monohydrate (TPMH), Ruc-ethyl maltol, and Ruc-maltol. All prepared cocrystals dissolved more rapidly than Ruc-camsylate salt at pH 2 and 4.5, with Ruc-TPMH having the maximum apparent solubility at pH 4.5 and 6.8. Pharmacokinetic studies on rats found that Ruc-TPMH had 2.4 times the C_{max} and 1.4 times the $AUC_{0\text{-}24h}$ at a single dosage compared to Ruc-Cam [66]. George and coworkers designed cocrystals of olaparib (OLA) and kaempferol (KAE) for sustained release. Intrinsic dissolution rate (IDR) tests indicated that OLA-KAE cocrystals had 0.40 times higher IDR values than pure OLA. Dissolution tests lowered the dissolution rate to 21.4% of the original drugs. After 3 months of accelerated conditions, OLA-KAE cocrystals demonstrated improved physical stability. OLA-KAE cocrystals inhibited ovarian cancer cells more effectively than OLA alone. The OLA-KAE cocrystals were observed to be a viable strategy for generating prolonged-release formulations with increased pharmacological efficacy [67]. Yin and colleagues formulated cocrystals of oxaliplatin (OXA) by employing baicalein and naringenin as cocrystal formers. Compared to OXA, both cocrystals displayed a slower release rate and prolonged hydrolysis. Cell toxicity was investigated utilizing the CCK-8 assay, which established that OXA-naringenin was safer, while OXA-baicalein had more substantial effects on suppressing cancer cells [68]. Table 7.5 lists the reported cocrystals of anticancer drugs along with a method of preparations and key findings. Therefore, cocrystals are an effective strategy for elevating both the efficacy and safety profiles of anticancer drugs, including potential enrichment in solubility, dissolution rates, bioavailability, and, ultimately, therapeutic results.

7.4 Reported Drug-Drug Cocrystals for Antituberculosis Drugs

Tuberculosis (TB) is a contagious disorder triggered by *Mycobacterium Tuberculosis* (MTB), predominantly affects the lungs and is characterized by symptoms such as chronic coughing, fever, and chest pain [82]. It has emerged as a major global health concern in the previous few years because development of multidrug resistance strains and rising poverty [83]. For the management of TB, DDCs are gaining interest as they offer a promising, cost-effective strategy for developing enhanced drug delivery with minimal risk. These cocrystals have the potential to enhance both physicochemical and biopharmaceutical properties without chemical alterations [84]. In this context, Liu *et al.* fabricated isoniazid (INH)-QCT cocrystals to address the severe hepatotoxicity associated with the isoniazid. The study outcomes revealed that the dissolution rate of QCT significantly increased by 51.67 times. Furthermore, oral bioavailability in rats was enhanced by 28.91 times. The INH-QCT cocrystals substantially decreased hepatotoxicity related to INH. The findings highlight potential routes for reducing the toxicity of anti-TB drugs using the crystallization process [85]. Similarly, Yadav *et al.* developed cocrystals of INH with hepatoprotective coformers such as syringic acid and QCT to tackle hepatotoxicity. The investigation indicated elevated levels of alanine aminotransferase, aspartate aminotransferase, and alkaline phosphatase enzymes among the INH-treated group, while the levels found in the INH-synergic acid and INH-QCT cocrystal–treated groups were similar to those in the untreated group. Research showed favorable evidence for the efficacy of nutraceutical coformers in alleviating hepatotoxicity linked with diverse drugs [86]. Salem *et al.* fabricated cocrystals of 4-aminosalicylic acid and sulfamethazine *via* high-power ultrasound and high-pressure homogenization (HPH). Both procedures yielded grade I cocrystals and exhibited high crystallinity. HPH generated nanosized cocrystals, whereas high-power ultrasound established microsized. At higher pressures, HPH resulted in smaller, narrow-sized cocrystals with restricted distribution. HPH cocrystals seemed needle-like, whereas high-power ultrasound showed a distinct morphology. Both nano- and microsized cocrystals improved the sulfamethazine dissolution rate of both drugs [87]. Simultaneously, Sakamoto *et al.* synthesized two cocrystals of quabodepistat, one with 2,5-dihydroxybenzoic acid (2,5-DHBA) and another with 2-hydroxybenzoic acid (2-HBA). The prepared cocrystals had higher bioavailability in beagle dogs, while the 2-HBA cocrystals had 1.4 times higher bioavailability. Dissolution studies indicated similar dissolution profiles for both cocrystals, although the 2-HBA cocrystals dissolved more slowly at pH 1.2. The findings suggest its potential for improving the quabodepistat oral bioavailability [88]. Likewise, Li and coworkers synthesized the cocrystals of ethionamide and nutraceutical chrysin to reduce the hepatotoxic effect. The research found that chrysin had an excellent hepatoprotective effect, nearly eradicating the liver toxicity of ETA. This could boost ETA's therapeutic efficacy, potentially alleviating the TB epidemic [89]. Additionally, Table 7.6 outlined the identified cocrystals of antituberculosis drugs and their key consequences.

Table 7.6 Cocrystals of anti-TB drugs and their enhanced properties.

Drug	Coformers	Synthesis method	Salient outcomes	Ref.
Prothionamide	Salicylic and fumaric acid	Neat/solvent-drop grinding and slow evaporation	Fivefold increased aqueous solubility compared to the crystalline form of prothionamide.	[90]
Isoniazid	Glutaric acid	Solid-state and solvent-assisted grinding	Isoniazid-Glutaric acid Enhanced solubility in different pH environments and in Pyrazinamide-glutaric acid improved solubility in different pH environments.	[91]
Pyrazinamide	Diflunisal	Liquid-assisted grinding	Improved the physicochemical properties and therapeutic activity of a drug	[92]
Rifampicin + pyrazinamide + ethambutol dihydrochloride	Isoniazid-caffeic and vanillic acids	---	Enhanced stability of drugs	[93]
4-aminosalicylic acid	Isonicotinamide and caffeine	Solvent-drop grinding	Improved solubility and stability	[94]
Isoniazid	Curcumin	Rapid solvent evaporation	Sustained release of isoniazid	[95]
Isoniazid	Protocatechuic acid	Slow evaporation	Enhanced stability but reduced solubility at pH 7.2	[96]

(*Continued*)

Table 7.6 Cocrystals of anti-TB drugs and their enhanced properties. (*Continued*)

Drug	Coformers	Synthesis method	Salient outcomes	Ref.
Piperazine-ferulate	Pyrazinamide	Liquid-assisted grinding	Reduced dissolution and solubility rate and sustained drug release	[97]
Isoniazid	Resveratrol	---	Decreased skin permeation by reducing solubility	[98]
5-fluorocytosine	Isoniazid	Slow evaporation	Improved stability of drug without impairing solubility of drugs	[99]

Therefore, cocrystals of anti-TB drugs represent a promising method, offering improved drug properties as well as reduced associated side effects.

7.5 Reported Drug-Drug Cocrystals for Diuretics

Diuretics are drugs designed to promote urine production, commonly employed in the management of disorders such as edema, hypertension, and heart failure [100]. Diuretics with limited aqueous solubility and permeability have led to the development of cocrystals for improving their properties. Zhang *et al.* synthesized two novel solid forms of acetazolamide (ACZ) to elevate its solubility and permeability. These forms include a salt containing piperazine and a crystal having theophylline. According to dissolution and *in vitro* permeation assays, the ACZ solubility and permeability were improved by cocrystal and salt. In conclusion, salt production was more helpful in enhancing solubility while cocrystallization might dramatically enhance the permeability [101]. Likewise, Acebedo-Martinez *et al.* developed cocrystals using furosemide and two popular NSAIDs, ethenzamide and piroxicam. The prepared samples were subjected to XRD, infrared spectroscopy, solubility, thermal analysis, and stability. The physicochemical properties were assessed, and the finding showed that the solubility improved in NSAIDs but not in furosemide. Additional research is needed to assess drug-drug interaction in these unique multicomponent solids [102]. In another study, Narala *et al.* fabricated hydrochlorothiazide (HCT) cocrystals by using resorcinol (RSL), nicotinamide (NIC), and catechol (CAT) as coformer *via* liquid-assisted grinding and hot-melt extrusion approach. Drug-coformer interactions resulted in greater solubility and higher diffusion/permeability for HCT-CAT and HCT-RSL. According to the study, hot-melt extrusion is an attractive method for synthesizing cocrystals within industrial scale-up [103]. Furosemide (FSM) is an effective diuretic prescribed for the management of congestive heart failure and edema, but it has low oral bioavailability. To tackle this issue, researchers synthesized salt and cocrystals of FSM with imidazole (IMZ) and 5-fluorocytosine (5-FCN) coformers. The new solid form's structural properties were examined through XRD, FTIR, and NMR spectroscopy. FSM-IMZ and FSM–5-FCN cocrystals showed substantial improvement in solubility, intrinsic dissolution, and permeability when compared to FSM alone. These findings suggest a potential approach to enhance FSM bioavailability [15]. The HCT molecular cocrystals by Sanphui and coworkers were produced utilizing nicotinic acid, nicotinamide-aminobenzoic acid, succinamide, resorcinol, and pyrogallol as coformer. Solubility and stability tests were carried out in water under ambient conditions. The cocrystal with p-aminobenzoic acid had a sixfold improvement in solubility over HCT and was stable in an aqueous solution for 24 hours. The cocrystals containing resorcinol, nicotinamide, and pyrogallol had a 1.5- to 2-fold rise in solubility and converted into HCT within 1 hour of dissolution [104]. Similarly, Ranjan and associates utilized a cocrystallization approach to fabricate the HCT cocrystals with water-soluble coformers phenazine (PHEN), 4-dimethylaminopyridine (DMAP), and picolinamide (PICA). It was found that HCT-DMAP and HCT-PHEN cocrystals were 4 and 1.4 times more soluble than the free drug. The study suggests the choice of coformer with considerable effect on the physicochemical characteristics of HCT. Additionally, it recommends cocrystallization as an attractive strategy for enhancing the solubility of the HCT in water [105]. Table 7.7 summarizes the cocrystal formulation of diuretics along with salient outcomes.

In summary, cocrystals of diuretics represent a significant advancement in pharmaceutical sciences, to revolutionize treatment paradigms by offering optimized therapeutic formulations with enhanced efficacy and safety profiles.

Table 7.7 Enhanced therapeutic efficacy of diuretics with cocrystals.

Drug	Coformers	Synthesis method	Salient outcomes	Ref.
ACZ	Theophylline	Liquid-assisted grinding and solvent evaporation	Enhanced solubility and permeability	[101]
Ethenzamide + Piroxicam	Furosemide	Mechanochemical	Improved solubility of ethenzamide and piroxicam	[102]
HCT	Nicotinamide	Hot-melt extrusion	Improved the twofold solubility of HCT	[103]
Spironolactone	Tryptophan	---	Threefold enhancement in the dissolution rate in comparison to the pure drug spironolactone	[106]
Acetazolamide	4,4-bipyridine-N,N-dioxide	---	Increased solubility and permeability.	[107]
HCT	Metformin hydrochloride	Mechanochemical	Increased 1.21-fold solubility of HCT.	[108]
FSM	Cytosine	Liquid-assisted grinding	11-fold higher solubility than pure FSM.	[109]
FSM	Nicotinamide	Slurry	High stability at 25°C	[110]

(*Continued*)

Table 7.7 Enhanced therapeutic efficacy of diuretics with cocrystals. (*Continued*)

Drug	Coformers	Synthesis method	Salient outcomes	Ref.
FSM	Piperazine	Liquid-assisted grinding	Three times higher solubility as compared to FZM-piperazine salt.	[111]
FSM	Isnicotinamide	Solvent-assisted mechanochemical	FSM-is nicotinamide has higher apparent solubility than FSM.	[112]

7.6 Reported Drug-Drug Cocrystals for Miscellaneous Drugs

Various DDCs have been developed for a wide range of drugs beyond those mentioned above. Therefore, miscellaneous DDCs are summarized in Table 7.8.

Overall, the aforementioned investigations indicated that DDCs are an effective strategy for enhancing the performance of diverse drugs, with

Table 7.8 Cocrystals of various drugs for the management of numerous disorders.

Drug	Coformers	Diseases	Preparation methods	Key findings	Ref.
Genistein	Piperazine	Diabetes, tumors, cardiovascular diseases, and menopausal symptoms	Grinding-assisted and solvent-based	Enhanced solubility and bioavailability	[113]
Ibrutinib	Ketoconazole	Lymphoproliferative	Liquid-assisted grinding	Elevated dissolution rate, solubility, and oral bioavailability of both drugs	[114]

(*Continued*)

Table 7.8 Cocrystals of various drugs for the management of numerous disorders. (*Continued*)

Drug	Coformers	Diseases	Preparation methods	Key findings	Ref.
Favipiravir	Theophylline	Respiratory viral infections	Rotary evaporation and spray drying	Enhanced dissolution rate	[115]
Apixaban	Quercetin	Thrombosis	Slurry	Improved solubility, bioavailability, and stability	[116]
Azithromycin	Paracetamol	Microbial infection	Slow solvent evaporation	Increased solubility and antimicrobial activity	[117]
Metaxalone	Nicotinamide	Muscle pain	---	Elevated solubility, bioavailability, and mitigated side effects	[118]
Nitrofurantoin	Trimethoprim	Urinary infection	---	Increased antibacterial activity against gram-negative at minimum concentration	[119]
Levofloxacin	Metacetamol	Bacterial infection	Grinding	Improved photostability and reduced hygroscopicity of levofloxacin	[120]
Milrinone	Ferulic acid	Cardiovascular diseases	---	Improved solubility and permeability, extended half-life, and bioavailability of drugs	[121]
Nicorandil	Suberic acid	Angina pectoris	---	Increased solubility and dissolution rate	[122]
Levofloxacin	Metacetamol	Bacterial infection	Liquid-assisted grinding	Improved photostability of levofloxacin	[123]
Efavirenz	Nevirapine	Viral infection	Solvent-assisted grinding	Showed a synergistic effect	[124]

potential advantages including increased solubility, stability, and bioavailability. Continued research and development in the field have the potential to generate advanced pharmaceutical formulations that improve patient health.

7.7 Patent Perspectives

DDCs are being developed to improve the pharmacokinetic and pharmacodynamic characteristics of drugs for the successful treatment of numerous disorders. Cocrystals have shown a wide range of benefits including improved dissolution, solubility, bioavailability, and stability. As a result, to accomplish the advantages mentioned above, there has been a progressive rise in patent filings for cocrystals, as seen in Table 7.9.

Table 7.9 Patents related to drug-drug cocrystals.

Patent table				
S. no.	**Patent no./ application no.**	**Title**	**Description**	**Reference**
1	US20120258170A1	Pharmaceutical cocrystals.	Improved solubility, IDR, stability, and great therapeutic potential were obtained. Its composition comprises quercetin and metformin.	[125]
2	EP1608339B1	Pharmaceutical cocrystal of celecoxib-nicotinamide.	Increased aqueous solubility and stability. It consists of a combination of celecoxib and nicotinamide.	[126]
3	WO2017191539 A1	Process for the preparation of dl-proline cocrystal of dapagliflozin.	Enhanced bioavailability, enhanced galenic properties, or higher purity and used in the treatment of patients with type 2 diabetes.	[127]

(*Continued*)

Table 7.9 Patents related to drug-drug cocrystals. (*Continued*)

Patent table				
S. no.	**Patent no./ application no.**	**Title**	**Description**	**Reference**
4	US8124603B2	*In vivo* studies of crystalline forms of meloxicam.	Improved bioavailability enhanced the rate of dissolution and shorter time to Cmax in blood, as compared to pure meloxicam.	[128]
5	US6001996A	Complexes of cephalosporins and carbacephalosporins with parabens	Maximize overall yield. This invention relates to novel cephalosporin and carbacephalosporin/ paraben complexes.	[129]
6	US7625910B2	Cocrystal	Less needs to control the humidity levels in the manufacturing process. Useful in the treatment of hyperproliferative diseases such as cancer.	[130]
7	EP3210975 A1	Cocrystals of lorcaserin	Promote weight loss through satiety. Its composition consists of (R)-lorcaserin hydrochloride with (+)-di-p-toluoyl-D-tartaric.	[131]
8	WO2017172811 A1	Novel cocrystals	The pharmaceutical composition is in the form of an injectable depot for extended release.	[132]

(*Continued*)

Table 7.9 Patents related to drug-drug cocrystals. (*Continued*)

Patent table				
S. no.	**Patent no./ application no.**	**Title**	**Description**	**Reference**
9	EP2300472B1	Glucocorticoid mimetics, methods of making them, pharmaceutical compositions, and uses thereof.	Providing therapeutic agents with a reduced side effect profile. Treating disease states or conditions mediated by the glucocorticoid receptor function in a patient in need of such treatment, and other uses.	[133]
10	US11478488B2	Cocrystals of tramadol and coxibs	Optimized solubility and IDR, as well as improved Cmax and bioavailability of celecoxib.	[134]
11	**2. US10150729B2**	Tranilast compositions and cocrystals	Improved solubility and photostability. Tranilast was found to inhibit the proliferation of fibroblasts and suppress collagen synthesis	[135]
12	US10689352B2	Solid-state forms of trisodium valsartan: sacubitril.	Chemical purity, flowability, solubility, dissolution rate, morphology or crystal habit, and stability—such as chemical stability as well as thermal and mechanical stability with respect to polymorphic conversion.	[136]
13	WO2012090225A2	Novel cocrystals/ molecular salts of metformin with oleoyl ethanolamide as an effective antidiabetic anti-obesity agent	Improved residence time thereby increasing the bioavailability and efficacy, as an antidiabetic agent.	[137]

Table 7.10 Drug-drug cocrystals in the market, brands, combinations, and companies.

S. no.	Brand name	Drug-drug combinations	Therapeutic category/ indication	Company	Year
1	Beta-Chlor®	Chloral betaine with chloral hydrate.	Sedative-hypnotic drug.	Mead Johnson	1963
2	Dimenhydrinate	Diphenhydramine and 8-chlorotheophylline.	Motion sickness.	Watson Laboratories	1982
3	Cafcit®	Caffeine citrate with citric acid.	Breathing problems in premature babies	HIKMA Pharmaceuticals	1999
4	Depakote®	Valproic acid with valproate sodium.	Anti-epileptic agent.	Abbott Laboratories	2002
5	Abilify®	Aripiprazole with fumaric acid	Schizophrenia Otsuka.	Otsuka Pharmaceuticals	2002
6	Lamivudine/ Zidovudine Teva®	Lamivudine with zidovudine	Human immunodeficiency virus (HIV) infection.	Teva Pharmaceuticals	2011
7	Suglat®	Ipragliflozin with L-proline	Antidiabetic.	Astellas Pharma	2014
8	Odomzo®	Sonidegib with phosphoric acid	Basal cell carcinoma.	Sun Pharmaceutical Industries Ltd	2015
9	Zafatek®	Trelagliptin and succinic acid	Antidiabetic.	Takeda Pharmaceutical Company	2015
10	Entresto®	Sacubitril and valsartan	Heart failure and to reduce the risk of cardiovascular death.	Novartis	2015
11	Steglatro®	Ertugliflozin with L-pyroglutamic	Antidiabetic.	Pfizer	2017
12	Mayzent®	Siponimod with fumaric acid	Sclerosis.	Novartis	2019
13	Seglentis®	Celecoxib and tramadol	Treatment of acute pain.	Kowa Pharmaceuticals	2021

7.8 Marketed Product

Market-ready DDC products mark a transformative rise in pharmaceutical innovation, offering improved solubility, bioavailability, and physicochemical properties, elevating patient outcomes. Their commercialization underscores the industry's dedication to overcoming traditional drug delivery challenges. Regulatory acceptance fuels an increasing market for DDC-based therapeutics, nurturing collaborations between academia and industry giants. This exemplary shift in drug development promises more targeted and efficient therapies to address medical needs effectively. With their demonstrated efficacy, DDC products are reshaping the pharmaceutical landscape, indicating a new era of precision in drug discovery and delivery. A summary of notable pharmaceutical cocrystal drug formulations with brand names, drug combinations, therapeutic indications, companies, and approval years is given in Table 7.10 [138–140].

7.9 Conclusion

Although thousands of pharmaceutical cocrystals have been described, drug-drug cocrystals are still relatively rare. This scarcity is due to the difficulty of selecting suitable drug combinations, including solubility variations, molecular interactions, packing patterns, dosage compatibility, therapeutic uses, and potential drug-drug interactions. The formation of drug-drug cocrystals is quite challenging. Hence, pharmaceutical scientists need to develop robust models that could anticipate DDC formation. In order to address these challenges, we can develop new crystallization techniques, could improve the screening processes. Moreover, broaden the reserch over cocrystals to encompass drug-nutraceutical combinations which could be new drug combination therapy and novel uses for existing pharmaceuticals.

Acknowledgment

The authors thankfully acknowledge the support and institutional facilities provided by Chitkara College of Pharmacy, Chitkara University Punjab, Institute of Pharmaceutical Sciences, Kurukshetra University, and Pt. C.L.S. Government College, Karnal, India, for providing the infrastructure and library facilities.

References

1. Bolla, G., Sarma, B., Nangia, A.K., Crystal engineering of pharmaceutical cocrystals in the discovery and development of improved drugs. *Chem. Rev.*, 122, 13, 11514–11603, 2022.
2. Srivastava, D., Fatima, Z., Kaur, D.C., Multicomponent pharmaceutical cocrystals: a novel approach for combination therapy. *Mini-Rev. Med. Chem.*, 18, 14, 1160–1167, 2018.
3. Cao, F., Rodriguez-Hornedo, N., Amidon, E.G., Mechanistic analysis of cocrystal dissolution, surface pH, and dissolution advantage as a guide for rational selection. *J. Pharm. Sci.*, 1, 243–251, 2019.
4. Thipparaboina, R., Kumar, D., Chavan, B.R., Shastri, R.N., Multidrug co-crystals: towards the development of effective therapeutic hybrids. *Drug Discov. Today*, 21, 3, 481–490, 2016.
5. Kaur, R., Cavanagh, L.K., Rodríguez-Hornedo, N., Matzger, J.A., Multidrug cocrystal of anticonvulsants: influence of strong intermolecular interactions on physiochemical properties. *Cryst. Growth design*, 17, 10, 5012–5016, 2017.
6. Almansa, C., Frampton, S.C., Vela, M.J., Whitelock, S., Plata-Salaman, R.C., Co-crystals as a new approach to multimodal analgesia and the treatment of pain. *J. Pain Res.*, 12, 2679–2689, 2019.
7. Shaikh, R., Singh, R., Walker, M.G., Croker, M.D., Pharmaceutical cocrystal drug products: an outlook on product development. *Trends Pharmacol. Sci.*, 39, 12, 1033–1048, 2018.
8. Jones, W., Motherwell, S.W., Trask, V.A., Pharmaceutical cocrystals: An emerging approach to physical property enhancement. *MRS Bull.*, 31, 11, 875–879, 2006.
9. Wang, N., Xie, C., Lu, H., Guo, N., Lou, Y., Su, W., Hao, H., Cocrystal and its application in the field of active pharmaceutical ingredients and food ingredients. *Curr. Pharm. Des.*, 24, 21, 2339–2348, 2018.
10. Goud, R.N., Suresh, K., Sanphui, P., Nangia, A., Fast dissolving eutectic compositions of curcumin. *Int. J. Pharm.*, 439, 1-2, 63–72, 2012.
11. Yadav, V.A., Shete, S.A., Dabke, P.A., Kulkarni, V.P., Sakhare, S.S., Co-crystals: a novel approach to modify physicochemical properties of active pharmaceutical ingredients. *Indian J. Pharm. Sci.*, 71, 4, 359, 2009.
12. Chavan, B.R., Thipparaboina, R., Yadav, B., Shastri, R.N., Continuous manufacturing of co-crystals: challenges and prospects. *Drug Deliv. Transl. Res.*, 8, 1726–1739, 2018.
13. Gadade, D.D. and Pekamwar, S.S., Pharmaceutical cocrystals: regulatory and strategic aspects, design and development. *Adv. Pharm. Bull.*, 6, 4, 479, 2016.
14. Nascimento, L.A., Fernandes, P.R., Charpentier, D.M., ter Horst, J.H., Caires, F.J., Chorilli, M., Co-crystals of non-steroidal anti-inflammatory drugs (NSAIDs): Insight toward formation, methods, and drug enhancement. *Particuology*, 58, 227–241, 2021.

15. Diniz, F.L., Carvalho Jr., S.P., Pena, A.S., Gonçalves, E.J., Souza, A.M., Filho de Souza, J.D., Fernandes, C., Enhancing the solubility and permeability of the diuretic drug furosemide via multicomponent crystal forms. *Int. J. Pharm.*, 587, 119694, 2020.
16. Sanphui, P. and Bolla, G., Curcumin, a biological wonder molecule: A crystal engineering point of view. Crystal Growth & Design, in: *Recent advances in improving oral drug bioavailability by cocrystals*, vol. 188, S. Emami, M. Siahi-Shadbad, K. Adibkia, M. Barzegar-Jalali (Eds.), pp. 5690–5711, p. 305, BIOIMPACTS, BI, 2018.
17. Kavanagh, N.O., Croker, M.D., Walker, M.G., Zaworotko, J.M., Pharmaceutical cocrystals: from serendipity to design to application. *Drug Discov. Today*, 24, 3, 796–804, 2019.
18. Guo, M., Sun, X., Chen, J., Cai, T., Pharmaceutical cocrystals: A review of preparations, physicochemical properties, and applications. *Acta Pharm. Sin. B.*, 11, 8, 2537–2564, 2021.
19. Peltonen, L., Design Space and QbD Approach for Production of Drug Nanocrystals by Wet Media Milling Techniques. *Pharmaceutics*, 10, 3, 104, 2018.
20. Clarke, H.D., *Crystal Engineering of Multi-Component Crystal Forms: The Opportunities and Challenges in Design*, University of South Florida, USA, 2012.
21. Banerjee, M., Nimkar, K., Naik, S., Patravale, V., Unlocking the potential of drug-drug cocrystals: A comprehensive review. *J. Control Release*, 348, 456–469, 2022.
22. Barnes, P.J., Theophylline. *Am. J. Respir. Crit. Care Med.*, 188, 8, 901–906, 2013.
23. Bhatt, P.M., Azim, Y., Thakur, T.S., Desiraju, G.R., Co-crystals of the anti-HIV drugs lamivudine and zidovudine. *Cryst. Growth Des.*, 9, 2, 951–957, 2009.
24. Rathi, R., Kushwaha, R., Goyal, A., Singh, I., Oxaliplatin-flavone pharmaceutical co-crystal-CN111205332A: patent spotlight. *Pharm. Pat. Anal.*, 11, 5, 147–154, 2022.
25. Kumar, S. and Nanda, A., Pharmaceutical cocrystals: an overview. *Indian J. Pharm. Sci.*, 79, 6, 858–871, 2017.
26. Shankar, S.J., Gowda, J.B.H., Akshatha, R.S., Metikurki, B., Cocrystallization: a novel approach to enhance the dissolution of poorly soluble drugs. *Indo. Am. J. Pharm. Res.*, 9, 3132–44, 2019.
27. Chabalenge, B., Korde, S., Kelly, A.L., Neagu, D., Paradkar, A., Understanding matrix-assisted continuous co-crystallization using a data mining approach in quality by design (QbD). *Cryst. Growth Des.*, 20, 7, 4540–4549, 2020.
28. Pawar, N., Saha, A., Nandan, N., Parambil, J.V., Solution co-crystallization: A scalable approach for co-crystal production. *Crystals*, 11, 3, 303, 2021.
29. Rathi, R. and Singh., I., Multicomponent crystal compromising dasatinib and selected co-crystals formers: a patent evaluation of EP2861589B1. *Pharm. Pat. Anal.*, 11, 1, 15–21, 2021.

30. Saindane, A.R. and Thombre, A.N., Drug-Drug Cocrystals: A Promising Approach to Overcome Barriers in Pain Management. *Cryst. Res. Technol.*, 58, 7, 2300029, 2023.
31. Maniruzzaman, M., Boateng, J.S., Snowden, M.J., Douroumis, D., A review of hot-melt extrusion: process technology to pharmaceutical products. *Int. Scholarly Res. Not.*, 2012, 1, 436763, 2012.
32. Madusanka, N., Eddleston, M.D., Arhangelskis, M., Jones, W., Polymorphs, hydrates and solvates of a co-crystal of caffeine with anthranilic acid. *Acta. Crystallography. B. Struct. Sci. Cryst. Eng. Mater.*, 70, 1, 72–80, 2014.
33. Ghosh, S., Mondal, A., Kiran, M.S.R.N., Ramamurty, U., Malla Reddy, C., The role of weak interactions in the phase transition and distinct mechanical behavior of two structurally similar caffeine co-crystal polymorphs studied by nanoindentation. *Cryst. Growth Des.*, 13, 10, 4435–4441, 2013.
34. Sharma, A., Puri, V., Kumar, P., Singh, I., Rifampicin-loaded alginate-gelatin fibers incorporated within transdermal films as a fiber-in-film system for wound healing applications. *Membranes*, 11, 1, 7, 2020.
35. Rathi, R. and Sanshita, I.S., Hot Melt Extrusion Technique for Developing Pharmaceutical Co-crystals: A Review. *Drug Deliv. Letters*, 14, 2, 97–108, 2024.
36. Malik, J., Khatkar, A., Nanda, A., A Comprehensive Insight on Pharmaceutical Co-crystals for Improvement of Aqueous Solubility. *Curr. Drug Targets*, 24, 2, 157–170, 2023.
37. Aramini, A., Bianchini, G., Lillini, S., Tomassetti, M., Pacchiarotti, N., Canestrari, D., Cocchiaro, P., Novelli, R., Dragani, M.C., Palmerio, F. and Mattioli, S., Ketoprofen, lysine and gabapentin co-crystal magnifies synergistic efficacy and tolerability of the constituent drugs: Pre-clinical evidences towards an innovative therapeutic approach for neuroinflammatory pain. *Biomed. Pharmacother.*, 163, 114845, 2023.
38. Castelli, V., d'Egidio, F., Maione, S., Luongo, L., Boccella, S., Cimini, A., Brandolini, L., Chierotti, M.R., Allegretti, M., Ketoprofen, lysine and gabapentin co-crystal magnifies synergistic efficacy and tolerability of the constituent drugs: Pre-clinical evidence towards an innovative therapeutic approach for neuroinflammatory pain. *Biomed. Pharmacother.*, 163, 114845, 2023.
39. Chen, A., Cai, P., Luo, M., Guo, M., Cai, T., Melt crystallization of celecoxib-carbamazepine cocrystals with the synchronized release of drugs. *Pharm. Res.*, 40, 567–577, 2023.
40. Sharma, M., Rathi, R., Kaur, S., Singh, I., Abd Kadir, E., Chahardehi, A.M., Lim, V., Antiinflammatory activity of herbal bioactive-based formulations for topical administration, in: *Recent Developments in Anti-Inflammatory Therapy*, pp. 245–277, Academic Press, USA, 2023.
41. Garg, U. and Azim, Y., Challenges and opportunities of pharmaceutical cocrystals: a focused review on non-steroidal anti-inflammatory drugs. *RSC Med. Chem.*, 12, 5, 705–721, 2021.

42. Charoenchaitrakool, M., Roubroum, T., Sudsakorn, K., Processing of a novel mefenamic acid-paracetamol-nicotinamide cocrystal using gas anti-solvent process. *J. CO2 Util.*, 62, 102080, 2022.
43. Ma, N., Liu, Y., Ling, G., Zhang, P., Preparation of meloxicam-salicylic acid co-crystal and its application in the treatment of rheumatoid arthritis. *J. Drug Deliv. Sci. Technol.*, 74, 103542, 2022.
44. Boccella, S., De Filippis, L., Giorgio, C., Brandolini, L., Jones, M., Novelli, R., Amorizzo, E., Leoni, M.L.G., Terranova, G., Maione, S., Luongo, L., Combination drug therapy for the management of chronic neuropathic pain. *Biomolecules*, 13, 12, 1802, 2023.
45. Nechipadappu, S.K., Tekuri, V., Trivedi, D.R., Pharmaceutical co-crystal of flufenamic Acid: Synthesis and characterization of two novel drug-drug co-crystal. *J. Pharm. Sci.*, 106, 1384–1390, 2017.
46. Devi, S., Kumar, A., Kapoor, A., Verma, V., Yadav, S., Bhatia, M., Ketoprofen–FA Co-crystal: *In vitro* and *in vivo* investigation for the solubility enhancement of drug by Design of Expert. *AAPS Pharm. SciTech.*, 23, 101, 2022.
47. Acebedo-Martínez, F.J., Domínguez-Martín, A., Alarcón-Payer, C., Sevillano-Páez, A., Verdugo-Escamilla, C., González-Pérez, J.M., Martínez-Checa, F., Choquesillo-Lazarte, D., Enhanced NSAIDs solubility in drug–drug formulations with ciprofloxacin. *Int. J. Mol. Sci.*, 24, 3305, 2023.
48. Ahirrao, S.P., Sonawane, M.P., Bhambere, D.S., Udavant, P.B., Ahire, E.D., Kanade, R., Kuber, D., Cocrystal formulation: A novel approach to enhance solubility and dissolution of etodolac. *Biosci. Biotechnol. Res. Asia*, 19, 111–119, 2022.
49. Rai, S.K., Allu, S., Nangia, A.K., Salts and cocrystal of etodolac: Advantage of solubility, dissolution, and permeability. *Cryst. Growth Des.*, 20, 4512–4522, 2020.
50. Xing, C., Chen, T., Wang, L., An, Q., Jin, Y., Yang, D., Zhang, L., Du, G., Lu, Y., Two novel co-crystals of naproxen: Comparison of stability, solubility, and intermolecular interaction. *Pharmaceuticals*, 15, 807, 2022.
51. Acebedo-Martínez, F.J., Alarcón-Payer, C., Frontera, A., Barbas, R., Prohens, R., Di Crisci, M., Domínguez-Martín, A., Gómez-Morales, J., Choquesillo-Lazarte, D., Novel polymorphic cocrystals of the non-steroidal anti-inflammatory drug niflumic acid: Expanding the pharmaceutical landscape. *Pharmaceutics*, 13, 2140, 2021.
52. Latif, S., Ijaz, Q.A., Hameed, M., ul ain Shoaib, Q., Fatima, K., Hussain, A., Arshad, M.S., Abbas, N., Improvement of physico-mechanical and pharmacokinetic attributes of naproxen by cocrystallization with L-alanine. *J. Drug Deliv. Sci. Technol.*, 61, 102236, 2021.
53. Modani, S., Gunnam, A., Yadav, B., Nangia, A.K., Nangia, A.K., Shastri, N.R., Generation and evaluation of pharmacologically relevant drug-drug cocrystal for gout therapy. *Cryst. Growth Des.*, 20, 3577–3583, 2020.

54. Yuliandra, Y., Zaini, E., Syofyan, S., Pratiwi, W., Putri, L.N., Pratiwi, Y.S., Arifin, H., Cocrystal of ibuprofen–nicotinamide: Solid-state characterization and *in vivo* analgesic activity evaluation. *Sci. Pharm.*, 86, 23, 2018.
55. Mittapalli, S., Chaitanya Mannava, M.K., Sahoo, R., Nangia, A., Cocrystals, salts, and supramolecular gels of nonsteroidal anti-inflammatory drug niflumic acid. *Cryst. Growth Des.*, 19, 219–230, 2019.
56. Almansa, C., Mercè, R., Tesson, N., Farran, J., Tomàs, J., Plata-Salamán, C.R., Co-crystal of tramadol hydrochloride-celecoxib (ctc): A novel API-API co-crystal for the treatment of pain. *Cryst. Growth Des.*, 17, 1884–1892, 2017.
57. Ando, S., Kikuchi, J., Fujimura, Y., Ida, Y., Higashi, K., Moribe, K., Yamamoto, K., Physicochemical characterization and structural evaluation of a specific 2:1 cocrystal of naproxen-nicotinamide. *J. Pharm. Sci.*, 101, 3214–3221, 2012.
58. Gascon, N., Almansa, C., Merlos, M., Miguel Vela, J., Encina, G., Morte, A., Smith, K., Plata-Salamán, C., Co-crystal of tramadol-celecoxib: preclinical and clinical evaluation of a novel analgesic. *Expert Opin. Investig. Drugs*, 28, 399–409, 2019.
59. Sung, H., Ferlay, J., Siegel, R.L., Laversanne, M., Soerjomataram, I., Jemal, A., Bray, F., Global cancer statistics 2020: GLOBOCAN estimates of incidence and mortality worldwide for 36 cancers in 185 countries. *CA Cancer J. Clin.*, 71, 209–249, 2021.
60. Kara, D.D. and Rathnanand, M., Cocrystals and drug–drug cocrystals of anticancer drugs: A perception towards screening techniques, preparation, and enhancement of drug properties. *Crystals*, 12, 1337, 2022.
61. Duan, C., Liu, W., Tao, Y., Liang, F., Chen, Y., Xiao, X., Zhang, G., Chen, Y., Hao, C., Two novel palbociclib-resorcinol and palbociclib-orcinol cocrystals with enhanced solubility and dissolution rate. *Pharmaceutics*, 14, 23, 2021.
62. Liu, L., Li, Y., Zhang, M., Zhang, Y., Lou, B., A drug-drug cocrystal of dihydromyricetin and pentoxifylline. *J. Pharm. Sci.*, 111, 82–87, 2022.
63. Yin, H.-M., Xie, J.-Y., Jiang, J.-Y., Hong, M., Zhu, B., Ren, G.-B., Qi, M.-H., Strategy to tune the performance of two drug components: Drug–drug cocrystals of lobaplatin with flavonoids. *Cryst. Growth Des.*, 22, 2602–2610, 2022.
64. Yu, Y.-M., Liu, L., Bu, F.-Z., Li, Y.-T., Yan, C.-W., Wu, Z.-Y., A novice cocrystal nanomicelle formulation of 5-fluorouracil with proline: The design, self-assembly and *in vitro/vivo* biopharmaceutical characteristics. *Int. J. Pharm.*, 617, 121635, 2022.
65. Hao, H., Zhang, Y., Hu, X., Guo, W., Yang, C., Wang, J., Cocrystallization of 5-fluorouracil with gallic acid: A novel 5-fluorouracil cocrystal displaying synergistic anti-tumor activity both in oral and intraperitoneal injection administration. *Eur. J. Pharm. Biopharm.*, 187, 12–23, 2023.
66. Xia, M., Jiang, Y., Cheng, Y., Dai, W., Rong, X., Zhu, B., Mei, X., Rucaparib cocrystal: Improved solubility and bioavailability over camsylate. *Int. J. Pharm.*, 631, 122461, 2023.

67. Duan, C., Chen, Y., Zhang, Y., Liang, F., Liu, W., Xiao, X., Xu, C., Zhuang, T., Hao, C., Chen, Y., Two cocrystals of olaparib with flavonoids toward sustained release: Structure, dissolution behavior, and anticancer activity analysis. *Cryst. Growth Des.*, 22, 4885–4894, 2022.
68. Yin, H.-M., Wu, N., Zhou, B.-J., Hong, M.-H., Zhu, B., Qi, M.-H., Ren, G.-B., Slow-release drug–drug cocrystals of oxaliplatin with flavonoids: delaying hydrolysis and reducing toxicity. *Cryst. Growth Des.*, 21, 75–85, 2021.
69. Rai, S.K., Marverti, G., Gunnam, A., Allu, S., Nangia, A.K., Dabrafenib-panobinostat salt: Improving the dissolution rate and inhibition of BRAF melanoma cells. *ACS Omega*, 8, 18255–18265, 2023.
70. Yan, T., Shu, B., Deng, X., Qian, K., Pan, R., Qiu, S.L., Yang, J., Fu, Q., Ma, Y., Antibacterial and anticancer activity, acute toxicity, and solubility of co-crystals of 5-fluorouracil and trimethoprim. *ACS Omega*, 8, 21522–21530, 2023.
71. Ding, Z., Li, F., Xie, L., Gu, M., Li, C., Liu, C., Peng, C., Li, W., Design and synthesis of novel phenylahistin derivatives based on co-crystal structures as potent microtubule inhibitors for anti-cancer therapy. *Mar. Drugs*, 20, 752, 2022.
72. Hong, M., Li, S., Ji, W., Qi, M.H., Bin Ren, G., Cocrystals of lenvatinib with sulfamerazine and salicylic acid: Crystal structure, equilibrium solubility, stability study, and anti-hepatoma activity. *Cryst. Growth Des.*, 21, 3714–3727, 2021.
73. Iqbal, U., Choudhary, M.I., Yousuf, S., Synthesis of co-crystals of anti-cancer nandrolone as a potential lead towards treatment of cancer. *J. Mol. Struct.*, 1224, 128981, 2021.
74. Pantwalawalkar, J., More, H., Bhange, D., Patil, U., Jadhav, N., Novel curcumin ascorbic acid cocrystal for improved solubility. *J. Drug Deliv. Sci. Technol.*, 61, 102233, 2021.
75. Luo, Y., Chen, S., Zhou, J., Chen, J., Tian, L., Gao, W., Zhang, Y., Ma, A., Li, L., Zhou, Z., Luteolin cocrystals: Characterization, evaluation of solubility, oral bioavailability, and theoretical calculation. *J. Drug Deliv. Sci. Technol.*, 50, 248–254, 2019.
76. Ray, E., Vaghasiya, K., Sharma, A., Shukla, R., Khan, R., Kumar, A., Verma, R.K., Autophagy-inducing inhalable co-crystal formulation of niclosamide-nicotinamide for lung cancer therapy. *AAPS PharmSciTech.*, 21, 1–4, 2020.
77. Kanagavel, M., Sparjan Samuvel, R.M., Ramalingam, V., Nechipadappu, S.K., Repurposing of antifungal drug flucytosine/flucytosine cocrystals for anti-cancer activity against prostate cancer targeting apoptosis and inflammatory signaling pathways. *Mol. Pharm.*, 21, 2577–2589, 2024.
78. Sathisaran, I., Devidas Bhatia, D., Vishvanath Dalvi, S., New curcumin-trimesic acid cocrystal and anti-invasion activity of curcumin multicomponent solids against 3D tumor models. *Int. J. Pharm.*, 587, 119667, 2020.

79. Li, P., Ramaiah, T., Zhang, M., Zhang, Y., Huang, Y., Lou, B., Two cocrystals of berberine chloride with myricetin and dihydromyricetin: crystal structures, characterization, and antitumor activities. *Cryst. Growth Des.*, 20, 157–166, 2020.
80. Allu, S., An, J.-H., Park, B.J., Kim, W.-S., Improving dissolution rate and solubility of palbociclib salts/cocrystal for anticancer efficacy. *J. Mol. Struct.*, 1305, 137756, 2024.
81. Inam, M., Yang, Y., Hu, J., Zheng, J., Deng, W., Zhou, Y., Qi, J., Xu, C., Chai, G., Dang, Y., Chen, W., Cocrystallization of gefitinib potentiate single-dose oral administration for lung tumor eradication via unbalancing the DNA damage/repair. *Pharmaceutics*, 15, 2713, 2023.
82. Rathi, R., Nagpal, M., Kaur, M., Ballouria, P., Dutta, R., Aggarwal, G., Arora, S., Treatment of tuberculosis in nano era: recent avenues. *Curr. Bioact. Compd.*, 18, 7, 13–26, 2022.
83. Neres, J., Hartkoorn, R.C., Chiarelli, L.R., Gadupudi, R., Pasca, M.R., Mori, G., Venturelli, A., Savina, S., Makarov, V., Kolly, G.S., Molteni, E., Binda, C., Dhar, N., Ferrari, S., Brodin, P., Delorme, V., Landry, V., De Jesus Lopes Ribeiro, A.L., Farina, D., Saxena, P., Pojer, F., Carta, A., Luciani, R., Porta, A., Zanoni, G., De Rossi, E., Costi, M.P., Riccardi, G., Cole, S.T., 2-carboxyquinoxalines kill mycobacterium tuberculosis through noncovalent inhibition of DprE1. *ACS Chem. Biol.*, 10, 705–714, 2015.
84. Wang, X., Du, S., Zhang, R., Jia, X., Yang, T., Zhang, X., Drug-drug cocrystals: Opportunities and challenges. *Asian J. Pharm. Sci.*, 16, 307–317, 2021.
85. Liu, F., Wang, L.Y., Yu, M.C., Li, Y.T., Wu, Z.Y., Yan, C.W., A new cocrystal of isoniazid-quercetin with hepatoprotective effect: The design, structure, and *in vitro/in vivo* performance evaluation. *Eur. J. Pharm. Sci.*, 144, 105216, 2020.
86. Yadav, B., Gunnam, A., Thipparaboina, R., Nangia, A.K., Shastri, N.R., Hepatoprotective cocrystals of isoniazid: Synthesis, solid state characterization, and hepatotoxicity studies. *Cryst. Growth Des.*, 19, 5161–5172, 2019.
87. Salem, A., Takácsi-Nagy, A., Nagy, S., Hagymási, A., Gősi, F., Vörös-Horváth, B., Balić, T., Pál, S., Széchenyi, A., Synthesis and characterization of nano-sized 4-aminosalicylic acid–sulfamethazine cocrystals. *Pharmaceutics*, 13, 1–13, 2021.
88. Sakamoto, N., Miyata, K., Fukami, T., Quabodepistat (OPC-167832), a novel antituberculosis drug candidate: Enhancing oral bioavailability via cocrystallization and mechanistic analysis of bioavailability in two cocrystals. *Mol. Pharm.*, 21, 358–369, 2024.
89. Li, J.-Y., Liu, F., Li, Y.-Y., Bao, X.-Y., Li, Y., Li, Y.-T., Liu, R.-M., Hepatoprotective cocrystal of ethionamide: A new attempt to refurbish old drug through crystal engineering. *J. Cryst. Growth*, 582, 126523, 2022.
90. Amrutha, S., Giri, L., Seethalekshmi, S., Varughese, S., Enhanced aqueous solubility of the solid forms of a BCS class-II anti-tuberculosis drug, prothionamide. *Cryst. Growth Des.*, 20, 5086–5096, 2020.

91. Ngilirabanga, J.B., Aucamp, M., Pires Rosa, P., Samsodien, H., Mechanochemical synthesis and physicochemical characterization of isoniazid and pyrazinamide co-crystals with glutaric acid. *Front. Chem.*, 8, 595908, 2020.
92. Wu, X., Wang, Y., Xue, J., Liu, J., Qin, J., Hong, Z., Du, Y., Solid phase drug-drug pharmaceutical co-crystal formed between pyrazinamide and diflunisal: Structural characterization based on terahertz/Raman spectroscopy combining with DFT calculation. *Spectrochim. Acta A Mol. Biomol. Spectrosc.*, 234, 118265, 2020.
93. Battini, S., Mannava, M.K.C., Nangia, A., Improved stability of tuberculosis drug fixed-dose combination using isoniazid-caffeic acid and vanillic acid cocrystal. *J. Pharm. Sci.*, 107, 1667–1679, 2018.
94. Churakov, A.V., Manin, A.N., Perlovich, G.L., Drug-drug cocrystals of antituberculous 4-aminosalicylic acid: Screening, crystal structures, thermochemical and solubility studies. *Eur. J. Pharm. Sci.*, 99, 228–239, 2017.
95. Xuan, B., Wong, S.N., Zhang, Y., Weng, J., Tong, H.H.Y., Wang, C., Sun, C.C., Chow, S.F., Extended release of highly water soluble isoniazid attained through cocrystallization with curcumin. *Cryst. Growth Des.*, 20, 1951–1960, 2020.
96. Mashhadi, S.M.A., Yunus, U., Bhatti, M.H., Ahmed, I., Tahir, M.N., Synthesis, characterization, solubility and stability studies of hydrate cocrystal of antitubercular isoniazid with antioxidant and anti-bacterial protocatechuic acid. *J. Mol. Struct.*, 1117, 17–21, 2016.
97. Yu, X.Z., Wang, L.Y., Liu, F., Li, Y.T., Wu, Z.Y., Yan, C.W., Sustained-Release Dual-drug ternary salt cocrystal of piperazine ferulate with pyrazinamide: Synthesis, structure, and hirshfeld surface analysis. *Cryst. Growth Des.*, 20, 2064–2073, 2020.
98. Rosa, J., Machado, T.C., Da Silva, A.K., Kuminek, G., Bortolluzzi, A.J., Caon, T., Cardoso, S.G., Isoniazid-resveratrol cocrystal: A novel alternative for topical treatment of cutaneous tuberculosis. *Cryst. Growth Des.*, 19, 5029–5036, 2019.
99. Souza, M.S., Diniz, L.F., Vogt, L., Carvalho Jr., P.S., D'vries, R.F., Ellena, J., Mechanochemical synthesis of a multicomponent solid form: The case of 5-fluorocytosine isoniazid codrug. *Cryst. Growth Des.*, 18, 5202–5209, 2018.
100. Kehrenberg, M.C.A. and Bachmann, H.S., Diuretics: A contemporary pharmacological classification. *Naunyn Schmiedebergs Arch Pharmacol.*, 395, 619–627, 2022.
101. Zhang, Y.X., Wang, L.Y., Dai, J.K., Liu, F., Li, Y.T., Wu, Z.Y., Yan, C.W., The comparative study of cocrystal/salt in simultaneously improving solubility and permeability of acetazolamide. *J. Mol. Struct.*, 1184, 225–232, 2019.
102. Acebedo Martínez, F.J., Alarcón Payer, C., Rodríguez Domingo, L., Domínguez Martín, A., Gómez Morales, J., Choquesillo Lazarte, D., Furosemide/non-steroidal anti-inflammatory drug–drug pharmaceutical solids: Novel opportunities in drug formulation. *Crystals*, 11, 1339, 2021.

103. Narala, S., Nyavanandi, D., Alzahrani, A., Bandari, S., Zhang, F., Repka, M.A., Creation of hydrochlorothiazide pharmaceutical cocrystals via hot-melt extrusion for enhanced solubility and permeability. *AAPS PharmSciTech.*, 23, 56, 2022.
104. Sanphui, P. and Rajput, L., Tuning solubility and stability of hydrochlorothiazide co-crystals. *Acta Crystallogr. B Struct. Sci. Cryst. Eng. Mater.*, 70, 81–90, 2014.
105. Ranjan, S., Devarapalli, R., Kundu, S., Vangala, V.R., Ghosh, A., Reddy, C.M., Three new hydrochlorothiazide cocrystals: structural analyses and solubility studies. *J. Mol. Struct.*, 1133, 405–410, 2017.
106. Li, M., Jin, Z., Han, D., Wu, S., Gong, J., Co-former selection for coamorphous amino acid/spironolactone formulations and exploration of the amorphization kinetics of systems. *Cryst. Growth Des.*, 23, 1511–1521, 2023.
107. Bai, G.-Y., Wu, Z.-Y., Xing, J.-L., Li, Y.-T., Comparative study of the cocrystals with layered/cavity structure in regulating *in vitro* pharmaceutical properties of diuretic acetazolamide. *J. Mol. Struct.*, 1271, 134035, 2023.
108. Coyote-Dotor, G., Páez-Franco, J.C., Canseco-González, D., Núñez-Pineda, A., Dorazco-González, A., Fuentes-Noriega, I., Vilchis-Néstor, A.R., Rodríguez-Hernández, J., Morales-Morales, D., Germán-Acacio, J.M., Synthesis, characterization, and intrinsic dissolution studies of drug–drug eutectic solid forms of metformin hydrochloride and thiazide diuretics. *Pharmaceutics*, 13, 1926, 2021.
109. Goud, N.R., Gangavaram, S., Suresh, K., Pal, S., Manjunatha, S.G., Nambiar, S., Nangia, A., Novel furosemide cocrystals and selection of high solubility drug forms. *J. Pharm. Sci.*, 101, 664–680, 2012.
110. Ueto, T., Takata, N., Muroyama, N., Nedu, A., Sasaki, A., Tanida, S., Terada, K., Polymorphs and a hydrate of furosemide-nicotinamide 1:1 cocrystal. *Cryst. Growth Des.*, 12, 485–494, 2012.
111. Banik, M., Gopi, S.P., Ganguly, S., Desiraju, G.R., Cocrystal and salt forms of furosemide: Solubility and diffusion variations. *Cryst. Growth Des.*, 16, 5418–5428, 2016.
112. Kerr, H.E., Softley, L.K., Suresh, K., Nangia, A., Hodgkinson, P., Evans, I.R., A furosemide-isonicotinamide cocrystal: an investigation of properties and extensive structural disorder. *CrystEngComm.*, 17, 6707–6715, 2015.
113. Wang, Z., Li, Q., An, Q., Gong, L., Yang, S., Zhang, B., Su, B., Yang, D., Zhang, L., Lu, Y., Du, G., Optimized solubility and bioavailability of genistein based on cocrystal engineering. *Nat. Prod. Bioprospect.*, 13, 30, 2023.
114. Kara, D.D., Bangera, P.D., Mehta, C.H., Tanvi, K., Rathnanand, M., In silico screening as a tool to prepare drug-drug cocrystals of ibrutinib-ketoconazole: A strategy to enhance their solubility profiles and oral bioavailability. *AAPS PharmSciTech.*, 24, 164, 2023.
115. Wong, S.N., Weng, J., Ip, I., Chen, R., Lakerveld, R., Telford, R., Blagden, N., Scowen, I.J., Chow, S.F., Rational development of a carrier-free dry powder inhalation formulation for respiratory viral infections via quality by design:

A drug-drug cocrystal of favipiravir and theophylline. *Pharmaceutics*, 14, 300, 2022.
116. Zhang, L., Kong, D., Wang, H., Jiao, L., Zhao, X., Song, J., Yang, D., Yang, H., Yang, S., Du, G., Lu, Y., Cocrystal of apixaban-quercetin: Improving solubility and bioavailability of drug combination of two poorly soluble drugs. *Molecules*, 26, 2677, 2021.
117. Islam, N.U., Khan, E., Umar, M.N., Shah, A., Zahoor, M., Ullah, R., Bari, A., Enhancing dissolution rate and antibacterial efficiency of azithromycin through drug-drug cocrystals with paracetamol. *Antibiotics*, 10, 939, 2021.
118. Gohel, S.K., Palanisamy, V., Sanphui, P., Prakash, M., Singh, G.P., Chernyshev, V., Isostructural cocrystals of metaxalone with improved dissolution characteristics. *RSC Adv.*, 11, 30689–30700, 2021.
119. Maity, D.K., Paul, R.K., Desiraju, G.R., Drug-drug binary solids of nitrofurantoin and trimethoprim: Crystal engineering and pharmaceutical properties. *Mol. Pharm.*, 17, 4435–4442, 2020.
120. Shinozaki, T., Ono, M., Higashi, K., Moribe, K., A novel drug-drug cocrystal of levofloxacin and metacetamol: Reduced hrygroscopicity and improved photostability of levofloxacin. *J. Pharm. Sci.*, 108, 2383–2390, 2019.
121. Liu, L., Yu, Y.-M., Yu, Y., Bu, F.-Z., Wu, Z.-Y., Yan, C.-W., Li, Y.-T., Drug-nutrient cocrystallization-driven strategy towards self-assembly of milrinone and ferulic acid provides an exemplification in perfecting *in vitro/vivo* characteristics of anti-heart failure drugs. *New J. Chem.*, 46, 10357–10367, 2022.
122. Mannava, M.K.C., Gunnam, A., Lodagekar, A., Shastri, N.R., Nangia, A.K., Solomon, K.A., Enhanced solubility, permeability, and tabletability of nicorandil by salt and cocrystal formation. *CrystEngComm.*, 23, 227–237, 2021.
123. Shinozaki, T., Ono, M., Higashi, K., Moribe, K., A novel drug-drug cocrystal of levofloxacin and metacetamol: reduced hygroscopicity and improved photostability of levofloxacin. *J. Pharm. Sci.*, 108, 2383–2390, 2019.
124. Ngilirabanga, J.B., Rosa, P.P., Aucamp, M., Kippie, Y., Samsodien, H., Dual-drug co-crystal synthesis for synergistic *in vitro* effect of three key first-line antiretroviral drugs. *J. Drug Deliv. Sci. Technol.*, 60, 101958, 2020.
125. KruthiventiIqbal, A.K., Reddy Jaggavarapu, J.S., Nagalapalli, R.K., Viswanadha, G.S., Anand, S.K., Pharmaceutical co-crystals of quercetin, US20120258170A1, assigned to Nutracryst Therapeutics Pvt Ltd, 2012.
126. ALMARSSON, O., Hickey, M.B., Peterson, M., Zaworotko, M.J., Moulton, B. Rodriguez-Hornedo, N., Pharmaceutical co-crystal of celecoxib-nicotinamide, EP1608339B1, assigned to University of South Florida, Johnson and Johnson Consumer Inc., University of Michigan, 2012.
127. NATRAJAN, S.K., Kishore, KARUMAN, CHIVittal, TANGIRALA, Sivakumaran, M., Process for the preparation dl-proline co-crystal of dapagliflozin, WO2017191539A1, assigned to Aurobindo Pharma Limited, 2017.

128. Hanna, M., Shan, N., Cheney, M.L., Weyna, D.R., *In vivo* studies of crystalline forms of meloxicam, US8124603B2, assigned to Thar Pharma LLC, 2012.
129. Amos, J.G., Indelicato, J.M., Pasini, C.E., Reutzel, S.M., US6001996A, assigned to Eli Lilly and Co, 1999.
130. Sependa, G.J. and Storey, R., Co-crystal, US7625910B2, assigned to Astra Zeneca AB, 2009.
131. Tesson, N. and CAMPÓN, E.G., Cocrystals of lorcaserin, EP3210975A1, assigned to Enantia SL, 2017.
132. Wennogle, L.P., Li, P., Aret, E., Novel co-crystals, WO2017172811A1, assigned to Intra-Cellular Therapies, Inc, 2017.
133. Ingelheim Pharma Gmbh&Co., Boehringer, K., Betageri, R., Bosanac, T., Jason Burke, M., Harcken, C., Kim, S., Kuzmich, D., Wai-Ho Lee, T., Li, Z., Liu, P., Lord, J., Razavi, H., Reeves, J.T., Thomson, D., Glucocorticoid mimetics, methods of making them, pharmaceutical compositions, and uses thereof, EP2300472B1, assigned to Boehringer Ingelheim International GmbH Boehringer Ingelheim Pharma GmbH and Co KG, 2012.
134. PLATA-SALAMAN, C.R. and Tesson, N., Co-crystals of tramadol and coxibs, US11478488B2, assigned to Esteve Pharmaceuticals SA, 2022.
135. Holland, J. and Frampton, C., Tranilast compositions and cocrystals, US10150729B2, assigned to Nuformix Technologies Ltd Nuformix Ltd, 2018.
136. Edislav, LEKSIC, Samec, D.S., Damir, SAHNIC, Kisicek, D., Hrkovac, M., Topic, E., Solid state forms of trisodium valsartan: sacubitril, US10689352B2, assigned to Teva Pharmaceuticals International GmbH Pliva Hrvatska doo, 2020.
137. Reddy, J.S, Dandela, R., Viswanadha, G.S., Nagalapalli, R., SOLOMON, A.K., Javed, I., Kruthiventi, A.K., Novel cocrystals / molecular salts of metformin with oleoyl ethanolamide as an effective anti-diabetic + anti-obesity agent, WO2012090225A2, assigned to Nutracryst Therapeutics Private Limited, 2016.
138. Kumar Bandaru, R., Rout, S.R., Kenguva, G., Gorain, B., Alhakamy, N.A., Kesharwani, P., Dandela, R., Recent Advances in Pharmaceutical Cocrystals: From Bench to Market. *Front Pharmacol.*, 12, 780582, 2021 Nov 11, doi: 10.3389/fphar.2021.780582.
139. Rathi, R., Guggal, Y., Twinkle, V., Sandhu, Singh, I., Exploring the Patent Landscape and Regulatory Prospective on Pharmaceutical Cocrystals. *J. Pharm. Technol. Res. Manag.*, 11, 49–60, 2023.
140. Singh, M., Harsh, B., Vaskuri, G.S., Jyothi, S., Madhukiran, R., Dhondale, A.G., Nambiar, A.K., Agrawal, P.K., Shastri, N.R., Kumar, D., Cocrystals by Design: A Rational Coformer Selection Approach for Tackling the API Problems. *Pharm.*, 15, 4, 1161, 2023.

8

Scaling Up Pharmaceutical Cocrystals: Exploring Commercial Opportunities and Industrial Prospective

Kampanart Huanbutta[1]*, Tanikan Sangnim[2], Ritu Rathi[3] and Inderbir Singh[3]

[1]Department of Manufacturing Pharmacy, College of Pharmacy, Rangsit University, Pathum Thani, Thailand

[2]Faculty of Pharmaceutical Sciences, Burapha University, Seansook, Muang, Chonburi, Thailand

[3]Chitkara College of Pharmacy, Chitkara University, Punjab, India

Abstract

Cocrystals are complex molecules showing great promise in improving drug properties. However, turning this promise into actual products faces significant challenges when scaling up for commercial use. This chapter discusses the balance between scientific creativity and practical needs in overcoming these challenges. It outlines the difficulties of scaling up cocrystallization for pharmaceuticals. Various methods for cocrystal scale-up, including established techniques like solution-mediated and mechanochemical cocrystallization, as well as newer methods like spray drying and continuous manufacturing, are also described and given examples in this chapter. The advantages and limitations of each method are discussed to provide information for researchers regarding further cocrystallization manufacturing on an industrial scale. Examples of commercial cocrystal pharmaceutical products and their benefit in cocrystal form are provided. The future and possibility of cocrystallization manufacturing from laboratory scale to large scale are concluded in the last part of this chapter.

Keywords: Scale-up, large-scale cocrystallization, continuous manufacturing, spray drying, commercial drug products

**Corresponding author*: kampanart.h@rsu.ac.th

Inderbir Singh, Subrata Mallick and Ritu Rathi (eds.) Cocrystals in Pharmaceutical Sciences: Design to Applications, (231–256) © 2025 Scrivener Publishing LLC

8.1 Introduction

Cocrystals are molecular complexes formed through noncovalent interactions between an active pharmaceutical ingredient (API) and a coformer offering several drug properties improvement from conventional drug forms [1]. Among several promising advantages, the solubility enhancement of hydrophobic drugs is the main purpose of this technique. Higher solubility from cocrystal drugs with suitable coformers exhibits remarkable improvements in terms of bioavailability and therapeutic efficacy [2–4]. Moreover, cocrystals offer better stability of drugs and lower hygroscopicity of conventional drug forms. These could extend drug potency over their shelf life [5]. Beyond stability and solubility benefits, cocrystals possess the capacity to refine additional desirable properties such as taste masking [6] and thermal stability [7]. Several benefits of cocrystallization resulted in numerous research articles concerning drug developments from this technique. For industrial-scale production of cocrystal drugs, the European Union EMA (2015) and the US FDA (2013) have issued respective guidance for industry and reflection papers that inform the significance of cocrystals as an emerging category of drug substances that may find utility within the pharmaceutical industry [8].

Nonetheless, producing modified API by means of cocrystal is limited to the industrial scale due to a number of production bottlenecks [9]. Firstly, it is impossible to form a cocrystal of an API in a reproducible way on a large scale due to the variety of factors of cocrystallization operations, such as interaction temperature, pressure, and solvent composition. Secondly, batch-to-batch performance reproducibility becomes a challenge for many cases when the size of synthesis is comparable to that of laboratory to industrial. Thirdly, there are drug regulatory requirements for manufacturing, standards of testing, control, and documentation implementation that require additional measures of validation and compliance. Finally, implementation of coformer and finding a proper source for it can be a large-scale problem, providing the cost and the conditions of production and hence the supply chain, being of the paramount cost part for the pharmaceutical industry [10]. To overcome the discussed limitations, new methodologies and techniques are required in various fields as well as more interdisciplinary research on the ways cocrystal technologies processes optimization for commercial usage and increase the sustainability of cocrystal-based psychiatric drugs.

This chapter has discussed the problems associated with scaling up the cocrystal production from lab to commercial scale. Furthermore, it has discussed the various methods by which cocrystal production is scaled up. The advantages and limitations of each cocrystallization method on a large scale are inquired into. The methods that can be used for producing the various cocrystal systems are presented and their suitability is discussed. The examples are given of the cocrystals already in the market or at the later stages of development. Finally, the future tendencies, the recent technology in cocrystal production, opportunities for commercial application, and the development of cocrystals are discussed.

8.2 Challenges and Considerations in Cocrystal Scale-Up

8.2.1 Process Control and Consistency

The manufacture of pharmaceutical cocrystals is a complex process in its nature because it consists of several components and is associated with complex kinetical patterns. The coexistence of multiple initial materials together with the cocrystal makes it complicated to learn this type of cocrystallization and its features, meaning that producing pharmaceutical cocrystal products involves significant scaling up. This perspective explains the necessity of using a multicomponent phase diagram to outline the best operation regions for solution cocrystallization. It covers how various operating conditions affect process kinetics and the complex relationship among these catalysts. It also includes an explanation of the clarification of intermolecular interactions, which are highly relevant for cocrystal formation, an overview of cocrystallization methods, and the significance of quality control, especially in the pharmaceutical field. Furthermore, this viewpoint clarifies the objective of the design of experiment (DoE) and quality-by-design (QbD) approaches that can be employed for optimizing the process of cocrystallization scale-up. These tools are valuable for improving the scalability of the processes because they offer systematic methods for simplifying the seemingly complex relationships involved in upscaling cocrystallization, which ultimately improves process and product quality [11].

8.2.2 Kinetics of Cocrystallization

Kinetics of cocrystallization implies the investigation of the velocity of cocrystal formation. It may be related to the numerous stages such as nucleation and crystal growth [12]. However, if the laboratory scale of cocrystal production is a relatively straightforward mission that may be easily conducted with the usage of different solid-based and solvent-based methods, cocrystallization on a large scale is a process of the kinetics of cocrystallization and nucleation. This is because crystallization kinetics and the thermodynamics of numerous cocrystals are very competitive, which results in the high complexity of the crystallization. Thus, as for the scale-up of solution cocrystals, it is a tremendously dull mission to achieve a high amount of yield, low average size, consistent morphology, and high purity. This scarcity of comprehensive data hampers both quantitative and qualitative analyses essential for in-depth process investigations necessary for large-scale manufacturing.

For instance, Xuan and his colleagues observed that isoniazid and curcumin cocrystals exhibited distinct drug dissolution kinetics at varying rates of solvent removal during the cocrystallization process [13]. Another illustration involves the carbamazepine-saccharin cocrystal, synthesized using an antisolvent process utilizing a volume of 4.5 L. In order to track the progress of the reaction over 120 min, cocrystallization was combined with offline characterization and in-line near-infrared (NIR) measurements. It was discovered that increasing the addition speed or agitation speed favored the formation of form-I while reversing the kinetic conditions led to the production of highly pure form-II [14]. Consequently, thorough investigation during technology transfer from laboratory-scale to plant-scale manufacturing becomes imperative. Developing a QbD strategy tailored to cocrystallization processes requires a broad understanding of how process parameters affect crucial product qualities related to cocrystallization kinetics [15].

8.2.3 Selection of Appropriate Scale-Up Method

Various techniques for cocrystallization found in the literature can be broadly classified into two main categories: solution-based and solid-state processes. Solid-state approaches typically involve minimal or no solvent usage, whereas solution-based methods rely on substantial solvent quantities, necessitating subsequent separation of the crystalline product from the mother liquor [5]. The selection of an appropriate cocrystallization method hinges upon the desired properties of the final crystal, such as size, morphology, and purity, which are considered critical attributes. These attributes are significantly influenced by the chosen cocrystallization approach, making the selection of the most suitable method a crucial initial step in the production process.

8.2.4 Regulatory Compliance

In terms of regulations, the US FDA released a draft of its guideline in 2013 about the regulatory classification of cocrystals. The final version of the guidance was published in February 2018 [16]. The purpose of this guidance is to provide planning to the applicants for submitting new drug applications (NDAs) and abbreviated new drug applications (ANDAs) with information on the appropriate regulatory classification of pharmaceutical cocrystal solid-state forms. Additionally, the guidance gives information on the data that applicants must submit in support of the classification as a cocrystal solid-state form as well as the regulatory implications of such classification. In Europe, the European Medicines Agency (EMA) released a reflective document in 2015, which describes the cocrystals in a similar manner as the salts of APIs [17]. This reflection paper, first of all, should reflect the current thinking of the Human Use and Committee for Veterinary Medicinal Products (CHMP/CVMP) on different aspects concerning the use of cocrystals of active substance in medicinal products, either for human use or veterinary use.

It was expected that the FDA and EMA guidelines would support the growth of a cocrystal drug market. However, the FDA and EMA have provided only initial guidance, and some regulatory aspects regarding cocrystals, like long-term stability data requirements, might still be evolving. Furthermore, the practical implementation of regulatory guidelines and the interpretation of data requirements by regulatory agencies may vary, leading to potential delays or uncertainties in the approval process and commercialization timeline for cocrystal-based drugs. These factors underscore the need for further refinement and harmonization of regulatory frameworks, as well as continued research and development efforts to address the complexities associated with cocrystal drug development and market entry.

8.3 Promising Methods for Cocrystal Scale-Up

8.3.1 Solution-Mediated Cocrystallization

The method of solution-mediated cocrystallization is extensively employed in the synthesis of cocrystals. The process entails the solution of the API (active pharmaceutical ingredient) and coformer (cocrystallizing agent) in an appropriate solvent. Subsequently, the formation of the cocrystal is induced by manipulating various conditions (e.g., temperature and solvent evaporation). This technique provides numerous benefits including uncomplicated in nature when compared to alternative approaches such as milling or melt cocrystallization. The final cocrystal properties can be manipulated by researchers through the adjustment of solvent selection, concentration, and temperature. Moreover, solution-mediated cocrystallization offers scalability which is a key advantage of this technique. It enables the production of cocrystals on a larger magnitude [12]. Generally, solvent evaporation is the most common synthesis technique for cocrystal discovery; however, it is not commonly used on a larger scale due to the slow process rate and difficulty in controlling homogeneity. Therefore, fast-drying scalable techniques such as spray drying and Rotary evaporation are widely used in industries.

a. Spray Drying

From a production perspective within the pharmaceutical industry, spray drying is increasingly proving a highly viable and effective method for scaling up cocrystal production. This is due, in large part, to the many advantages this method offers over conventional methods of manufacturing, serving as an extremely useful tool in the area of cocrystal scale-up. For one thing, it is a highly scalable method, with adjustments to parameters such as feed rate and atomization enabling the easy transition from small-scale experiment to large-scale production. Moreover, its capability for continuous processing also serves to significantly improve efficiency and ensure a greater degree of consistency than is achievable via the batch model. Additionally, it also allows for highly precise control over cocrystal particle parameters such as size and morphology, aiding in the improvement or tuning of properties such as dissolution rate and bioavailability and ensuring that the cocrystal will be as effective as possible. Lastly, cocrystal formation and drying can be combined, with the two processes carried out simultaneously in one step, obviating the need for separate cocrystal formation and drying steps and yielding significant cost benefits as a result [18].

Shao *et al.* used the spray-drying technique to spray drugs together with polymer to enhance cocrystal stability. Different polymers were used for the spray-drying process, and the resultant spray-dried products were assessed [18]. Walsh *et al.* provided a comparative assessment of hot-melt extrusion (HME) and spray drying as the cocrystal production processes. The model cocrystal was selected for the research, nicotinamide and ibuprofen mixed in a 1:1 molar ratio. To achieve the processing objectives, different pharmaceutical excipients were selected including xylitol, mannitol, Soluplus, and PVP K-15. The results of the experiment showed that both spray drying and HME demonstrate the ability to cocrystallize the formulation of an API and a coformer with the excipient. However, spray drying is more feasible for performing no additional unit operations to yield the final pharmaceutical product (cocrystals with the carrier excipient) than using HME [19]. In addition to traditional high-temperature drying approaches, spraying is employed in freeze-drying processes where heat-sensitive drugs may be dried. Tanaka *et al.* spray freeze-dried the theophylline and oxalic acid–based cocrystals developed to produce an inhaled dry powder dosage form. Analyzing the product sample, the presence of theophylline anhydrate and oxalic acid cocrystals with partial coamorphous features was confirmed. This observation indicates a perspective for the usage of the spray freeze-dried process for cocrystallization in the case of the offshoot of dry powder inhaler products being a method toward residual pharmaceutical dry powders [20]. The future of the spray-drying technique for cocrystal scale-up manufacturing is very promising, and advancements in process control, automation, and knowledge of cocrystal behavior will be the basis for the highly efficient and quality production of cocrystals in pharmaceutical use.

b. Continuous Manufacturing

Batch manufacturing is the most common process for the production of cocrystals since it is simple, and the cost investment is low, however, for large-scale production, batch manufacturing includes some limitations such as firmly defined batch size, multiple sequential steps, raw material wastage, long waiting for next step, contamination risk between processes, required extensive validation and difficult to scale-up procedure [21]. Because of the described limitations, the regulatory authorities are promoting the development of alternative pharmaceutical processes either based on continuous manufacturing or single-step processing. In detail, a single-step process means that the raw materials are fed consistently into the system each step of the procedure is well-connected to other processing steps, and the final drug product results at a constant rate. Table 8.1 summarizes the major differences between these two approaches. For instance, batch processing involves discrete batches processed sequentially, resulting in longer operation times and higher raw material requirements, while continuous manufacturing enables a steady flow of materials, reducing production time and inventory needs. Furthermore, continuous processes offer enhanced scalability, consistent product quality with real-time monitoring, and reduced waste generation. However, they often require higher initial capital investment compared to batch processes.

Continuous manufacturing offers several benefits for the production of cocrystals. Firstly, it allows process control advantages by providing real-time process monitoring by process analytical technology (PAT) and it makes it possible to adjust some critical process parameters (CPPs) such as the temperature, pressure, and flow rates to obtain designed cocrystal

Table 8.1 Comparison of batch and continuous process.

Feature	Batch processing	Continuous processing
Process flow	Discrete batches, processed sequentially	Continuous flow of materials, processed in interconnected stages
Production time	Longer operation due to batch processing and holding step	Potentially shorter operation time
Inventory management	Requires larger quantities of materials in one batch process	Requires smaller amount/ volume of initial substances
Scalability	Limited; depending on the size of instruments	More readily scalable to meet demand fluctuations
Product quality	Might be different between batches in case of serval cocrystal formation factors	Consistent product quality with real-time monitoring and control
Waste generation	Higher potential for waste of solvent from each batch processing	Lower potential for waste with potential for solvent recycling
Process control	Limited control during processing, relying on hold points for quality checks	Real-time monitoring and control of critical parameters throughout the process
Capital investment	Lower initial investment	Higher initial investment in specialized equipment and process development
Mixing efficiency	Need long time and energy for better mixing efficiency	Homogeneously mixing in molecular level interaction
Regulatory considerations	Established regulatory framework for batch processes	Evolving regulatory framework for continuous manufacturing

products. As a result, because of the consistency in cocrystal formation, batch-to-batch variations are reduced. Secondly, it efficiently uses time, thus eliminating idle time for the constantly processing tasks and may reduce the time of overall production and increase the entire output when compared to batch processing methods. Thirdly, continuous manufacturing minimizes solvent usage, and waste production is minimized by using continuous recycling and reusing strategies. Last but not least, continuous manufacturing systems are much more scalable organisms rather than batch processes, and this ensures better and more flexible scalability and is more responsive to market dynamics [22].

The research team led by Kara *et al.* has utilized two techniques, namely, fluid bed drying and continuous manufacturing to produce the industrial scale of the sulfadimidine-4-aminosalicylic acid cocrystals. The diagram of the setup and connection of the two instruments is presented in Figure 8.1. The 3D printed microfluidic device was used in this research to produce cocrystals including sulfadimidine and 4-aminosalicylic acid. The microfluidic system assisted in such a task as the continuous production of cocrystals by supplying the solutions of sulfadimidine and 4-aminosalicylic acid and an antisolvent system including polyvinylpyrrolidone and distilled water. The microfluidic chip was properly incorporated into the fluidized bed providing a possibility to achieve a single step for the generation of the cocrystal formulation. The process combines mixing, coating onto carrier beads, and drying the cocrystals [23].

Apart from research applications of continuous manufacturing of cocrystals, the number of pharmaceutical manufacturers that use the continuous production approach has increased. For example, Vertex Pharmaceuticals uses the continuous manufacturing approach to produce Orkambi, a combination pharmaceutical product used to treat cystic fibrosis [24, 25]. The first notification of the intention to switch production from batch to continuous

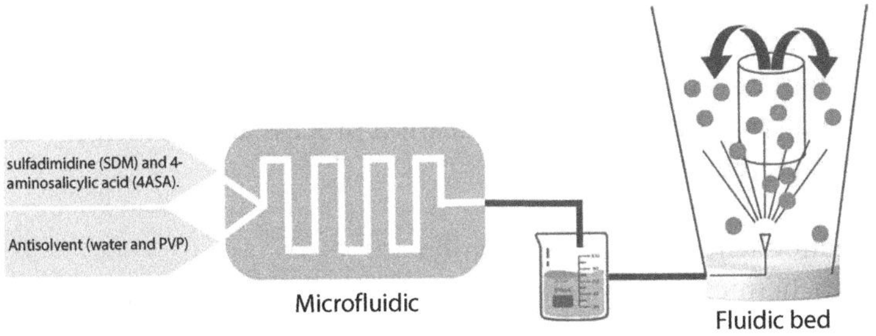

Figure 8.1 Schematic of cocrystal manufacturing by fluid bed drying and continuous manufacturing (microfluidic) modified from [23].

manufacturing was issued to Prezista by Janssen Pharmaceuticals, which produces this drug used in the treatment of HIV-1 [26]. The main advantage of this approach is that with the agility offered by continuous manufacturing, the company can quickly adjust its manufacturing volume to stabilize fluctuation in product demand. This way, the company successfully reduced inventory volume and switched to a demand-driven manufacturing approach, which improved overall operational performance.

8.3.2 Solid/Slurry Cocrystallization

Solid/slurry-mediated cocrystallization is generally considered to be potentially advantageous over solution-mediated cocrystallization in many ways. It consists of the following stages.

- The blend of solid formulations of both the API and the coformer.
- Mixing the said solid with another compound, i.e., the small volume of solvent, which translates solid formulation into a slurry, allowing one compound with superior solubility, typically in up to a minimum of 10% volume of solvent, to dissolve or disperse.
- The interaction of the saturated dissolved compounds creates the conditions for the formation and growth of the cocrystal, which is facilitated through the medium of unseparated solvent to allow the nuclei of the cocrystal to form and grow.
- The obtained cocrystal which is mixed with solvent requires a process of purification.

Solid/slurry cocrystallization methodology presents numerous benefits, such as avoiding high use of solvent, facilitating the effective evaluation of prospective coformers, and frequently producing cocrystals that are thermodynamically stable, a critical attribute for preserving consistent properties throughout storage and application. However, the long operation time and difficulty of the cocrystal purification process are the factors that need to be considered [27].

a. Mechanochemical Cocrystallization

Mechanochemical cocrystallization is the process of using mechanical energy for the cocrystal formation from solid starting materials. It is typically implemented by grinding in a ball mill or mortar and pestle. Such action causes high shear stress and frictional forces that, in their turn, result in a number of effects. First, the action breaks down the crystal lattices of the starting materials and therefore reduces their ordered structure. Second, it increases the surface area of both ground materials which, in turn, supports the contact and interaction of their molecules. Finally, the materials remain in close joint contact in the process of cocrystal formation because the action has broken down their surfaces; in this case, a large surface area of interaction facilitates the beginning of a new cocrystal stage due to the formation of new types of interactions, such as hydrogen bonding and van der Waals forces, between the molecules of the components. Eventually, the cocrystal is formed [28].

Mechanochemical cocrystallization has several positive aspects. First, it is a solid-based method, in contrast to solution-based cocrystallization, which uses a volatile solution. Therefore, it provides environmentally friendly and less expensive benefits. Second, the method is energy-efficient. It requires slight additional energy input compared to conventional solution-based methods. Third, it is applicable to a wide range of materials, such as pharmaceuticals, organic molecules, and metal-organic frameworks, providing versatility in usage. Fourth, the process is easy to scale up for industrial production [29, 30].

At the laboratory scale, mechanochemical cocrystallization usually utilizes a ball mill instrument for grinding. Ball milling offers several advantages. For example, it provides higher energy input, facilitating the formation of more diverse cocrystals. Additionally, ball milling allows for more precise control over reaction times, enhancing reproducibility. Finally, the ball milling process promotes superior mixing, which optimizes mass and energy transfer during cocrystallization. For industrial-scale production, twin-screw extrusion (TSE) emerges as a powerful alternative mechanochemical tool. TSE enables the large-scale preparation of cocrystals, a significant advantage for commercial applications. Notably, TSE has been successfully employed for the scale-up of various synthesis processes, including those for organic compounds, metal-organic frameworks (MOFs), and deep eutectic solvents [31].

b. Hot-Melt Extrusion

The HME method is an industrially feasible, solvent-free, scalable, and continuous cocrystal preparation technique [32]. The substance is incorporated in a molten polymeric matrix composed of starches, sugar alcohols, low-melting waxes, and thermoplastic polymers. The carrier systems are evaluated based on their favorable characteristics of solubility and stability, as presented in Figure 8.2. In order to attain effective processing, critical factors include the physical state, molecular weight, melting point, and polymer of the API.

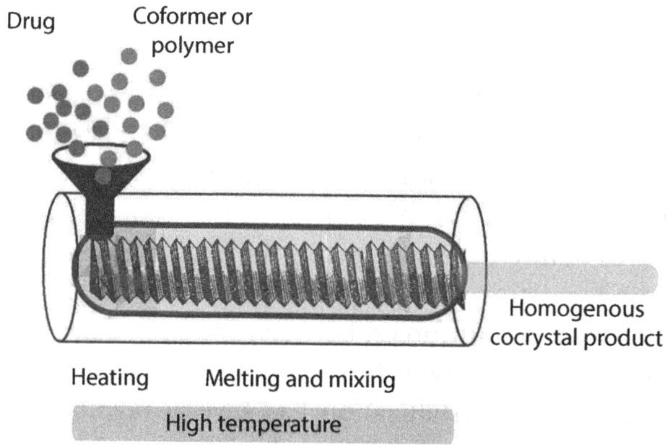

Figure 8.2 Hot-melt extrusion for cocrystal manufacturing.

Generally, the cocrystals prepared by hot-melt extruder are prepared by two methodologies. The first methodology involves the extrusion of the drug and coformer together and another is matrix-assisted cocrystallization, where polymer plays a crucial role in enhancing the cocrystallization process. During the cocrystal preparation by the HME technique, one should be aware of the critical quality attributes (CQAs) of the process. These CQAs are composed of the cocrystal, purity, thermal properties, particle size, and distribution. An optimal feed rate should be maintained to be able to successfully create cocrystals through extrusion. An excessive feed rate will prevent the drug-coformer mixture from having enough time to form cocrystals before they exit the extruder. The procedure will fail because the extruder nozzle will become blocked by the higher torque and pressure [33].

The utilization of HME has become a prevalent method for improving the solubility of medicinal goods. Leistritz and Thermo Fisher Scientific, producers of extruders, have been consistently creating more compact machinery to fulfill the pharmaceutical industry's requirements. Formulation and product development studies in early-stage development need the use of smaller-scale lab equipment due to the restricted availability of API. Nevertheless, it is imperative to ensure a smooth transition from laboratory testing to full-scale production. This guarantees that the product created in the laboratory can be manufactured on a bigger scale without encountering any problems. Twin-screw extruder manufacturers provide a range of different sizes for their twin-screw extruders [34].

8.4 Quality-By-Design Approaches (QbD) in Cocrystallization Manufacturing

QbD is a systematic approach to development that begins with predefined objectives and emphasizes understanding product, process, and control based on sound science and quality risk management [35]. The guidelines of the International Council for Harmonisation of Technical Requirements for Pharmaceuticals for Human Use (ICH) Q8 (R2) (Pharmaceutical Development), ICH Q9 (Quality Risk Management), and ICH Q10 (Pharmaceutical Quality System) have influenced the evolution of pharmaceutical QbD [36]. In order to achieve meaningful product quality specifications that are based on clinical performance, QbD can be implemented in cocrystallization manufacturing. This approach enhances product and process design, understanding, and control, thereby reducing product variability and defects and increasing process capability. Additionally, QbD has the potential to enhance the efficiency of manufacturing and product development. Lastly, QbD improves the management of post-approval change and the analysis of root causes.

As presented in Figure 8.3, a quality target product profile (QTPP) is a prospective summary of the quality characteristics of a cocrystal drug product that ideally will be achieved to ensure the desired quality, considering safety and efficacy. The main aim of QTPP in cocrystallization manufacturing is to improve the solubility and bioavailability of poorly soluble drugs [37]. CQAs are physical, chemical, biological, or microbiological properties or characteristics that need to be controlled to ensure product quality. In the production process of cocrystals, their CQAs are solubility, stability, purity, particle size, size distribution, and polymorphic form. CPPs are also very vital in the cocrystallization process. To ensure

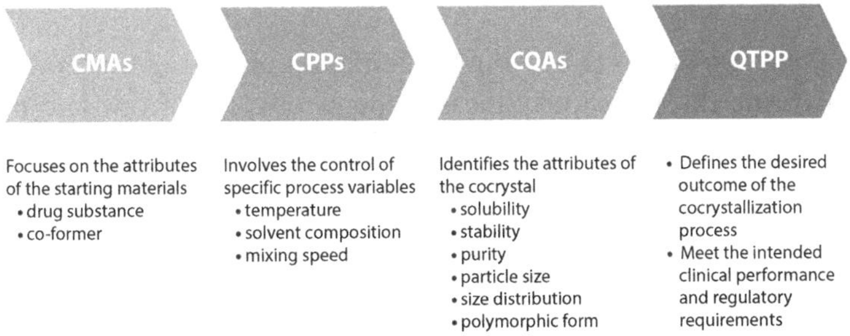

Figure 8.3 CMAs, CPPs, CQAs, and QTPP in cocrystallization manufacturing.

that the product meets its CQAs, it is crucial to identify and control CPPs within predefined limits during process development. Temperature control during the crystallization process, solvent-to-solid ratio, and mixing speed are general CPPs for the cocrystallization process [11]. Lastly, critical material attributes (CMAs) are the physical, chemical, biological, or microbiological properties or characteristics of input materials that can impact the final product quality. In the case of the cocrystallization process, the CMAs are the purity and particle size of the API and coformer which can influence the process and the quality of the resulting cocrystals.

QbD and DoE have been used in many studies to reveal the landscape of CPPs and also optimize the cocrystallization process. Xia and colleagues employed a QbD approach to optimize the preparation of carbamazepine and succinic acid cocrystals using slurry crystallization. A DoE approach in the central composite design model was used to optimize the formulation variables. The key variables examined were time, molar ratio, temperature, and concentration, with the goal, or QTPP, of maximizing the yield and purity of the cocrystals. The optimized formulation was characterized by powder X-ray diffraction [38]. Apart from solution cocrystallization, Ross and his team utilized a QbD approach to minimize material usage and optimize the selection process for continuous manufacturing via HME. The combination of computational and experimental methods effectively identified coformers that could form stable cocrystals [39]. QbD increases both the creation process of cocrystals and the understanding of the system for manufacturing drug nanocrystals using wet medium milling procedures. Several studies have indicated that QbD improves the scalability of production while ensuring the achievement of efficient and consistent nanocrystal manufacture [40].

8.5 Process Analytical Technology in Scale-Up Cocrystallization Process

PAT in pharmaceuticals is a system for designing, analyzing, and controlling manufacturing through timely measurements of CQA of raw and in-process materials and processes, with the goal of ensuring final product quality [41]. PAT involves the use of various tools and techniques to monitor and control the manufacturing process in real time, thereby improving the efficiency, consistency, and quality of pharmaceutical products. The application of QbD principles to the industrial crystallization process is made possible by the use of PAT. These important facilitators have a strong correlation with the cocrystallization process's scaling up for commercial production [12, 42].

Table 8.2 highlights the key PAT tools applied during the scale-up of pharmaceutical cocrystallization processes, illustrating their specific roles in monitoring critical quality attributes and ensuring process efficiency.

The focused beam reflectance measurement (FBRM), a sophisticated particle characterization technique, is commonly applied as a PAT tool in crystallization for *in situ* monitoring of particle size and count [43]. During the cocrystallization process, the FBRM probe is introduced into the process streams at an inclined angle. This positioning allows for smooth particle movement over the probe window, where the measurement is conducted. The tool is designed to monitor the formation and growth of particles in real time. It gives information on the number of particles and their distribution in terms of chord length, which is measured over time and in relation to temperature. This can also be utilized to ascertain the

Table 8.2 Process analytical technology used in the scale-up of cocrystallization processes in pharmaceutical applications.

PAT tool	Application	Key benefits	Reference
Focused beam reflectance measurement (FBRM)	Monitoring particle size and count in real time	Provides *in situ* measurements, noninvasive	[45] [46]
Process Raman spectroscopy	Identifying and quantifying cocrystal formation	Real-time chemical information, nondestructive	[47] [48]
Near-infrared (NIR) spectroscopy	Monitoring API and cocrystal content in blends	Rapid analysis, minimal sample preparation	[49] [50]
Particle vision and measurement (PVM)	Visualizing particle shape and size in real time	Provides image-based information, *in situ*	[51] [46]
Attenuated total reflectance–Fourier transform infrared (ATR-FTIR) spectroscopy	Analyzing molecular interactions during cocrystallization	Real-time chemical information, nondestructive	[52] [53]

solubility and the minimum solidification zone width (MSZW). Another PAT utilized in the crystallization process is particle vision and measurement (PVM), which allows for the observation and preservation of high-quality microscopy photographs of crystals, particles, and droplets as they naturally occur in the dynamic and real process concentration. It validates the information obtained from FBRM by offering real-time photos and also provides further understanding of the crystal structure, shape, production of secondary liquid or oiling out, monitoring of different crystal forms, and other aspects related to the process of crystallization [44]. Apart from crystal size and morphology monitoring, attenuated total reflectance–Fourier transform infrared (ATR-FTIR), ultraviolet/visible spectroscopy (ATR-UV/vis), Fourier transform infrared (FTIR) spectroscopy, and NIR spectroscopy can be used for monitoring the solution composition during the cocrystallization process [12].

8.6 Commercial Opportunities for Cocrystals

The prevalence of drug candidates with low solubility in water is increasing across the pharmaceutical industry. According to Takagi (2006), around 40% of the immediate-release oral products that are commercially available are classified as practically insoluble. Approximately 70% of the newly identified chemical entities (NCE) by combinatorial screening programs are estimated to be water-insoluble. This has led to the introduction of cocrystal technology in the pharmaceutical industry for commercial purposes. According to a report by Grand View Research Inc., the market size for cocrystal technology is estimated to be around US$200 million by 2025 [54]. The growth of this market is expected to be driven by the increasing demand for new and improved drug delivery systems, as well as the rising prevalence of chronic diseases. The examples of cocrystal drugs that are currently on the market are presented in Table 8.3.

8.7 Future Directions and Outlook

8.7.1 Machine Learning and AI

Various computational modeling techniques will be implemented in the development of cocrystal products, including lattice energy minimization, supramolecular recognition modeling, and machine learning techniques.

Table 8.3 Examples of cocrystal drugs that are currently on the market.

Commercial name	Indication	API	Coformer	Benefit of cocrystallization
Entresto®	Chronic heart failure	Valsartan	Sacubitril	Enhance bioavailability of valsartan
Tecfidera®	Multiple sclerosis	Dimethyl fumarate	Nicotinamide	Improve thermodynamic stability, solubility, and dissolution profile
Suglat®	Type 2 diabetes	Ipragliflozin	L-Proline	Enhance the stability of anhydrous form
Steglatro®	Type 2 diabetes	Ertugliflozin	Z-pyroglutamic acid	Improve stability
Depakote®	Acute bipolar mania and epilepsy	Valproic acid	Valproate sodium	Solid phase stability and less hygroscopicity
Lexapro®	Depression and anxiety	Escitalopram	Oxalate	Improve stability of API
Beta-Chlor®	Diagnosis or treatment of arthritis, allergic disorders, and atopic dermatitis	Chloral hydrate	Betaine	Improve thermal stability

Computational models have the capability to evaluate potential coformers, thereby aiding researchers in the selection of coformers that can generate a robust and stable cocrystal structure and forecast cocrystal formation and properties. This, consequently, can accelerate the process of cocrystal discovery and design. In a study by Mswahili and the team, they extracted descriptor values from the simplified molecular input line entry system of compounds and compared various machine learning models using data from 1476 instances. The results showed that the artificial neural network exhibited the highest accuracy, sensitivity, and F1 score. Additionally, the model performed comparably with only half of the descriptors selected through feature selection algorithms. These findings suggest that this approach can lead to faster and more accurate cocrystal development, enhancing the efficiency of the process [55].

8.7.2 Advanced Process Analytical Technology (PAT)

The primary application of PATs in product manufacturing has been the design, analysis, and control of diverse functional processes. These technologies are employed to provide essential data regarding a multitude of parameters that have the potential to substantially impact distinct CQA. The crystallization of pharmaceutical active ingredients involves an extensive series of processes. Such severe conditions may result in product failure. To implement PAT, each critical phase and parameter (whether in-line or out-of-line processing) can therefore be utilized to closely monitor crystal formation and development. This technology has emerged as an innovative instrument, and there are currently a multitude of academic and industrial investigations underway. The objective of PAT is to ensure superior quality by conducting accurate and timely assessments of crucial performance and quality of basic materials, in-process materials, and finished goods. PAT application entails the integration of scientifically grounded process design and optimization, suitable sensor technologies, information and statistical tools (chemometrics), and feedback process control strategies—all of which collaborate harmoniously to ensure the production of excellent products. A variety of *in situ* analytical techniques combined with chemometric instruments for the analysis of multivariate process information will serve as the foundation for future advancements in crystallization process modeling, simulation, and control [56–58].

8.7.3 Green Chemistry

Pharmaceutical cocrystallization is increasingly recognized for enhancing the solubility and bioavailability of poorly soluble drugs without altering their chemical identity or therapeutic effect. Scaling up the production of cocrystal products has traditionally required a high amount of solvent. Therefore, green and convenient techniques have been introduced and developed to minimize synthetic steps and solvent use. Recent studies have explored various methods to manufacture cocrystal drug products using green chemistry approaches, further promoting environmentally friendly and efficient drug development processes [59]. For example, Duarte and team used spray congealing for the large-scale production of cocrystals. The obtained cocrystal particles were compact and spherical consisting of aggregates of individual cocrystals entangled or adhered with each other. By varying atomization and cooling efficiency, cocrystal particle properties can be real time adjusted [60].

Furthermore, the utilization of PAT allows for accurate regulation of reaction parameters, immediate access to analytical data, and sophisticated data analysis techniques to promote sustainability objectives by comprehensively comprehending reaction kinetics, thermodynamics, and the direct impact of process variables on yield and waste production [61]. Merck scientists are developing biocatalytic-based syntheses for pharmaceutical intermediates, emphasizing fewer reaction steps, milder conditions, and better sustainability compared to classical methods. They successfully applied biocatalysis in the production of Belzutifan, addressing challenges with PAT tools to ensure precise control and overcome analytical difficulties in handling dense slurries and reaction monitoring [62].

Another approach to reduce solvent use is flow chemistry. Purdue University chemists and engineers modified the continuous flow process for lomustine to use green solvents, 2-MeTHF, and acetic acid, replacing dichloromethane. This new approach improved solubility, yield, and productivity while maintaining the process's environmental friendliness, demonstrating the importance of reactant solubility and reactor configuration in optimizing continuous flow processes [63].

8.8 Conclusion

Based on the preceding explanation, it can be inferred that cocrystallization is a superior option compared to conventional approaches such as salt production, solvates, and polymorphs for enhancing the physicochemical and processability properties of APIs. The utilization of solvates, salt, and polymorphs for producing pharmaceutically acceptable forms of APIs is limited due to a naturally restricted number of solvates and polymorphs, as well as the absence of ionizable groups in some APIs. Research conducted in the past decade has significantly advanced multiple aspects of cocrystallization, including screening, development, characterization, production, and formulations of cocrystals. The guidelines provided by regulatory agencies such as the US FDA and EMA confirm the recognized significance of pharmaceutical cocrystals in the advancement of better candidates with enhanced properties. However, this technology has not yet become a standard procedure for the manufacturing of medications. In order to effectively utilize cocrystals in the pharmaceutical industry, it is crucial to concentrate research efforts on establishing consistent synthetic procedures for cocrystals, determining and enhancing the factors that impact the quality of cocrystals (such as purity, yield, and reproducibility), and integrating cocrystal screening and development into the drug development process.

Acknowledgment

The authors wish to extend sincere gratitude to Rangsit University and Burapha University for providing access to the Journals database.

References

1. Vishweshwar, P., McMahon, J.A., Bis, J.A., Zaworotko, M.J., Pharmaceutical co-crystals. *J. Pharm. Sci.*, 95, 3, 499–516, 2006.
2. Naqvi, A., Ahmad, M., Minhas, M.U., Khan, K.U., Batool, F., Rizwan, A., Preparation and evaluation of pharmaceutical co-crystals for solubility enhancement of atorvastatin calcium. *Polym. Bull.*, 77, 6191–6211, 2020.
3. Emami, S., Siahi-Shadbad, M., Adibkia, K., Barzegar-Jalali, M., Recent advances in improving oral drug bioavailability by cocrystals. *BioImpacts*, 8, 4, 305–320, 2018.
4. Chadha, R., Saini, A., Arora, P., Bhandari, S., Pharmaceutical cocrystals: a novel approach for oral bioavailability enhancement of drugs. *Crit. Rev. Ther. Drug Carrier Syst.*, 29, 3, 183–218, 2012.
5. Dhondale, M.R., Thakor, P., Nambiar, A.G., Singh, M., Agrawal, A.K., Shastri, N.R., Kumar, D., Co-crystallization approach to enhance the stability of moisture-sensitive drugs. *Pharmaceutics*, 15, 1, 189, 2023.
6. Dhakate, C.S., Upadhye, K.P., Dixit, G.R., Bakhale, S.S., Umate, R.M., Taste masking by co-crystallization: A review. *World J. Pharm. Res.*, 6, 1531–1548, 2017.
7. Yadav, A., Shete, A., Dabke, A., Kulkarni, P., Sakhare, S., Co-crystals: a novel approach to modify physicochemical properties of active pharmaceutical ingredients. *Indian J. Pharm. Sci.*, 71, 4, 359–370, 2009.
8. Lin, S.-Y., Mechanochemical approaches to pharmaceutical cocrystal formation and stability analysis. *Curr. Pharm. Des.*, 22, 32, 5001–5018, 2016.
9. Lee, S.L., O'Connor, T.F., Yang, X., Cruz, C.N., Chatterjee, S., Madurawe, R.D., Moore, C.M., Yu, L.X., Woodcock, J., Modernizing pharmaceutical manufacturing: from batch to continuous production. *J. Pharm. Innov.*, 10, 191–199, 2015.

10. Ghadi, R., Ghuge, A., Ghumre, S., Waghmare, N., Kadam, V.J., Co-crystals: emerging approach in pharmaceutical design. *Indo Am. J. Pharm. Res.*, 4, 7, 3881–3892, 2014.
11. Saha, A., Ahangar, A.A., Dar, A.A., Thirunahari, S., Parambil, J.V., Pharmaceutical cocrystals: a perspective on development and scale-up of solution cocrystallization. *Cryst. Growth Des.*, 23, 11, 7558–7581, 2023.
12. Pawar, N., Saha, A., Nandan, N., Parambil, J.V., Solution cocrystallization: A scalable approach for cocrystal production. *Crystals*, 11, 3, 303, 2021.
13. Xuan, B., Wong, S.N., Zhang, Y., Weng, J., Tong, H.H., Wang, C., Sun, C.C., Chow, S.F., Extended release of highly water soluble isoniazid attained through cocrystallization with curcumin. *Cryst. Growth Des.*, 20, 3, 1951–1960, 2020.
14. Lee, M.-J., Wang, I.-C., Kim, M.-J., Kim, P., Song, K.-H., Chun, N.-H., Park, H.-G., Choi, G.J., Controlling the polymorphism of carbamazepine-saccharin cocrystals formed during antisolvent cocrystallization using kinetic parameters. *Korean J. Chem. Eng.*, 32, 1910–1917, 2015.
15. Bianchi, F., Fornari, F., Riboni, N., Spadini, C., Cabassi, C.S., Iannarelli, M., Carraro, C., Mazzeo, P.P., Bacchi, A., Orlandini, S., Development of novel cocrystal-based active food packaging by a Quality by Design approach. *Food Chem.*, 347, 129051, 2021.
16. US FDA, Guidance for industry. Regulatory classification of pharmaceutical co-crystals, 2018, https://www.fda.gov/regulatory-information/search-fda-guidance-documents/regulatory-classification-pharmaceutical-co-crystals.
17. European Medicines Agency, Reflection paper on the use of cocrystals of active substances in medicinal products, 2015, https://www.gmp-compliance.org/files/guidemgr/reflection-paper-use-cocrystals-active-substances-medicinal-products_en.pdf.
18. Shao, S., Stocker, M.W., Zarrella, S., Korter, T.M., Singh, A., Healy, A.M., *In Situ* Cocrystallization via Spray Drying with Polymer as a Strategy to Prevent Cocrystal Dissociation. *Mol. Pharmaceutics*, 20, 9, 4770–4785, 2023.
19. Walsh, D., Serrano, D.R., Worku, Z.A., Madi, A.M., O'Connell, P., Twamley, B., Healy, A.M., Engineering of pharmaceutical cocrystals in an excipient matrix: Spray drying versus hot melt extrusion. *Int. J. Pharm.*, 551, 1-2, 241–256, 2018.
20. Tanaka, R., Hattori, Y., Otsuka, M., Ashizawa, K., Application of spray freeze drying to theophylline-oxalic acid cocrystal engineering for inhaled dry powder technology. *Drug Dev. Ind. Pharm.*, 46, 2, 179–187, 2020.
21. Moradiya, H.G., Islam, M.T., Scoutaris, N., Halsey, S.A., Chowdhry, B.Z., Douroumis, D., Continuous manufacturing of high quality pharmaceutical cocrystals integrated with process analytical tools for in-line process control. *Cryst. Growth Des.*, 16, 6, 3425–3434, 2016.

22. Chavan, R.B., Thipparaboina, R., Yadav, B., Shastri, N.R., Continuous manufacturing of co-crystals: challenges and prospects. *Drug Deliv. Transl. Res.*, 8, 1726–1739, 2018.
23. Kara, A., Kumar, D., Healy, A.M., Lalatsa, A., Serrano, D.R., Continuous manufacturing of cocrystals using 3d-printed microfluidic chips coupled with spray coating. *Pharmaceuticals*, 16, 8, 1064, 2023.
24. Juneja, P., Kaur, B., Odeku, O.A., Singh, I., Development of Corn Starch-Neusilin UFL2 Conjugate as Tablet Superdisintegrant: Formulation and Evaluation of Fast Disintegrating Tablets. *J. Drug Deliv.*, 2014, 1, 827035, 2014.
25. Krogmeier, S., Pritchard, J., Dokou, E., Miles, S., Connelly, G., Medendorp, J., Bourland, M., Swinney, K., Orkambi: a continuous manufacturing approach to process development at Vertex, in: *How to Design and Implement Powder-To-tablet Continuous Manufacturing Systems*, pp. 383–396, Academic Press, Amsterdam, 2022.
26. Jain, S., Kaur, S., Rathi, R., Nagaich, U., Singh, I., Application of co-processed excipients for developing fast disintegrating tablets: A review. *Polym. Med.*, 53, 1, 59–68, 2023.
27. Rathi, R., Kaur, S., Singh, I., A Review on Co-crystals of Herbal Bioactives for Solubility Enhancement: Preparation Methods and Characterization Techniques. *Cryst. Growth Des.*, 22, 3, 2023–42, 2022.
28. Gröls, J.R. and Castro-Dominguez, B., Mechanochemical co-crystallization: Insights and predictions. *Comput. Chem. Eng.*, 153, 107416, 2021.
29. Braga, D., Maini, L., Grepioni, F., Mechanochemical preparation of co-crystals. *Chem. Soc. Rev.*, 42, 18, 7638–7648, 2013.
30. Pagola, S., Outstanding advantages, current drawbacks, and significant recent developments in mechanochemistry: a perspective view. *Crystals*, 13, 1, 124, 2023.
31. Solares-Briones, M., Coyote-Dotor, G., Páez-Franco, J.C., Zermeño-Ortega, M.R., de la O Contreras, C.M., Canseco-González, D., Avila-Sorrosa, A., Morales-Morales, D., Germán-Acacio, J.M., Mechanochemistry: A green approach in the preparation of pharmaceutical cocrystals. *Pharmaceutics*, 13, 6, 790, 2021.
32. Rathi, R., Sanshita, Singh, I., Hot Melt Extrusion Technique for Developing Pharmaceutical Co-crystals: A Review. *Drug Deliv. Lett.*, 14, 2, 97–108, 2024.
33. Nandi, U., Trivedi, V., Ross, S.A., Douroumis, D., Advances in twin-screw granulation processing. *Pharmaceutics*, 13, 5, 624, 2021.
34. Patil, H., Vemula, S.K., Narala, S., Lakkala, P., Munnangi, S.R., Narala, N., Jara, M.O., Williams III, R.O., Terefe, H., Repka, M.A., Hot-Melt Extrusion: from Theory to Application in Pharmaceutical Formulation—Where Are We Now? *AAPS PharmSciTech*, 25, 2, 37, 2024.
35. US FDA, Q8(R2) Pharmaceutical Developmen, 2009, https://www.fda.gov/regulatory-information/search-fda-guidance-documents/q8r2-pharmaceutical-development.

36. US FDA, Q8, Q9, & Q10 Questions and Answers – Appendix: Q&As from Training Sessions (Q8, Q9, & Q10 Points to Consider), 2012, https://www.fda.gov/regulatory-information/search-fda-guidance-documents/q8-q9-q10-questions-and-answers-appendix-qas-training-sessions-q8-q9-q10-points-consider.
37. Thakor, P., Yadav, B., Modani, S., Shastri, N.R., Preparation and optimization of nano-sized cocrystals using a quality by design approach. *CrystEngComm*, 22, 13, 2304–2314, 2020.
38. Xia, Y., Optimization of Carbamazepine-Succinic Acid Cocrystal Preparation Using Quality by Design Approach. *Open Access Lib. J.*, 8, 2, e7074, 2021.
39. Ross, S.A., Ward, A., Basford, P., McAllister, M., Douroumis, D., A Quality by Design strategy for cocrystal design based on novel computational and experimental screening strategies with the aim of rapid scale-up to continuous manufacture via Hot-Melt Extrusion: Part A, 2024, 16 June 2024, PREPRINT (Version 1), https://doi.org/10.21203/rs.3.rs-4477916/v1.
40. Peltonen, L., Design space and QbD approach for production of drug nanocrystals by wet media milling techniques. *Pharmaceutics*, 10, 3, 104, 2018.
41. US FDA, Guidance for industry, PAT-A framework for innovative pharmaceutical development, manufacturing and quality assurance, 2004, http://www. fda. gov/cder/guidance/published. html.
42. Aucamp, M. and Milne, M., The physical stability of drugs linked to quality-by-design (QbD) and in-process technology (PAT) perspectives. *Eur. J. Pharm. Sci.*, 139, 105057, 2019.
43. Holaň, J., Ridvan, L., Billot, P., Štěpánek, F., Design of co-crystallization processes with regard to particle size distribution. *Chem. Eng. Sci.*, 128, 36–43, 2015.
44. Sihorkar, V., How PAT tools can help improve understanding of the crystallisation process, 2024, https://www.sailife.com/how-can-pat-tools-help-improve-understanding-of-the-crystallization-process/.
45. Wang, X., Du, S., Zhang, R., Jia, X., Yang, T., Zhang, X., Drug-drug cocrystals: Opportunities and challenges. *Asian J. Pharm. Sci.*, 16, 3, 307–17, 2021.
46. Kim, P., Cho, M.-Y., Choi, G.J., Understanding the Mechanism of Indomethacin-Saccharin Co-crystal Formation Using In-line Monitoring System based on PVM and FBRM. *Korean Chem. Eng. Res.*, 55, 2, 180–189, 2017.
47. Powell, K., Croker, D., Rielly, C., Nagy, Z., PAT-based design of agrochemical co-crystallization processes: A case-study for the selective crystallization of 1: 1 and 3: 2 co-crystals of p-toluenesulfonamide/triphenylphosphine oxide. *Chem. Eng. Sci.*, 152, 95–108, 2016.
48. Otaki, T., Tanabe, Y., Kojima, T., Miura, M., Ikeda, Y., Koide, T., Fukami, T., *In situ* monitoring of cocrystals in formulation development using low-frequency Raman spectroscopy. *Int. J. Pharm.*, 542, 1-2, 56–65, 2018.
49. Kelly, A.L., Gough, T., Dhumal, R.S., Halsey, S., Paradkar, A., Monitoring ibuprofen–nicotinamide cocrystal formation during solvent free continuous

cocrystallization (SFCC) using near infrared spectroscopy as a PAT tool. *Int. J. Pharm.*, 426, 1-2, 15–20, 2012.
50. Sarraguça, M.C., Ribeiro, P.R., Santos, A.O., Silva, M.C., Lopes, J.A., A PAT approach for the on-line monitoring of pharmaceutical co-crystals formation with near infrared spectroscopy. *Int. J. Pharm.*, 471, 1-2, 478–484, 2014.
51. Simon, L.L., Pataki, H., Marosi, G.R., Meemken, F., Hungerbühler, K., Baiker, A., Tummala, S., Glennon, B., Kuentz, M., Steele, G., Assessment of recent process analytical technology (PAT) trends: a multiauthor review. *Org. Process Res. Dev.*, 19, 1, 3–62, 2015.
52. Powell, K.A., Bartolini, G., Wittering, K.E., Saleemi, A.N., Wilson, C.C., Rielly, C.D., Nagy, Z.K., Toward continuous crystallization of urea-barbituric acid: a polymorphic co-crystal system. *Cryst. Growth Des.*, 15, 10, 4821–4836, 2015.
53. Barmpalexis, P., Karagianni, A., Nikolakakis, I., Kachrimanis, K., Artificial neural networks (ANNs) and partial least squares (PLS) regression in the quantitative analysis of cocrystal formulations by Raman and ATR-FTIR spectroscopy. *J. Pharm. Biomed. Anal.*, 158, 214–224, 2018.
54. Kumar Bandaru, R., Rout, S.R., Kenguva, G., Gorain, B., Alhakamy, N.A., Kesharwani, P., Dandela, R., Recent advances in pharmaceutical cocrystals: From bench to market. *Front. Pharmacol.*, 12, 780582, 2021.
55. Mswahili, M.E., Lee, M.-J., Martin, G.L., Kim, J., Kim, P., Choi, G.J., Jeong, Y.-S., Cocrystal prediction using machine learning models and descriptors. *Appl. Sci.*, 11, 3, 1323, 2021.
56. Das, S.S., Alkahtani, S., Nayak, A.K., Hasnain, M.S., Chapter 8 - Process analytical technology (PAT) tools: Uses in pharmaceutical manufacturing, in: *Advances and Challenges in Pharmaceutical Technology*, A.K. Nayak (Ed.), pp. 243–259, Academic Press, Amsterdam, 2021.
57. Rathi, R., Guggal, Y., Twinkle, Sandhu, V., Singh, I., Exploring the Patent Landscape and Regulatory Prospective on Pharmaceutical Cocrystals. *J. Pharm. Technol. Res. Manag.*, 11, 49–60, 2023.
58. Lawrence, X.Y., Lionberger, R.A., Raw, A.S., D'Costa, R., Wu, H., Hussain, A.S., Applications of process analytical technology to crystallization processes. *Adv. Drug Deliv. Rev.*, 56, 3, 349–369, 2004.
59. Rathi, R. and Singh, I., Multicomponent crystal compromising dasatinib and selected co-crystals formers: a patent evaluation of EP2861589B1. *Pharm. Pat. Anal.*, 11, 1, 15–21, 2021.
60. Duarte, Í., Andrade, R., Pinto, J.F., Temtem, M., Green production of cocrystals using a new solvent-free approach by spray congealing. *Int. J. Pharm.*, 506, 1-2, 68–78, 2016.
61. Mettler Toledo®, PAT for Green Chemistry in Pharmaceutical and Chemical Industries, 2024, https://www.mt.com/th/th/home/library/stories/automated-reactors/examples-of-green-chemistry-in-industry.html.

62. Qin, Y., Mattern, K.A., Zhang, V., Abe, K., Kim, J., Zheng, M., Gangam, R., Kalinin, A., Kolev, J.N., Axnanda, S., Evolution of a Green and Sustainable Manufacturing Process for Belzutifan: Part 4— Applications of Process Analytical Technology in Heterogeneous Biocatalytic Hydroxylation. *Org. Process Res. Dev.*, 28, 2, 432–440, 2024.
63. Murbach-Oliveira, G., Akturk, I., Nagy, Z., Thompson, D.H., Process Development and Kinetic Analysis in the Green Synthesis of Lomustine. *Org. Process Res. Dev.*, 28, 5, 1910–1916, 2024.

9

Pharmaceutical Cocrystals for Solubility Enhancement of Drugs

Mouli Das, Sk Habibullah, Tanisha Das and Subrata Mallick*

Department of Pharmaceutics, School of Pharmaceutical Sciences, Siksha 'O' Anusandhan (Deemed to be University), Bhubaneswar, Odisha, India

Abstract

Cocrystallization is considered an ecologically and possibly active method to improve the ability of a low water-soluble drug to be dissolved in an aqueous medium and consequently bioavailability without changing the physico-pharmacological properties of the drug. In the field of drug development and research, cocrystallization is an appropriate, convenient, and green practice with a small amount of solvent or without any solvent. In the world of chemistry, solubility plays a vital role in understanding the behavior of substances in various solutions. Solubility denotes the capability of a material to dissolve in a solvent and form a homogeneous mixture which remains influenced by various factors like solute-solvent nature, temperature, pressure and presence of another solute, particle size, molecular size, and polymorphism. The first key aspect to consider is the nature of the solvent and solute involved in the dissolution process. Solvents can be broadly classified into polar and nonpolar. Polar solvents, such as water, have a positive and negative end, allowing them to interact with polar substances. Weakly basic drugs with acidic coformers can improve solubility by altering the pH. Cocrystallization involves combining multiple molecular components in precise ratios to form solid-phase materials through intermolecular interactions of dispersion forces, dipole-dipole forces, dipole-induced dipole forces, and hydrogen bonds. Cocrystallization could also be utilized in pharmaceutical formulation development of taste masking, controlled release dosage forms, targeted delivery systems, etc. It can exhibit specific physical, chemical, and mechanical properties of various material science applications. Furthermore, cocrystallization finds application beyond pharmaceuticals, extending to areas like nutraceuticals, food science, and material engineering. Its versatility enables tailored solutions for taste

**Corresponding author*: profsmallick@gmail.com; subratamallick@soa.ac.in

Inderbir Singh, Subrata Mallick and Ritu Rathi (eds.) Cocrystals in Pharmaceutical Sciences: Design to Applications, (257–292) © 2025 Scrivener Publishing LLC

masking, controlled-release formulations, and targeted drug delivery systems. As research progresses, the refinement of cocrystallization strategies promises to address the escalating demand for improved drug solubility.

Keywords: Bioavailability, cocrystallization, dissolution rate, polymorphism, solubility

9.1 Introduction

The manufacturing of effective drug products influences the successful delivery of any active pharmaceutical ingredient (API) to patients. Generally, the challenges faced in the successful delivery of drugs and manufacturing include specific property deficits, such as solubility, stability, organoleptic properties, bioavailability, and mechanical characteristics [1]. Lack of solubility in pharmaceutical drug development is a major challenge for developing any dosage form as well as any other formulations. Different solid pharmaceutical multi-components produced from API are shown in Figure 9.1. This section will comprise the various techniques related to cocrystallization in order to improve poor drug solubility thereby enhancing their bioavailability. Consecutive administration of oral drugs, bioavailability studies, and complexed inter-associated drug absorption deals with crucial factors related to dosage forms. The primary factor includes the release of API from the formulation which significantly affects pharmacological actions. The dissolution profile followed by the solubility studies of the drug enclosed in the physiological environment of GIT also serves as another key factor influencing bioavailability. Any API transition from solid to dissolved state is an important factor to be considered in the dissolution process for effective absorption. Additionally, permeability also plays a significant role in absorption at GIT. The drug's ability to cross the mucosal barrier across the GI tract and enter into the general circulation is highly influenced by the permeability factors depending upon specific characteristics of the site of absorption. Fundamentally, the interplay of drug release, dissolution profile, and permeability jointly rules the overall bioavailability, determining the rate and extent at which the administered API reaches the systemic circulation and exerts its therapeutic activity. Cocrystallization is considered likely to be a successful approach in improving the aqueous solubility of weakly soluble drugs. It encompasses the formation of a multicomponent solid crystalline substance of a drug component and an appropriate conformer in a specific molar ratio to form a crystal lattice. Cocrystals can enhance the physical characteristics of the drug like solubility without hampering its pharmacological

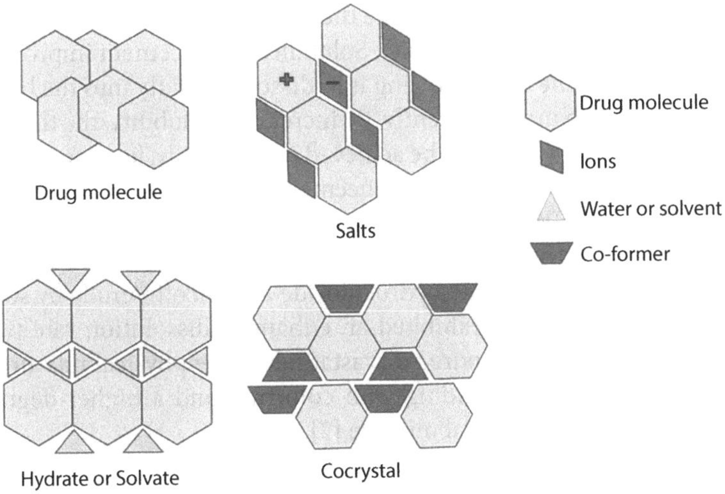

Figure 9.1 Different solid pharmaceutical multi-components produced from API.

effectiveness [2–5]. Cocrystals of fluoxetine hydrochloride with benzoic acid and fumaric acid dissolve more sluggishly, while cocrystals with succinic acid dissolve more rapidly compared to the pure drug because of the hydrogen bond formation between neutral carboxylic acid and chloride ion of fluoxetine hydrochloride [6]. The selection of a suitable coformer is a key challenge in cocrystal development and can be done through trial and error, supramolecular synthon, or virtual screening approaches. Drug material and coformer are used to prepare cocrystals employing solvent vaporization technique, solvent-drop grinding, and solution cooling crystallization. Overall, cocrystallization offers a promising strategy for enhancing not only the solubility of drugs but also improving their pharmaceutical properties.

9.2 Why Solubility Enhancement is Important?

Drug solubility plays a crucial role in the field of pharmacology as it has a direct influence on the bioavailability of the drug. Bioavailability denotes the degree of availability of the drug in the bloodstream via the absorption process. Therefore, enhancing the solubility of the material could be of utmost importance as it determines how efficiently the drug can be absorbed and available in the body. When a drug is soluble, it can readily dissolve in a solvent, typically water, and will be ready for absorption at the absorption site. When a drug is poorly soluble, it is not readily available at the absorption site which reduces the therapeutic efficacy due to long residence time near and around the site. Solubility enhancement improves the dissolution rate of the drug allowing it to dissolve rapidly into the biological fluid and absorb more efficiently. By increasing solubility, the therapeutic concentration of a drug can be achieved more readily, leading to a faster onset of action. The solubility enhancement method of cocrystallization proposes to modify the solid-state properties of a drug without altering its pharmacological activity. For example, in a study, cocrystal of atorvastatin calcium with glucosamine hydrochloride and nicotinamide by solvent vaporization method also exhibited an enhanced dissolution rate two to three times faster than the pure atorvastatin and the physical mixture due to weak inter-particular bonding with coformer and a higher degree of drug crystallinity in a physical mixture [7].

9.3 Factors Affecting Solubility

Understanding the factors that influence the capability of a material to dissolve in an assumed solvent is vital in enhancing solubility. Crucial variables that influence solubility are discussed below:

9.3.1 Nature of Solute and Solvent

The major aspect is the nature of the solvent and solute participating in the dissolution process. The two broad categories of solvents are polar solvents and nonpolar solvents. Polar solvents, such as water, have a positive and negative end, which allows them to interact with polar substances. On the other hand, nonpolar solvents lack such polarity and interact more readily with nonpolar substances. Understanding the nature of the solute is equally important. Different substances have varying levels of solubility based on their physicochemical properties. For example, polar substances have a tendency to be soluble readily in polar liquids, whereas nonpolar substances have greater solubility in nonpolar liquids.

9.3.2 Particle Size

The size of the solute particle and their surface area can influence solubility. Decreased particle size increases surface area and facilitates drug dissolution. Active surface area is directly related to the dissolution rate rather than the entire external area of the material. Enhanced intimate contact of the active solid external area with the aqueous medium enables augmented dissolution. The association between the material size and the solubility is governed by the principle from physical chemistry specifically the Noyes-Whitney equation. It also exhibits different thermodynamic properties compared to the larger particle size. It influences solubility by affecting factors like Gibbs free energy and enthalpy of the same compound. Particle agglomeration of hydrophobic drugs may form larger particles due to high surface free energy resulting in a fall of dissolution. The use of a wetting agent reduces interfacial tension and converts the total surface area of the hydrophobic drug to active surface area and improves dissolution. The reduction of size and crystallinity is effective in solubility as well as dissolution profile [8]. The preparation of nano-cocrystal formulation is still in the investigative stage. The wet particulate milling crystallization technique was found more robust for the preparation of nano-cocrystals. In a recent study, prepared nano-cocrystals (300 to 450 nm) exhibited better physical stability and improved kinetic solubility by 6.6 and 51.5 times than the solubility of itraconazole–succinic acid cocrystal and pure itraconazole respectively [9].

9.3.3 Temperature

Temperature is a critical factor that influences the solubility of solid drug substances. Usually, a rise in temperature hints at increased solubility for most solid solutes. This is because elevated temperatures extend additional energy for the particulate materials to break down into individual pieces and interact with the liquid molecules. However, the ability of the gas solutes to be dissolved tends to decrease with increasing temperature. This is due to the inverse relationship between temperature and the solubility of gasses described by Henry's law. In some cases, solubility may be relatively constant with temperature changes. This is common for solutes where the dissolution process is not significantly affected by temperature or where the exothermic and endothermic effects balance out each other.

9.3.4 Pressure

The pressure exerted on a gas solute can impact its solubility in a liquid solvent. According to Henry's law, the "solubility" of a gas is directly related to the fractional pressure of the gas above the fluid. Therefore, an increase in pressure leads to higher gas solubility, while a decrease in pressure reduces solubility. This principle is vital in various applications such as carbonation of beverages and gas absorption processes.

9.3.5 Molecular Size

Smaller molecules often have higher solubility. This is because smaller molecule experiences weaker intermolecular forces and can more easily fit between solvent molecules. Small molecules are more soluble in polar solvents as they form favorable interactions with polar functional groups in the solvent. The presence of polar or charged functional groups in a molecule can improve solubility in charged solvents. Bonding like hydrogen bonding, dipole-dipole interaction, and ion-dipole interactions between solute and solvent molecules can significantly impact solubility.

9.3.6 Polymorphism

Polymorphs (having different molecular arrangements) can have a notable impact on the solubility of a molecule. Variations in the crystal structure, packing arrangement, and surface properties can affect the solubility of any compound. There are some factors about polymorphs that can affect solubility. Different polymorphs can have distinct crystal packing arrangements which lead to various molecular densities within the crystal lattice that can influence the exposure of functional groups and affect the solubility in a given solvent. Polymorphs having looser packing may have higher solubility because solvent molecules can penetrate the crystal lattice very easily. It is also seen that thermodynamically stable polymorphs play a crucial role because they have lower solubility leads to less dissolution.

9.3.7 Polarity

Polarity is a fundamental factor that strongly influences the solubility of a substance. Polarity plays a crucial role in the interaction between molecules of the solvent and solute. In a general rule, a polar solvent tends to dissolve in a polar solvent, i.e., "like dissolves like." The polarity of the solvent facilitates the breaking down of the intermolecular forces within the solute. Similarly, nonpolar solutes tend to dissolve in nonpolar solvents where van der Waals forces play a significant role. For ionic compounds, the interaction between ions and polar solvents is very crucial. Polar solvents like water can enhance the dissolution of an ionic compound by surrounding and stabilizing it through ion-dipole interaction.

9.3.8 Common-Ion Effect

The presence of a common ion in a solution can affect the solubility of a given solute due to a shift in chemical equilibrium via the application of Le

Chatelier's principle of equilibrium reaction. A drug in salt form decreases solubility when dissolved in a solution comprising one of its ions [10].

9.4 Solubility and Bioavailability

The amount and rate of any drug entering the systemic circulation is known as bioavailability. When it comes to the development of any formulation, a significant challenge comes to its low solubility and oral bioavailability. A cellular membrane is considered a combination of biological and physicochemical barriers to drug transport. The solid drug present in solution form is only readily available for permeation through the gastrointestinal membrane. A limited factor in drug absorption is solubility in the early stage of drug design. A common cause of poor solubility of any drug compound is incomplete and erratic absorption of the drug after oral administration because of extreme lipophilicity.

The biopharmaceutical classification system (how well a drug dissolves in aqueous media and how readily a drug diffuses across the biomembrane) categorizes four groups, but it has been claimed that the primary properties are not permeability and solubility but bonding (especially hydrogen bonding) and molecular size [11]. Synthesizing and designing any pharmaceutical cocrystal to increase aqueous solubility is the crucial work of crystal engineering for possible improvement of oral bioavailability. One report of a pharmacokinetic study conducted using apixaban-cocrystal demonstrated more than 20-fold improvement in oral bioavailability when compared with apixaban alone [12]. Another example of a cocrystal of baicalein with coformer nicotinamide showed 2.80- and 2.49-fold enhanced area under the curve (AUC) and peak plasma concentration (Cmax) in rats respectively as compared to pure API [13]. When tested in rats, meloxicam cocrystal demonstrated significantly higher bioavailability relative to the drug alone [14]. Additionally, the anticancer compound 6-mercaptopurine cocrystals exhibited 168.7% boosted oral bioavailability in rats when compared to pure drugs [15]. Cocrystallization, a powerful technique that can modify the drug's crystal lattice can significantly enhance the solubility and stability of various substances influencing their physicochemical properties and finally improving the bioavailability of the drug. By forming complexes with two or more compounds, cocrystallization offers a unique approach to overcoming challenges in drug formulation, material science, and other fields including solubility properties leading to improved therapeutic efficacy. Poorly soluble drugs often face challenges in effective absorption and consequent bioavailability.

9.5 Permeability and Cocrystallization

One of the key factors of absorption and distribution of API depends on the drug's permeability through the cellular membrane. The permeability of any pharmaceutical compound is mostly dependent on the "octanol-water partition coefficient" ($K_{o/w}$) measuring its lipophilicity/hydrophilicity of the unmodified API. One study investigated the permeability of hydrochlorothiazide and cocrystals with various coformers using Franz diffusion cells and it showed higher drug flow levels as compared to pure drugs. Permeability enhanced due to the development of hetero-synthon formation between API and the coformer [16]. In a cocrystal formation between drug and coformer, the properties and structure of the coformer play an important role. Coformer selection is based on its hydrophilicity and hydrophobicity to prepare cocrystals with higher solubility and penetrability [17]. Likewise, API and coformer interaction, molecular packing, and interaction with solvent play an important part in modifying penetrability to select an optimal formulation [18]. Previous literature suggests that cocrystallization can alter the polarity of the relevant chemical entity which results in altering the permeability of the drug compound [16]. Coformers like glutaric acid, lactic acid, tartaric acid, and fumaric acid are used to improve the permeability, particularly of highly lipophilic drugs.

Membrane permeability and/or diffusion could potentially be improved by cocrystallization without modifying the molecule's structure or disturbing membrane integrity. Pharmaceutical cocrystallization is known as a novel and modified method to improve penetrability as it is well reproducible, free of tissue harm, and can be shaped on a big scale related to other methods [16, 19, 20]. Ferretti and colleagues conducted the study on indomethacin for solubility as well as permeability enhancement with three different coformers, i.e., 2-methoxy-5-nitroaniline, 2-hydroxy-4-methyl-pyridine, and saccharine. Their result demonstrated that there is a 2-fold and 2.5-fold increase in permeability with 2-hydroxy-4-methyl-pyridine and saccharine and no remarkable change with 2-methoxy-5-nitroaniline. This difference in results may be due to the affinity of API for protein being induced by specific weak interaction [20]. BCS class IV drug hydrochlorothiazide with coformers piperazine and picolinamide gave 6.6-fold and 3-fold higher solubility along with permeability respectively using cellulose nitrate membrane, and cocrystals using TMP, isoniazid, and malonamide also contributed higher cumulative permeation [21]. A similar study on hydrochlorothiazide with succinic acid (4.7-fold lower solubility, lower permeability), nicotinic acid (1.4-fold decreased solubility, two-time increased penetrability), resorcinol (2.4-folds enhanced solubility, 1.6-fold enhanced penetrability), nicotinamide (solubility enhanced by 1.3-fold, penetrability enhanced by 1.8 times), and 4-aminobenzoic acid (increased solubility by 2.4 times and increased permeability by 1.3-fold), and all cocrystal exhibit altered solubility as well as permeability [16].

9.6 Enhancing Solubility through Cocrystallization

Several methods to prepare cocrystals have been investigated like solvent vaporization method, reaction crystallization, fusion, and grinding method, etc. Drug solubility increases by the preparation of cocrystals by altering original molecular arrangements or by reducing the lattice energy of the packing of drug molecules [22]. Nevertheless, supersaturation is achieved on the surface of the cocrystal diffusion layer due to the increase in the drug concentration. Crystal growth and nucleation coincide with the dissolution of the cocrystal and that transforms into the real crystal structure which precipitates on "the cocrystal surface," consequently avoiding the fast procedure of dissolution and losing the benefit of the solubility of the cocrystal [23]. Several techniques have been introduced for the improvement of solubility of API such as hot-melt extrusion (HME), SE, ultrasonic preparation, high-pressure homogenization, supercritical fluid process, freeze drying, ultrasonic precipitation, and spray drying. Apart from the preparation method, the nature of the materials is another factor in developing a formulation, having improved solubility leading to improved bioavailability [24–27].

9.6.1 Hot-Melt Extrusion Technique of Cocrystallization to Enhance Solubility

HME is a widely used technology not only in the food and polymer industry but also in cocrystallization for the development of various types of drug delivery systems. HME is advantageous because it is a solvent-free technique, a readily scalable process, and a continuous or single-step method [28–30]. Cocrystal agglomerates prepared with ibuprofen and nicotinamide (1:1) using HME showed an enhanced dissolution rate compared to ibuprofen alone [31]. In another study, Soluplus matrix-assisted cocrystallization of carbamazepine using nicotinamide as a coformer by the HME method presented a very fast dissolution profile [32]. Drugs with compatible coformers are used to create a "supramolecular synthon"-type cocrystal.

There are two categories of "supramolecular synthon," i.e., "homosynthon" having the "same" functional group and "heterosynthon" having a "different" functional group [33]. Combining more than two drugs in a single dose has developed a popular drug-development approach and also in the growth of cocrystals. Mixing more than two drugs in the preparation of a cocrystal is called a drug-drug cocrystal [34]. In recent years, meloxicam-aspirin, tramadol hydrochloride–celecoxib, metformin-glipizide, and many more as drug-drug cocrystals have already been reported [33].

9.6.2 Amino Acid in Cocrystal to Enhance Solubility

It is considered that amino acids are the construction blocks of polypeptides and protein, which play crucial roles in food as well as fitness. The functions of the amino acid include "cell signaling," osmolyte synthesis, gene expression, regulation of hormone secretion, and protein synthesis. They are widely available in wheat, corn, rice, sorghum, etc. and they are considered as commonly regarded as harmless compounds. Since, they have a great propensity to develop a possible bond like "hydrogen bonds" with several groups like pyridyl, phenolic hydroxyl, hydroxyl, and carboxyl. In addition, they are known to have low toxicity [34]. For all those reasons, the amino acid is used as a coformer in the formation of a cocrystal with additional benefits. Liquid-assisted grinding (LAG), SE, and neat grinding are well-developed reported techniques for preparing cocrystals using amino acids. Myricetin-proline cocrystals exhibited an enhanced drug release when compared with the physical combination and pure myricetin [35]. A similar study conducted to prepare a cocrystal of acetazolamide–L-Proline exhibited higher solubility than pure acetazolamide with an increasing value of pH [36]. Other studies also revealed improved solubility as well as dissolution in cocrystals of diclofenac sodium with L-proline trihydrate as well as L-proline monohydrate [37]. LAG techniques were also used in indomethacin [20], diclofenac with L-proline [38], and chlorothiazide with a number of coformers [39], and all studies presented a higher solubility than the pure drug. Overall, it can be explained that the use of amino acid as a coformer is superior in developing a safer, simple, environmentally friendly formulation for drug solubility enhancement.

9.6.3 Polymeric Effect on Cocrystallization

Polymeric excipients, which can improve drug delivery and act as progressive coformers rather than small molecules, have not been explored extensively previously. Nevertheless, in efforts to enhance the effectiveness of poorly solvable medications, a range of polymers have been employed in solid dispersion techniques. These typically involve drug-polymer systems comprising two components, with the molecularly dispersed drug within an amorphous polymer matrix. In addition to these methods, limited instances of drug-polymer crystalline inclusion complexes have been documented.

Until now, the majority of pharmaceutical cocrystals have consisted of API paired with small molecules of coformers like saccharin, nicotinamide, and succinic acid. Conversely, instances of cocrystals among API and polymers have been relatively limited, with examples such as hydroxybenzenes-PEG cocrystals [40, 41].

9.6.4 Cocrystals of Herbal Bioactive to Enhance Solubility

Herbal bioactives as the natural compounds sourced from plants are usually chosen widely because of their synergistic effect. Direct administration of these compounds comes across several problems like low water solubility, least permeability through the membrane, and limited stability leading to low bioavailability. To boost the aqueous solubility of herbal bioactives like curcumin, naringenin, berberine, quercetin, resveratrol, and baicalein, the cocrystallization technique was used in several previous studies [42]. Coformers like resorcinol and pyrogallol were used with curcumin by using the LAG method and the study demonstrated 5- and 12-time increases in solubility with resorcinol and pyrogallol, respectively [43]. Several other coformers like salicylic acid [44], N-acetylcysteine [45], hydroxyquinoline [44], nicotinamide, tartaric acid, p-hydroxybenzoic acid [46], ascorbic acid [47], hydroxyquinol, trimesic acid, succinic acid, paracetamol, salicylic acid, and folic acid dihydrate [47] were used, and in every study, it exhibited enhanced solubility of curcumin.

In the case of quercetin, various coformers were used, including caffeine, isonicotinamide, isonicotinic acid, nicotinamide, theobromine dihydrate, malonic acid, succinic acid, and picolinic acid. Various methods like slow evaporation, grinding, and LAG methods are used, and in each study, the increased solubility was observed by several folds [48–54]. For instance, quercetin-caffeine cocrystals enhanced (14 times) the solubility [48], and quercetin–picolinic acid cocrystals demonstrated a 10-time increase in solubility [51]. There are several other herbal bioactives where cocrystallization techniques were used to enhance the solubility as listed in Table 9.1.

Table 9.1 Solubility enhancement of herbal drugs by cocrystallization technique.

Herbal bioactive	Description	Coformer	Method used	Result	Reference
Berberine	Isoquinoline alkaloid; biological activities like antibacterial, anti-inflammatory, anticarcinoma, and antidiabetic; poor oral absorption and low bioavailability	Fumaric acid	Recrystallization and slurry method	The dissolution rate increased by three times	[55]
		Citric acid	Physical mixture	Enhanced solubility	[56]
		Myricetin and dihydromyricetin	SE	Enhanced solubility for both the cocrystal	[57]
		L (+)-lactic acid	Physical mixing	Increased solubility	[58]
		Pimelic acid, adipic acid, succinic acid, and glutaric acid	Slow-evaporation technique	Improved solubility, cocrystal with pimelic acid showed maximum dissolution profile	[59]
		3-methylcinnamic acid	SE	Solubility enhanced by 13.9, 1.5, 4.7, and 15.8 times in water, methanol, ethanol, and isopropanol, respectively	[60]

(*Continued*)

Table 9.1 Solubility enhancement of herbal drugs by cocrystallization technique. (*Continued*)

Herbal bioactive	Description	Coformer	Method used	Result	Reference
Resveratrol	Polyphenolic stilbenoid; activity as antiviral, antimicrobial, cardioprotective, estrogenic, immunoprotective, and neuroprotective; low solubility and <1% bioavailability	Isoniazid and nicotinamide	Gas antisolvent technique	Enhanced dissolution profile	[61]
		Isoniazid and 4-aminobenzamide	LAG and SE	Increased solubility and tabletability	[62]
		Ascorbic acid	Evaporation method	Bioavailability enhanced by 3.73-folds and increased AUC, C_{max}	[63]
		L-proline	Recrystallization method	15.8% increase in solubility and dissolution	[64]
Naringenin	Flavonoids; possess anticancer, anti-bacterial, antioxidant, antiviral, anti-inflammatory, and cardioprotective activity	Isonicotinamide, picolinic acid, and betaine	Evaporation technique	Improved solubility and dissolution rate	[65]

(*Continued*)

Table 9.1 Solubility enhancement of herbal drugs by cocrystallization technique. (*Continued*)

Herbal bioactive	Description	Coformer	Method used	Result	Reference
Baicalein	Flavonoids; have activities like anti-inflammatory, antitumor, antioxidant, and cardioprotective	Nicotinamide	High-pressure homogenization	2.17 times increased solubility	[66]
Myricetin	Flavonoids; have anticancer, antioxidant, and anti-inflammatory activity	Caffeine	SE	17-time increase in dissolution profile	[67]
Theobromine	Bitter alkaloid; antioxidant, anti-inflammatory activity	"2,3-dihydroxybenzoic acid, 3,4-dihydroxybenzoic acid, 2,5-dihydroxybenzoic acid, 2,4-dihydroxybenzoic acid, 2,6-dihydroxybenzoic acid, 3,5-dihydroxybenzoic acid"	LAG and slow evaporation technique	4.4 to 26 times enhanced solubility	[68]

(*Continued*)

Table 9.1 Solubility enhancement of herbal drugs by cocrystallization technique. (*Continued*)

Herbal bioactive	Description	Coformer	Method used	Result	Reference
Chrysin	Flavone compound; activity against inflammation, erectile dysfunction, gout, anxiety	Thiamine and cytosine	Solvent-drop grinding method	Enhanced solubility by means of three to four times	[69]
Fisetin	Flavonoid, active against inflammation and cancer	Caffeine and nicotinamide	Cooling crystallization	Twofold increase in solubility	[70]
Piperine	Alkaloid-based compound	Saccharin	SE	Enhanced dissolution	[71]
Hesperetin	Flavonoids found in citrus fruits; activity against tumors, and inflammation, and act as an antioxidant	Picolinic acid, caffeine, and nicotinamide	Solvent-drop grinding	Four- to fivefolds solubility enhanced	[72]

9.6.5 Solubility Enhancement of Anticancer Drug via Cocrystallization

Because of the huge global concern for public health, solubility enhancement of anticancer drugs is very important. Approximately 19 million new cases and 10 million cases of death were reported in 2020 according to the Global Cancer Observatory (GCO). In this conflict on cancer, it has been projected that, possibly, there will be 16 million death cases reported in the next decade. Over the past few decades, it has been noticed that science and technology trying to fill the gaps related to cancer drug delivery and most treatments are solely through the parenteral route. However, parenteral routes are easy to administer, but looking at the frequency and patient compliance, the "oral" administration involves a lot of friendly consideration and delivers the drug conveniently due to its comfort and safety, as well as acceptance. However, lower water-solubility, low permeability, first-pass metabolism, and higher efflux ratio hamper the bioavailability, particularly the chemotherapeutic compound. Previously, many efforts have been applied like nano-formulations, complexation, crystal engineering, and lipid-based systems to overcome the solubility problem and improve the properties of the drug compound. Physiological parameters like solubility, hygroscopicity, dissolution, and stability can affect a compound's therapeutic effectiveness. To escape from this issue, cocrystallization is best suited for preparing a stable form of any drug without changing the chemical form at the molecular level. Most anticancer APIs are weakly basic or acidic and have similar pKa values which is an advantage for choosing coformers such as amides, carboxylic amides, polyphenols, and carboxylic acid. Very few anticancer drugs are zwitterion like paclitaxel. The most commonly used coformer is nicotinamide and its derivative for the cocrystallization of anticancer drugs. Having pyridine and amide groups, they act like primary synthons [73]. Most of the studies focused on the improvement of solubility of any poorly water-soluble compound by using various techniques of cocrystallization. Even though solubility is the rate-limiting factor in oral bioavailability, metabolism may be a significant factor in a drug's bioavailability. Sometimes, cocrystal preparation increases the dissolution profile rather than improving solubility. Faster dissolution enhances the faster absorption in the GI tract by decreasing Tmax.

Apatinib cocrystallization (low oral bioavailability) with three various coformers, i.e., sebacic, mandelic, and adipic acid, in a recent report, improved both solubility and bioavailability [74]. This study suggested that apatinib–sebacic acid cocrystals give a 7.5-fold increase in solubility along

Table 9.2 Cocrystallization of the anticancer drug for improved activity.

Anticancer drug	Coformers	Method used	Result	References
Ethylene bipyridine	Flavone compounds	Vaporization of solvent	Increased solubility and also enhanced effect against HCT-8 and Caco-2 cancer cells	[74]
5-Fluorouracil (5-FU)	Urea, thiourea, and pyrazinamide	Supercritical antisolvent (for crystallization) and supercritical solvent (for co-crystallization)	Improved solubility and anticancer activity	[75]
	Acetanilide, thiourea, urea, aspirin	Solid-state grinding, non-grinding method	Enhanced solubility and percent growth inhibition were observed through MTT assay against HCT 116 colorectal cell line study by all the cocrystals	[76]
	3,4-dihydroxybenzoic acid, gentisic acid, 4-aminopyridine,	solution crystallization and solvent-assisted grinding	Enhanced permeation and 2-3-folds improved in relative bioavailability studied in rat skin, cell viability dropped by cocrystals in MCF-7, Hela, and Caco-2 cell line which indicate improved antitumor activity	[77]

(Continued)

Table 9.2 Cocrystallization of the anticancer drug for improved activity. (*Continued*)

Anticancer drug	Coformers	Method used	Result	References
	2,5- Dihydroxybenzoic acid (25DHBA), 34DHBA, 35DHBA	Slurry and liquid assisted grinding method	5FU-34DHBA and 5FU-34DHBA.2H_2O showed increased solubility (34% and 11%), while 5FU-35DHBA and 5FU-25DHBA resulted in decreased solubility (27% and 51%). The pharmacokinetic parameters, including Cmax, tmax, and $t_{1/2}$, were also altered *in vivo* pharmacokinetic study	[78]
Mercaptopurine	Zinc triflate	SE	Enhanced the ability to be dissolved in aqueous solvent	[79]
Axitinib	Cinnamic acid and fumaric acid	Slurry method and LAG	Improved solubility and dissolution without compromising stability	[80]
	Glutaric acid	Slurry method	Enhanced solubility and stability	[81]
Betulinic acid	Ascorbic acid	Slow evaporation technique	Rise in solubility and bioavailability and showed superior cytotoxic activity on tumor cell lines, i.e., HaCat, B164A5, B16F0, MCF7, and MDA-MB-231	[82]
Baicalein	Nicotinamide	Slurry method	Improved dissolution, solubility, and bioavailability	[66]

(*Continued*)

Table 9.2 Cocrystallization of the anticancer drug for improved activity. (*Continued*)

Anticancer drug	Coformers	Method used	Result	References
Artemisinin	Acetylene dicarboxylic acid	Co-grinding without solvent	Enhancement of solubility	[83]
Ceritinib	Quercitin and nicotinamide	Neat grinding	Improved solubility and stability	[84]
Dasatinib	Vanillin, ethyl gallate, methyl gallate, propyl gallate, nicotinamide, and N-methyl-4-hydroxybenzoate	SE	Better solubility, bioavailability, and stability	[85]
Doxorubicin	Maraviroc	Slow SE technique	Solubility improved	[86]
Docetaxel	Nicotinamide	Slow SE technique	Rise in dissolution, solubility, and penetration properties	[87]
Emodin	Nicotinamide	SE	Oral absorption, improved solubility, and bioavailability	[88]

(*Continued*)

Table 9.2 Cocrystallization of the anticancer drug for improved activity. (*Continued*)

Anticancer drug	Coformers	Method used	Result	References
Lenvatinib	Salicylic acid	Slow-SE	Enhanced solubility and stability and improved anti-hepatoma activity	[89]
	Sulfamerazine	Slow-SE	Enhanced solubility, and stability	
	Salicylic acid and sulfamerazine	Slurry reaction crystallization	Higher dissolution, solubility, and stability	
Lenalidomide	Gallic acid	LAG	Rise in solubility, dissolution, and stability	[90]
	Urea, 3,5-dihydroxybenzoic acid	SE	Enhance solubility	[91]
Ibrutinib	Succinic acid, hydroxybenzoic acid, hydroxy naphthoic acid	Slurry and ultrasonic technique	Solubility and dissolution were enhanced by 6.64- and 1.85-folds, respectively	[92]
Lapachone	Resorcinol	Power driven crushing technique	Enhanced solubility and bioavailability (oral)	[93]
Imatinib	Syringic acid	Slurry method	Enhanced dissolution, solubility, and penetrability	[94]

(*Continued*)

Table 9.2 Cocrystallization of the anticancer drug for improved activity. (*Continued*)

Anticancer drug	Coformers	Method used	Result	References
Luteolin	Caffeine	Rapid solvent removal method and LAG	Enhanced solubility, bioavailability, and thermal stability	[95]
Palbociclib	Orcinol	Solvent evaporation	Enhanced dissolution, solubility, and stability	[96]
	Resorcinol	Solvent evaporation	Improved solubility, bioavailability (1.24 folds), and stability and increased IC50 value (*in vitro* cytotoxicity assay)	
Resveratrol	Nicotinamide	CO_2 antisolvent recrystallization (GAS technique)	Higher rate of dissolution and improved bioavailability	[61]
Zoledronic acid	Nicotinamide, tartaric acid	Slurry method, dry grinding, and solvent vaporization	Improved solubility	[97]
Regorafenib	Glutaric acid, malonic acid, and pimelic acid	LAG and slurry method	Enhanced solubility and dissolution	[98]
Methyltestosterone	Thiosalicylic acid and salicylic acid	Solid-state grinding method	Improved solubility	[99]

(*Continued*)

Table 9.2 Cocrystallization of the anticancer drug for improved activity. (*Continued*)

Anticancer drug	Coformers	Method used	Result	References
Oridonin	Nicotinamide, gallic acid, fumaric acid, and salicylic acid	Rotary evaporation method	Improved solubility and bioavailability by 1.34 and 1.18 times and showed cytotoxicity activity on HepG2 cells	[100]
Cytarabine	Uracil	Antisolvent method	Increased permeability and improved antitumor activity against EL-4, K562, and HL-60 cells cultured in DMEM medium	[101]

Table 9.3 Drug-drug cocrystal for enhancement of activity.

Drug 1	Drug 2	Therapeutic activity	Method used	Effect on solubility	Other activity	Reference
5-FU	**Regorafenib**	Anticancer	Hetero-seeding method	Improved	Enhanced stability and improved antitumor activity	[104]
5-FU	**Sorafenib**	Anticancer	Slurry method	Improved	Enhanced stability and improved antitumor effect	
5-FU	**Trimethoprim**	Anticancer and antibiotic	Slow evaporation method	Improved solubility	Higher anticancer activity against human breast cancer cell	[105]
5-FU	**Kaempferol**	Anticancer and flavonoids	Slurry method	Improved	Improved dissolution	[106]
5-FU	**Aspirin**	Anticancer and NSAIDs	Solid-state grinding method	Improved	-	[76]
Oxaliplatin	**Naringenin**	Anticancer and flavonoids	SE	-	Release profile reduced	[107]
Oxaliplatin	**Baicalein**				Delayed hydrolysis	
Erlotinib	**Furosemide**	Anticancer and diuretic	LAG	Improved	Improved dissolution and thermal stability	[108]

(Continued)

Table 9.3 Drug-drug cocrystal for enhancement of activity. (*Continued*)

Drug 1	Drug 2	Therapeutic activity	Method used	Effect on solubility	Other activity	Reference
Luteolin	**Isoniazid**	Anticancer and anti-tuberculosis	Rapid solvent removal method and LAG	Enhanced	Improved bioavailability and thermal stability	[95]
Gefitinib	**Mefenamic**	Anticancer and NSAIDs	LAG	Improved	Higher dissolution	[108]
Emodin	**Berberine chloride**	Anticancer drugs	Slurry method	Increased solubility of emodin and sustained release of berberine chloride	Higher Cmax and AUC of emodin	[109]
Naftopidil	**Aspirin, tolfenamic acid, mefenamic acid, and niflumic acid**	Antti-cancer and NSAIDs	SE	Eightfolds higher improved solubility	-	[110]
Apigenin	**Piperazine hexahydrate**	Flavonoid and anthelmintic	SE	Enhanced solubility	Higher dissolution rate	[111]
Imatinib mesylate	**5-FU**	Antidrugs	SE	Improved solubility	Improved stability	[112]
Temozolomide	**Hesperidin**	Anticancer and flavonoids	LAG	Rise in solubility	Improved dissolution as well as stability	[113]

with greater aqueous solubility [74]. Several other anticancer drugs are recorded in Table 9.2.

9.6.6 Drug-Drug Cocrystallization to Enhance Solubility

A drug-drug cocrystal is created when two or more drugs are combined to form a crystal. This is a trending approach and gained more attention in the pharmaceutical industry due to low-cost experiments, less risk, and improved biopharmaceutical and physicochemical properties of the drugs [57]. It denotes an encouraging area of research in combination therapies for better treatment and a large number of clinical conditions. It is pre-assumed that cocrystal and multiple APIs can overcome the limitations associated with combinational therapy. Drug-drug combinational cocrystallization can offer advantages of additive and synergistic effects, reduction of dose, and cost-effectiveness. There are several previous studies where more than one drug has been used to prepare drug-drug cocrystals. Recently, Eesam and colleagues reported the preparation of cocrystal of carvedilol (poor aqueous solubility and high permeability and is used for treating high blood pressure and congestive heart failure) with hydrochlorothiazide (diuretic) as a coformer. This study confirmed that prepared cocrystal showed 7.3 times enhanced solubility, 2.7-fold improved dissolution rather than carvedilol alone, and remained stable for 3 months [102].

There are several other studies conducted on anticancer drugs to enhance solubility, dissolution, and bioavailability. For example, Shridhar *et al.* published drug-drug crystallization of gefitinib with furosemide as a coformer applying solution-crystallization procedure. Hydrogen bonding between the components creates compact crystal packing with enhanced stability as compared to pure gefitinib [103]. Other studies on drug-drug cocrystals as tabulated in Table 9.3.

9.6.7 Cocrystallization for Food Encapsulation

Sustainable food production is the greatest challenge in the food industry to maintain the structural integrity of any compound. It is important to manage the food-bioactive until their administration and consumption. It is crucial to improve the dissolving capability in water, taste masking, as well as its bioavailability. High-quality food products can boost functionality in cardiovascular disease, cancer, diabetes, obesity, stroke, etc. Recently, the encapsulation technique has been used to enhance the shelf life as well as the stability of food bioactive compounds. Encapsulation is the technique where the active component is covered within a secondary compound. Researchers are concentrating on the cocrystallization technique to encapsulate the food bioactive components because of its cost-effectiveness and the method's simplicity to overcome the limitations related to food bioactives. Cocrystallization using sucrose as a primary component is preferred due to the improvement of solubility, dispersibility, wettability, and stability of food materials [114, 115]. Preparation of cocrystals in sucrose solution encapsulating culinary bananas (pulp and peel) exhibited enhanced dissolution rate as well as high solubility, very good flowability, and low hygroscopicity [116]. The result of the butterfly pea flower drink using the cocrystallization technique developed by Marpaung *et al.* revealed stable antioxidant activity and ultimately represented a potentially functional drink [117]. Cocrystallization with sucrose in another study displayed an improved encapsulation efficiency of β-carotene (extracted from carrot) and also showed low hygroscopicity, low moisture content, and greater encapsulation efficacy [118]. Future studies can fill the gap between laboratory-scale and large-scale applications without compromising stability and efficacy.

9.7 Conclusion

Cocrystallization is a process that does not involve any type of tedious reaction and eliminates a large amount of solvent, making it green and sustainable research even after preparation is made by employing a common grinding method to fulfill the idea of "green chemistry." Pharmaceutical cocrystallization is an advanced technique in which polished physicochemical properties like solubility, dissolution rate, and physical, and chemical stability turn the compound into a superior compound to the pure drug molecule. The unique crystal structure contains more than one molecular compound held together by noncovalent bonds resulting in an enhancement of performance when compared to the individual compound. Furthermore, cocrystallization plays an important role in pharmaceutical industries but also faces various challenges [119]. In addition, cocrystal can streamline the intelligent property security of APIs, offering a method to encompass patent uniqueness. In the field of material sciences, physicochemical properties can be improved by cocrystallization, thus exploring a new application of cocrystals in several fields like enhancement of solubility of a drug molecule, physical as well as chemical stability, etc. However, there are several challenges that came across during the making of cocrystals. For the synthesis as well as characterization of cocrystals, it is important to have expertise and specialized techniques for development and manufacturing on an industrial scale. Moreover, cocrystals need to be well-defined from regulatory aspects to ensure proper approval of procedure and evaluation. Additionally, understanding the fundamental principles leading to the formation of the cocrystals as well as their stability is still under study. Various simulations and computational approaches can significantly enhance the design and discovery of cocrystals by predicting their properties. Overall, cocrystal gives a talented opportunity in different areas of material science as well as pharmaceuticals. Trials associated with fundamental consideration, synthesis, engineering, and regulation are very important for understanding the complete potential of cocrystals. Ongoing research and associated teamwork among scientific experts, engineers, and regulatory authorities will propel the progress of the use of cocrystals in the future.

Acknowledgment

The authors are thankful to the School of Pharmaceutical Sciences, Siksha 'O' Anusandhan (Deemed to be University) for their support and for providing other facilities.

References

1. Sun, C.C., Cocrystallization for successful drug delivery. *Expert Opin. Drug Discovery*, 10, 201–213, 2013.
2. Alatas, F., Ratih, H., Kurnia, H., Soewandhi, S.N., Solubility enhancement of clozapine through cocrystal formation with isonicotinamide. *Indones. J. Pharm.*, 2, 1–6, 2019.
3. Rathi, R., Kushwaha, R., Goyal, A., Singh, I., Oxaliplatin-flavone pharmaceutical cocrystal-CN111205332A: patent spotlight. *Pharm. Pat. Anal.*, 11, 5, 147–154, 2022.
4. Wicaksono, Y., Barikah, K.Z., Della Yudatama, A., Maulia, H., Nuri, Setyawan, D., Increasing the solubility and anti-inflammatory activity of curcumin by cocrystallization, *Sci. Technol. Indones.*, 8, 501–508, 2023.
5. Narala, S., Nyavanandi, D., Srinivasan, P., Mandati, P., Bandari, S., Repka, M.A., Pharmaceutical cocrystals, salts, and co-amorphous systems: A novel opportunity of hot melt extrusion. *J. Drug Delivery Sci. Technol.*, 61, 102209, 2021.

6. Parmar, V.K. and Shah, S.A., Hydrochloride salt co-crystals: Preparation, characterization and physicochemical studies. *Pharm. Dev. Technol.*, 18, 2, 443–453, 2013.
7. Al-Kazemi, R., Al-Basarah, Y., Nada, A., Dissolution enhancement of atorvastatin calcium by cocrystallization. *Adv. Pharm. Bull.*, 9, 559, 2019.
8. Lipert, M.P. and Rodríguez-Hornedo, N., Cocrystal transition points: role of cocrystal solubility, drug solubility, and solubilizing agents. *Mol. Pharm.*, 12, 3535–3546, 2015.
9. Huang, Z., Staufenbiel, S., Bodmeier, R., Combination of cocrystal and nanocrystal techniques to improve the solubility and dissolution rate of poorly soluble drugs. *Pharm. Res.*, 39, 949–961, 2022.
10. Ayogu, P., Ezugwu, M., Eze, F., Principle of Common-ion Effect and its Application in Chemistry: a Review. *J. Chem. Lett.*, 1, 77–83, 2020.
11. Mannhold, R., Kubinyi, H., Folkers, G., *Drug bioavailability: estimation of solubility, permeability, absorption and bioavailability*, John Wiley & Sons, Weinheim, Germany, 2009.
12. Madan, J.R., Waghmare, S.V., Patil, R.B., Awasthi, R., Dua, K., Cocrystals of apixaban with improved solubility and permeability: formulation, physicochemical characterization, pharmacokinetic evaluation, and computational studies. *ASSAY Drug Dev. Technol.*, 19, 2, 124–38, 2021.
13. Huang, Y., Zhang, B., Gao, Y., Zhang, J., Shi, L., Baicalein–nicotinamide cocrystal with enhanced solubility, dissolution, and oral bioavailability. *J. Pharm. Sci.*, 103, 2330–2337, 2014.
14. Weyna, D.R., Cheney, M.L., Shan, N., Hanna, M., Zaworotko, M.J., Sava, V., Song, S., Sanchez-Ramos, J.R., Improving solubility and pharmacokinetics of meloxicam via multiple-component crystal formation. *Mol. Pharm.*, 9, 7, 2094–2102, 2012.
15. Wang, J.-R., Yu, X., Zhou, C., Lin, Y., Chen, C., Pan, G., Mei, X., Improving the dissolution and bioavailability of 6-mercaptopurine via cocrystallization with isonicotinamide. *Bioorg. Med. Chem. Lett.*, 25, 1036–1039, 2015.
16. Sanphui, P., Devi, V.K., Clara, D., Malviya, N., Ganguly, S., Desiraju, G.R., Cocrystals of hydrochlorothiazide: solubility and diffusion/permeability enhancements through drug–coformer interactions. *Mol. Pharm.*, 12, 1615–1622, 2015.
17. Banik, M., Gopi, S.P., Ganguly, S., Desiraju, G.R., Cocrystal and salt forms of furosemide: solubility and diffusion variations. *Cryst. Growth Des.*, 16, 5418–5428, 2016.
18. Saikia, B., Bora, P., Khatioda, R., Sarma, B., Hydrogen bond synthons in the interplay of solubility and membrane permeability/diffusion in variable stoichiometry drug cocrystals. *Cryst. Growth Des.*, 15, 5593–5603, 2015.
19. Yan, Y., Chen, J.-M., Lu, T.-B., Simultaneously enhancing the solubility and permeability of acyclovir by crystal engineering approach. *CrystEngComm*, 15, 6457–6460, 2013.

20. Ferretti, V., Dalpiaz, A., Bertolasi, V., Ferraro, L., Beggiato, S., Spizzo, F., Spisni, E., Pavan, B., Indomethacin cocrystals and their parent mixtures: does the intestinal barrier recognize them differently? *Mol. Pharm.*, 12, 1501–1511, 2015.
21. Gopi, S.P., Banik, M., Desiraju, G.R., New cocrystals of hydrochlorothiazide: optimizing solubility and membrane diffusivity. *Cryst. Growth Des.*, 17, 308–316, 2017.
22. Elder, D.P., Holm, R., De Diego, H.L., Use of pharmaceutical salts and cocrystals to address the issue of poor solubility. *Int. J. Pharm.*, 453, 88–100, 2013.
23. Almeida e Sousa, L., Reutzel-Edens, S.M., Stephenson, G.A., Taylor, L.S., Supersaturation potential of salt, cocrystal, and amorphous forms of a model weak base. *Cryst. Growth Des.*, 16, 737–748, 2016.
24. Savjani, K.T., Gajjar, A.K., Savjani, J.K., Drug solubility: importance and enhancement techniques. *Int. Sch. Res. Not.*, 2012, 1–10, 2012.
25. Sareen, S., Mathew, G., Joseph, L., Improvement in solubility of poor water-soluble drugs by solid dispersion. *Int. J. Pharm. Investig.*, 2, 12, 2012.
26. Singh, D., Bedi, N., Tiwary, A.K., Enhancing solubility of poorly aqueous soluble drugs: Critical appraisal of techniques. *J. Pharm. Investig.*, 48, 509–526, 2018.
27. Eedara, B.B., Kallakunta, V.R., Bandari, S., Self-nanoemulsifying powders for improved oral delivery of poorly water-soluble drugs. *Ther. Deliv.*, 6, 899–901, 2015.
28. Dash, R., Sahoo, R.N., Si, S.C., Mallick, S., Effect of eutectic formulation on photodegradation kinetics of ornidazole in aqueous state: predicted interaction and molecular binding in docking studies. *Chem. Pap.*, 76, 2823–2832, 2022.
29. Rathi, R., Sanshita, Singh, I., Hot Melt Extrusion Technique for Developing Pharmaceutical Cocrystals: A Review. *Drug Deliv. Lett.*, 14, 2, 97–108, 2024.
30. Kshirsagar, S.M., Chatale, B.C., Amin, P.D., Comparative evaluation of ibuprofen co-crystals prepared by solvent evaporation and hot melt extrusion technology. *J. Drug Delivery Sci. Technol.*, 67, 103003, 2022.
31. Dhumal, R.S., Kelly, A.L., York, P., Coates, P.D., Paradkar, A., Cocrystalization and simultaneous agglomeration using extrusion. *Pharm. Res.*, 27, 2725–2733, 2010.
32. Boksa, K., Otte, A., Pinal, R., Matrix-assisted cocrystallization (MAC) simultaneous production and formulation of pharmaceutical cocrystals by hot-melt extrusion. *J. Pharm. Sci.*, 103, 2904–2910, 2014.
33. Aitipamula, S. and Tan, B.H., Pharmaceutical cocrystals: Crystal engineering and applications, in: *Multi-Component Cryst. Synth. Concepts, Funct*, E.R. Tiekink and J.Z. Schpector (Eds.), pp. 1–27, 2018.
34. Othman, M.F., Jamburi, N., Anuar, N., Rahim, S.A., Rohalim, N.H., Ibuprofen-amino acids cocrystal screening via co-grinding methods, in: *MATEC Web of Conferences*, EDP Sciences, 2016.

35. Liu, M., Hong, C., Yao, Y., Shen, H., Ji, G., Li, G., Xie, Y., Development of a pharmaceutical cocrystal with solution crystallization technology: Preparation, characterization, and evaluation of myricetin-proline cocrystals. *Eur. J. Pharm. Biopharm.*, 107, 151–159, 2016.
36. Dengale, S.J., Grohganz, H., Rades, T., Löbmann, K., Recent advances in co-amorphous drug formulations. *Adv. Drug Delivery Rev.*, 100, 116–125, 2016.
37. Nugrahani, I., Kumalasari, R.A., Auli, W.N., Horikawa, A., Uekusa, H., Salt cocrystal of diclofenac sodium-l-proline: Structural, pseudopolymorphism, and pharmaceutics performance study. *Pharmaceutics*, 12, 690, 2020.
38. Nugrahani, I., Utami, D., Ibrahim, S., Nugraha, Y.P., Uekusa, H., Zwitterionic cocrystal of diclofenac and L-proline: Structure determination, solubility, kinetics of cocrystallization, and stability study. *Eur. J. Pharm. Sci.*, 117, 168–176, 2018.
39. Aljohani, M., Pallipurath, A.R., McArdle, P., Erxleben, A., A comprehensive cocrystal screening study of chlorothiazide. *Cryst. Growth Des.*, 17, 10, 5223–5232, 2017.
40. Dosiere, M., Structure and crystallization of molecular complexes between poly (ethylene oxide) and hydroxybenzenes. *J. Macromol. Sci. Part B Phys.*, 35, 303–328, 1996.
41. Spěváček, J., Paternostre, L., Damman, P., Draye, A.C., Dosiere, M., Solid state 13C NMR study of molecular complexes of poly (ethylene oxide) and hydroxybenzenes. *Macromolecules*, 31, 3612–3616, 1998.
42. Haskins, M.M. and Zaworotko, M.J., Screening and preparation of cocrystals: A comparative study of mechanochemistry vs slurry methods. *Cryst. Growth Des.*, 21, 7, 4141–4150, 2021.
43. Sanphui, P., Goud, N.R., Khandavilli, U.B.R., Nangia, A., Fast dissolving curcumin cocrystals. *Cryst. Growth Des.*, 11, 4135–4145, 2011.
44. Sathisaran, I. and Dalvi, S.V., Crystal engineering of curcumin with salicylic acid and hydroxyquinol as coformers. *Cryst. Growth Des.*, 17, 3974–3988, 2017.
45. Ribas, M.M., Sakata, G.S., Santos, A.E., Dal Magro, C., Aguiar, G.P., Lanza, M., Oliveira, J.V., Curcumin cocrystals using supercritical fluid technology. *J. Supercrit. Fluids*, 152, 104564, 2019.
46. Goud, N.R., Suresh, K., Sanphui, P., Nangia, A., Fast dissolving eutectic compositions of curcumin. *Int. J. Pharm.*, 439, 63–72, 2012.
47. Pantwalawalkar, J., More, H., Bhange, D., Patil, U., Jadhav, N., Novel curcumin ascorbic acid cocrystal for improved solubility. *J. Drug Delivery Sci. Technol.*, 61, 102233, 2021.
48. Kelly, G.S., Quercetin. *Altern. Med. Rev.*, 16, 172–195, 2011.
49. Smith, A.J., Kavuru, P., Wojtas, L., Zaworotko, M.J., Shytle, R.D., Cocrystals of quercetin with improved solubility and oral bioavailability. *Mol. Pharm.*, 8, 1867–1876, 2011.

50. Wu, N., Zhang, Y., Ren, J., Zeng, A., Liu, J., Preparation of quercetin–nicotinamide cocrystals and their evaluation under *in vivo* and *in vitro* conditions. *RSC Adv.*, 10, 21852–21859, 2020.
51. Vasisht, K., Chadha, K., Karan, M., Bhalla, Y., Jena, A.K., Chadha, R., Enhancing biopharmaceutical parameters of bioflavonoid quercetin by cocrystallization. *CrystEngComm*, 18, 1403–1415, 2016.
52. Setyawan, D., Permata, S.A., Zainul, A., Lestari, M.L.A.D., Improvement *in vitro* dissolution rate of quercetin using cocrystallization of quercetin-malonic acid. *Indones. J. Chem.*, 18, 531–536, 2018.
53. Athiyah, U., Kusuma, P.A., Tutik, T., Lestari, M.L., Isadiartuti, D., Paramita, D.P., Setyawan, D., Crystal engineering of quercetin by liquid assisted grinding method. *J. Teknol.*, 81, 39–45, 2019.
54. Setyawan, D., Oktavia, I.P., Farizka, R., Retno, S., Physicochemical characterization and *in vitro* dissolution test of quercetin-succinic acid cocrystals prepared using solvent evaporation. *Turk. J. Pharm. Sci.*, 14, 280, 2017.
55. Yang, D., Cao, J., Jiao, L., Yang, S., Zhang, L., Lu, Y., Du, G., Solubility and stability advantages of a new cocrystal of berberine chloride with fumaric acid. *ACS Omega*, 5, 8283–8292, 2020.
56. Lu, Q., Dun, J., Chen, J.-M., Liu, S., Sun, C.C., Improving solid-state properties of berberine chloride through forming a salt cocrystal with citric acid. *Int. J. Pharm.*, 554, 14–20, 2019.
57. Li, P., Ramaiah, T., Zhang, M., Zhang, Y., Huang, Y., Lou, B., Two cocrystals of berberine chloride with myricetin and dihydromyricetin: crystal structures, characterization, and antitumor activities. *Cryst. Growth Des.*, 20, 157–166, 2019.
58. Wang, L., Liu, S., Gao, Z., Crystal structure, dissolution and hygroscopicity of a novel cocrystal hydrate of berberine hydrochloride with L (+)-lactic acid. *Die Pharm. Int. J. Pharm. Sci.*, 75, 483–487, 2020.
59. Wang, L., Liu, S., Chen, J., Wang, Y., Sun, C.C., Novel Salt-Cocrystals of Berberine Hydrochloride with Aliphatic Dicarboxylic Acids: Odd–Even Alternation in Physicochemical Properties. *Mol. Pharm.*, 18, 1758–1767, 2021.
60. He Chen S, Y., Li, M., Gao, Y., Feng, H., Umar, Q., Yin, D., Feng, Y., Novel cocrystal of 3-methylcinnamic acid with berberine (1: 1): synthesis, characterization, and intestinal absorption property. *Drug Dev. Ind. Pharm.*, 49, 617–627, 2023.
61. Pessoa, A.S., Aguiar, G.P.S., Oliveira, J.V., Bortoluzzi, A.J., Paulino, A., Lanza, M., Precipitation of resveratrol-isoniazid and resveratrol-nicotinamide cocrystals by gas antisolvent. *J. Supercrit. Fluids*, 145, 93–102, 2019.
62. Zhou, Z., Li, W., Sun, W.J., Lu, T., Tong, H.H., Sun, C.C., Zheng, Y., Resveratrol cocrystals with enhanced solubility and tabletability. *Int. J. Pharm.*, 509, 391–399, 2016.

63. Neerati, P. and Gudimandula, R., A System for Synthesizing Resveratrol Cocrystals Using Cocrystallization Techniques to Improve the Bio Availability of Resveratrol, 2024.
64. Lou, Y., Yu, K., Wu, X., Wang, Z., Cui, Y., Bao, H., Wang, J., Hu, X., Ji, Y., Tang, G., Cocrystals of resveratrol and polydatin with L-proline: crystal structures, dissolution properties, and *in vitro* cytotoxicities. *Molecules*, 26, 5722, 2021.
65. Luo, C., Liang, W., Chen, X., Wang, J., Deng, Z., Zhang, H., Pharmaceutical cocrystals of naringenin with improved dissolution performance. *CrystEngComm*, 20, 3025–3033, 2018.
66. Pi, J., Wang, S., Li, W., Kebebe, D., Zhang, Y., Zhang, B., Qi, D., Guo, P., Li, N., Liu, Z., A nano-cocrystal strategy to improve the dissolution rate and oral bioavailability of baicalein. *Asian J. Pharm. Sci.*, 14, 154–164, 2019.
67. Xu, J., Wei, Y., Qian, S., Zhang, J., Preparation of myricetin-caffeine cocrystal and its single crystal analysis. *J. China Pharm. Univ.*, 47, 324–328, 2016.
68. Gołdyn, M., Larowska, D., Nowak, W., Bartoszak-Adamska, E., Synthon hierarchy in theobromine cocrystals with hydroxybenzoic acids as coformers. *CrystEngComm*, 21, 7373–7388, 2019.
69. Sa, R., Zhang, Y., Deng, Y., Huang, Y., Zhang, M., Lou, B., Novel salt cocrystal of chrysin with berberine: preparation, characterization, and oral bioavailability. *Cryst. Growth Des.*, 18, 4724–4730, 2018.
70. Mohite, R., Mehta, P., Arulmozhi, S., Kamble, R., Pawar, A., Bothiraja, C., Synthesis of fisetin cocrystals with caffeine and nicotinamide using the cooling crystallization technique: biopharmaceutical studies. *New J. Chem.*, 43, 13471–13479, 2019.
71. Zaini, E., Riska, D., Oktavia, M.D., Ismed, F., Fitriani, L., Improving dissolution rate of piperine by multicomponent crystal formation with saccharin. *Res. J. Pharm. Technol.*, 13, 1928–1932, 2020.
72. Chadha, K., Karan, M., Bhalla, Y., Chadha, R., Khullar, S., Mandal, S., Vasisht, K., Cocrystals of hesperetin: structural, pharmacokinetic, and pharmacodynamic evaluation. *Cryst. Growth Des.*, 17, 2386–2405, 2017.
73. Kara, D.D. and Rathnanand, M., Cocrystals and drug–drug cocrystals of anticancer drugs: A perception towards screening techniques, preparation, and enhancement of drug properties. *Crystals*, 12, 1337, 2022.
74. Zhu, B., Wang, J.R., Zhang, Q., Li, M., Guo, C., Ren, G., Mei, X., Stable cocrystals and salts of the antineoplastic drug apatinib with improved solubility in aqueous solution. *Cryst. Growth Des.*, 18, 4701–4714, 2018.
75. Cuadra, I.A., Cabañas, A., Cheda, J.A.R., Türk, M., Pando, C., Cocrystallization of the anticancer drug 5-fluorouracil and coformers urea, thiourea or pyrazinamide using supercritical CO2 as an antisolvent (SAS) and as a solvent (CSS). *J. Supercrit. Fluids*, 160, 104813, 2020.
76. Jubeen, F., Liaqat, A., Sultan, M., Iqbal, S.Z., Sajid, I., Sher, F., Green synthesis and biological evaluation of novel 5-fluorouracil derivatives as potent anticancer agents. *Saudi Pharm. J.*, 27, 1164–1173, 2019.

77. Gautam, M.K., Besan, M., Pandit, D., Mandal, S., Chadha, R., Cocrystal of 5-fluorouracil: characterization and evaluation of biopharmaceutical parameters. *AAPS PharmSciTech*, 20, 1–17, 2019.
78. Dai, X.L., Wu, C., Li, J.H., Liu, L.C., He, X., Lu, T.B., Chen, J.M., Modulating the solubility and pharmacokinetic properties of 5-fluorouracil via cocrystallization. *CrystEngComm*, 22, 21, 3670–3682, 2020.
79. Yao, J., Chen, J.-M., Xu, Y.-B., Lu, T.-B., Enhancing the solubility of 6-mercaptopurine by formation of ionic cocrystal with zinc trifluoromethanesulfonate: single-crystal-to-single-crystal transformation. *Cryst. Growth Des.*, 14, 5019–5025, 2014.
80. Ren, B.-Y., Dai, X.-L., Wang, J., Wu, C., Lu, T.-B., Chen, J.-M., Cocrystallization of axitinib with carboxylic acids: preparation, crystal structures and dissolution behavior. *CrystEngComm*, 23, 5504–5515, 2021.
81. Ren, B.-Y., Dai, X.-L., Chen, J.-M., Lu, T.-B., Two anhydrous forms and one monohydrate of a cocrystal of axitinib and glutaric acid: characterization, property evaluation and phase transition study. *CrystEngComm*, 24, 2138–2148, 2022.
82. Nicolov, M., Ghiulai, R.M., Voicu, M., Mioc, M., Duse, A.O., Roman, R., Ambrus, R., Zupko, I., Moaca, E.A., Coricovac, D.E., Farcas, C., Cocrystal formation of betulinic acid and ascorbic acid: Synthesis, physico-chemical assessment, antioxidant, and antiproliferative activity. *Front. Chem.*, 7, 92, 2019.
83. Makadia, J., Madu, S.J., Arroo, R., Seaton, C.C., Li, M., Artemisinin-acetylenedicarboxylic acid cocrystal: screening, structure determination, and physicochemical property characterisation. *CrystEngComm*, 24, 1056–1067, 2022.
84. Awasthi, A., Dheeraj, H.M., Birangal, S., Pai, A., Pai, G., Sathyanarayana, M.B., Fabrication of ceritinib cocrystals with improved solubility: Preparation, solid-state characterization, solubility studies, and molecular docking studies. *Rasayan J. Chem.*, 14, 905–913, 2021.
85. T. Chiodo, A. Hafner, T. Hintermann, B. Salvador, M. Szelagiewicz, F. Blatter, B. Siebenhaar, M. Vossen, Multicomponent crystals comprising dasatinib and selected cocrystal formers. Google Patents, 2016.
86. Dheeraj, H.M., Awasthi, A., Pai, A., Birangal, S., Vasanthraju, S.G., Pai, G., Sathyanarayana, M.B., Designing of stable cocrystals of cytarabine and doxorubicin hcl using suitable coformers. *Lat. Am. J. Pharm.*, 40, 39–48, 2021.
87. Fan, X., Chen, J., Shen, Q., Docetaxel–nicotinamide complex-loaded nanostructured lipid carriers for transdermal delivery. *Int. J. Pharm.*, 458, 296–304, 2013.
88. Ban, E., An, S.H., Park, B., Park, M., Yoon, N.E., Jung, B.H., Kim, A., Improved solubility and oral absorption of emodin-nicotinamide cocrystal over emodin with PVP as a solubility enhancer and crystallization inhibitor. *J. Pharm. Sci.*, 109, 3660–3667, 2020.

89. Hong, M., Li, S., Ji, W., Qi, M.-H., Ren, G., Cocrystals of lenvatinib with sulfamerazine and salicylic acid: Crystal structure, equilibrium solubility, stability study, and anti-hepatoma activity. *Cryst. Growth Des.*, 21, 3714–3727, 2021.
90. Song, J.-X., Chen, J.-M., Lu, T.-B., Lenalidomide–gallic acid cocrystals with constant high solubility. *Cryst. Growth Des.*, 15, 4869–4875, 2015.
91. Song, J.-X., Yan, Y., Yao, J., Chen, J.-M., Lu, T.-B., Improving the solubility of lenalidomide via cocrystals. *Cryst. Growth Des.*, 14, 3069–3077, 2014.
92. Shi, X., Wang, C., Chen, Q., Shen, S., Song, S., Zhou, X., Improving physicochemical properties of Ibrutinib with cocrystal strategy based on structures and natures of the carboxylic acid coformers. *J. Drug Delivery Sci. Technol.*, 63, 102554, 2021.
93. Liu, C., Liu, Z., Chen, Y., Chen, Z., Chen, H., Pui, Y., Qian, F., Oral bioavailability enhancement of β-lapachone, a poorly soluble fast crystallizer, by cocrystal, amorphous solid dispersion, and crystalline solid dispersion. *Eur. J. Pharm. Biopharm.*, 124, 73–81, 2018.
94. Reggane, M., Wiest, J., Saedtler, M., Harlacher, C., Gutmann, M., Zottnick, S.H., Piechon, P., Dix, I., Müller-Buschbaum, K., Holzgrabe, U., Meinel, L., Bioinspired cocrystals of Imatinib providing enhanced kinetic solubility. *Eur. J. Pharm. Biopharm.*, 128, 290–299, 2018.
95. Luo, Y., Chen, S., Zhou, J., Chen, J., Tian, L., Gao, W., Zhang, Y., Ma, A., Li, L., Zhou, Z., Luteolin cocrystals: Characterization, evaluation of solubility, oral bioavailability and theoretical calculation. *J. Drug Deliv. Sci. Technol.*, 50, 248–254, 2019.
96. Duan, C., Liu, W., Tao, Y., Liang, F., Chen, Y., Xiao, X., Zhang, G., Chen, Y., Hao, C., Two novel palbociclib-resorcinol and palbociclib-orcinol cocrystals with enhanced solubility and dissolution rate. *Pharmaceutics*, 14, 23, 2021.
97. Varma, A., Laxmi, P., Pai, A., Pai, G., Vasantharaju, S.G., Sathyanarayana, M.B., Designing of stable cocrystals of zoledronic acid using suitable coformers. *Chem. Pharm. Bull.*, 67, 816–823, 2019.
98. Jia, J.-L., Dai, X.L., Che, H.J., Li, M.T., Zhuang, X.M., Lu, T.B., Chen, J.M., Cocrystals of regorafenib with dicarboxylic acids: Synthesis, characterization and property evaluation. *CrystEngComm*, 23, 653–662, 2021.
99. Fatima, R., Yousuf, S., Choudhary, M.I., Cocrystallization of methyltestosterone with salicylic acid and 2, 2′-disulfanediyldibenzoic acid-Effect on solubility of methyltestosterone and growth of leishmania parasite (amastigotes and promastigotes). *J. Mol. Struct.*, 1295, 136741, 2024.
100. Jia, X., Hao, H., Zhang, Q., Yang, M.X., Wang, N., Sun SL, S.L., Yang, Z.N., Jin, Y.R., Wang, J., Du, Y.F., The bioavailability enhancement and insight into the action mechanism of poorly soluble natural compounds from cocrystals preparation: Oridonin as an example. *Phytomedicine*, 122, 155179, 2024.
101. Yu, Y.-M., Bu, F.-Z., Meng, S.-S., Yan, C.-W., Wu, Z.-Y., Li, Y.-T., The first nano-cocrystal formulation of marine drug cytarabine with uracil based

on cocrystal nanonization strategy for long-acting injection exhibiting enhanced antitumor activity. *Int. J. Pharm.*, 644, 123300, 2023.
102. A.R., Eesam, S., Bhandaru, J.S., Naliganti, C., Bobbala, R.K., Solubility enhancement of carvedilol using drug–drug cocrystallization with hydrochlorothiazide. *J. Pharm. Sci.*, 6, 1–3, 2020.
103. Thorat, S.H., Sahu, S.K., Patwadkar, M.V., Badiger, M.V., Gonnade, R.G., Drug–drug molecular salt hydrate of an anticancer drug gefitinib and a loop diuretic drug furosemide: an alternative for multidrug treatment. *J. Pharm. Sci.*, 104, 12, 4207–4216, 2015.
104. Liu, C., Wang, C., Wan, S., Liu, L., Sun, C.C., Qian, F., An elusive drug–drug cocrystal prepared using a heteroseeding strategy. *Cryst. Growth Des.*, 21, 5659–5668, 2021.
105. Yan, T., Shu, B., Deng, X., Qian, K., Pan, R., Qiu, S., Yang, J., Fu, Q., Ma, Y., Antibacterial and Anticancer Activity, Acute Toxicity, and Solubility of Cocrystals of 5-Fluorouracil and Trimethoprim. *ACS Omega*, 8, 21522–21530, 2023.
106. Lv, W.-T., Liu, X.-X., Dai, X.-L., Long, X.-T., Chen, J.-M., A 5-fluorouracil–kaempferol drug–drug cocrystal: a ternary phase diagram, characterization and property evaluation. *CrystEngComm*, 22, 8127–8135, 2020.
107. Yin, H.-M., Wu, N., Zhou, B.J., Hong, M.H., Zhu, B., Qi, M.H., Ren, G.B., Slow-release drug–drug cocrystals of oxaliplatin with flavonoids: delaying hydrolysis and reducing toxicity. *Cryst. Growth Des.*, 21, 75–85, 2020.
108. George, C.P., Thorat, S.H., Shaligram, P.S., Suresha, P.R., Gonnade, R.G., Drug–drug cocrystals of anticancer drugs erlotinib–furosemide and gefitinib–mefenamic acid for alternative multi-drug treatment. *CrystEngComm*, 22, 6137–6151, 2020.
109. Deng, Y., Zhang, Y., Huang, Y., Zhang, M., Lou, B., Preparation, crystal structures, and oral bioavailability of two cocrystals of emodin with berberine chloride. *Cryst. Growth Des.*, 18, 7481–7488, 2018.
110. Bandaru, R.K., Giri, L., Kesharwani, P., Dandela, R., Drug-drug salts of Naftopidil with non-steroidal anti-inflammatory drugs for potential multidrug therapy. *Colloids Surf. A Physicochem. Eng. Asp.*, 682, 132835, 2024.
111. Zhang, J., Shen, R., Zhang, X., Li, G., Wang, X., A pharmaceutical cocrystal of apigenin with piperazine: Preparation, structural characterization, and dissolution performance. *J. Mol. Struct.*, 1298, 137027, 2024.
112. Veverka, M., Šimon, P., Gallovič, J., Jorík, V., Veverková, E., Dubaj, T., Imatinib mesylate cocrystals: synthesis, screening, and preliminary characterization. *Monatsh. Chem.-Chem. Mon.*, 143, 1405–1415, 2012.
113. Wang, J., Dai, X.-L., Lu, T.-B., Chen, J.-M., Temozolomide–hesperetin drug–drug cocrystal with optimized performance in stability, dissolution, and tabletability. *Cryst. Growth Des.*, 21, 838–846, 2021.
114. Chezanoglou, E. and Goula, A.M., Cocrystallization in sucrose: A promising method for encapsulation of food bioactive components. *Trends Food Sci. Technol.*, 114, 262–274, 2021.

115. López-Córdoba, A., Deladino, L., Agudelo-Mesa, L., Martino, M., Yerba mate antioxidant powders obtained by cocrystallization: Stability during storage. *J. Food Eng.*, 124, 158–165, 2014.
116. Khawas, P. and Deka, S.C., Encapsulation of natural antioxidant compounds from culinary banana by Cocrystallization. *J. Food Process. Preserv.*, 41, e13033, 2017.
117. Marpaung, A.M., Lee, M., Kartawiria, I.S., The development of butterfly pea (Clitoria ternatea) flower powder drink by cocrystallization. *Indones. Food Sci. Technol. J.*, 3, 34–37, 2020.
118. Kaur, P., Elsayed, A., Subramanian, J., Singh, A., Encapsulation of carotenoids with sucrose by cocrystallization: Physicochemical properties, characterization and thermal stability of pigments. *LWT*, 140, 110810, 2021.
119. Rathi, R., Guggal, Y., Twinkle, Sandhu, V., Singh, I., Exploring the Patent Landscape and Regulatory Prospective on Pharmaceutical Cocrystals. *J. Pharm. Technol. Res. Manag.*, 11, 49–60, 2023.

10

Pharmaceutical Cocrystals for Improving Physicochemical Properties of Drugs: Enhancement of Therapeutic Efficacy

Debabrata Ghosh Dastidar*, Arpan Datta, Sanchari Bhattacharya and Moumita Chowdhury

Guru Nanak Institute of Pharmaceutical Science and Technology, Panihati, Kolkata, West Bengal, India

Abstract

Cocrystallization has emerged as a crucial technique in pharmaceutical sciences for enhancing the physicochemical and therapeutic properties of drugs. A cocrystal, defined as a crystalline complex of two or more molecules in a non-ionic association, held together by weaker noncovalent intermolecular forces such as hydrogen bonding, offers opportunities for modulating the important properties of active pharmaceutical ingredients (APIs). Many APIs face challenges like poor solubility, dissolution rate, bioavailability, and stability, which impact drug development and commercialization. Traditional techniques have limitations, but cocrystallization provides a unique opportunity to enhance properties like solubility, dissolution, permeability, and stability without altering the chemical identities of the molecules. By optimizing physicochemical properties like crystallinity, particle size, and crystal habit early in drug development, cocrystals can accelerate development timelines and lower costs. This chapter provides an overview of the science of cocrystallization and its applications in enhancing the therapeutic efficacy of drugs. It covers topics such as cocrystal formation fundamentals, characterization methods, and the influence of cocrystals on pharmacokinetic properties. Additionally, industrially relevant cocrystallization techniques and examples of marketed cocrystal-based pharmaceutical products are discussed, showcasing the potential and impact of cocrystals in drug development.

**Corresponding author*: debabrata.ghoshdastidar@gnipst.ac.in

Inderbir Singh, Subrata Mallick and Ritu Rathi (eds.) Cocrystals in Pharmaceutical Sciences: Design to Applications, (293–328) © 2025 Scrivener Publishing LLC

***Keywords*:** Solubility enhancement, crystal engineering, biopharmaceutical properties, combinatorial approaches, therapeutic efficacy

10.1 Introduction

Cocrystallization is emerging as an important technique in pharmaceutical sciences for improving the physicochemical and therapeutic properties of drugs. A cocrystal is defined as a crystalline complex comprising two or more different molecules in a non-ionic association held together by weaker noncovalent intermolecular forces such as hydrogen bonding. In cocrystals, the component molecules interact through supramolecular synthons resulting in new crystalline forms with properties distinct from their individual components [1]. Many pharmaceutical active pharmaceutical ingredients (APIs) suffer from poor solubility, dissolution rate, bioavailability, and stability issues, which pose challenges for drug development and commercialization. Traditionally, techniques such as salt formation, polymorphism, and solvates/hydrates have been utilized to modify the physicochemical properties of APIs. However, these approaches have limitations and cannot be universally applied. Cocrystallization provides a unique opportunity to modulate important properties like solubility, dissolution, permeability, and stability of APIs without altering their chemical identities [2].

The ultimate goal of drug development is to deliver safe and effective therapeutics to patients. However, many promising drug candidates fail due to poor solubility, dissolution, and bioavailability issues leading to a lack of therapeutic efficacy. It is estimated that over 40% of new drug candidates exhibit poor aqueous solubility, a key determinant of oral bioavailability. Dissolution of the drug is the rate-limiting step for systemic absorption and attainment of pharmacological concentrations at the site of action. Low bioavailability translates to difficulties in achieving the intended pharmacological response even at higher doses, increasing toxicity risks [3]. Enhancing solubility, dissolution and oral bioavailability of drug compounds during early development phases is therefore critical for improving therapeutic efficacy and enabling successful commercialization [3]. This requires optimization of physicochemical properties like crystallinity, particle size, and habit without altering the molecular structure or pharmacological activity of the drug molecule. Strategic modifications at the preformulation stage can significantly accelerate development timelines and lower costs by eliminating candidates with inherent poor solubility or dissolution properties early on [4]. Properties like stability also impact the treatment efficacy and safety over time. Drugs with stability issues may degrade during manufacturing, storage, or *in vivo*, reducing potency and consistency of therapeutic outcome [5]. Unstable forms also face challenges in commercial scale-up due to difficulties in process optimization and validation. Designing stable solid forms of compounds early provides an advantage in manufacturing.

Over the past two decades, significant advances have been made in the fundamental understanding of cocrystal formation and characterization. A wide range of pharmaceutical cocrystals have been reported with improved solubility, dissolution, and stability profiles relative to their parent molecules [1]. Recent industrial and regulatory acceptance of cocrystals has validated their potential for commercial applications [6]. This chapter aims to provide an overview of cocrystallization science and its pharmaceutical applications in enhancing the therapeutic efficacy of drugs. Key topics that will be covered include fundamentals of cocrystal formation, screening and characterization methods, the influence of cocrystals on pharmacokinetic properties, and industrially relevant cocrystallization techniques. Representative examples of marketed cocrystal-based pharmaceutical products will also be discussed.

10.2 Preparation and Characterization Methods for Cocrystals

Several experimental and computational techniques have been developed for the synthesis and screening of pharmaceutical cocrystals (Figure 10.1). The selection of an appropriate method is governed by the intrinsic properties of the components as well as objectives such as laboratory-scale synthesis versus industrial-scale production [7]. The commonly used cocrystal formation methods include solution-mediated methods involving solvent evaporation, spray drying, slurry conversion, and thermal treatment. Solutions of components are employed for cocrystal nucleation and growth. While effective for smaller-scale synthesis, issues arise with solvent residuals and hydrate/solvate formation during scale-up [8]. The solvent-free methods for cocrystal preparation include neat/grinding cocrystallization, melt-mediated methods, mechanical activation, and gas antisolvent technique. Neat/grinding cocrystallization involves grinding or grinding with little or no solvent. This technique mimics the solution process on the surfaces. It is simple, requiring only mortar/pestle [9] but may induce undesired phases and lack reproducibility at the

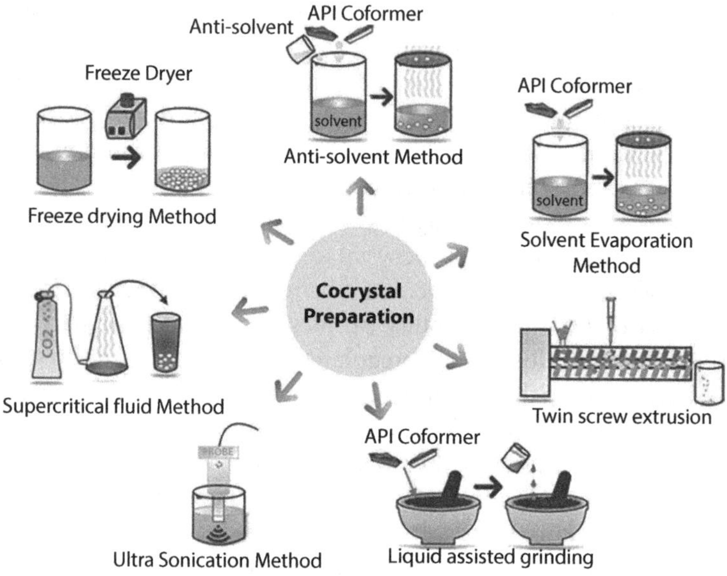

Figure 10.1 Cocrystal preparation methods. Various techniques are used for the preparation of cocrystals. The most used methods include: (1) freeze-drying, (2) antisolvent, (3) solvent evaporation, (4) twin-screw extrusion, (5) liquid-assisted grinding, (6) ultrasonication, and (7) supercritical fluid. The figure is adapted from (Singh *et al.*, 2023; https://doi.org/10.3390/pharmaceutics15041161) published under a CC4 license.

industrial scale [10]. Hot melt extrusion is a solvent-free process utilizing elevated temperatures. They enable continuous processing but are limited by the thermal stability of components [11]. Mechanical activation techniques like grinding, vibration-milling, ultrasound, and sonication utilize mechanical energy for cocrystal formation [12]. While providing better control, scale-up remains a challenge. Gas antisolvent techniques involve rapid expansion of supercritical solutions or aerosol solvent extraction employs high-pressure CO_2-based processing [13]. However, specialized equipment requirements limit widespread application.

A multitude of analytical techniques have been developed to characterize cocrystals and ascertain critical structure-property relationships. The selection of appropriate techniques provides a comprehensive insight into cocrystal composition, crystal structure, and phase identification. Solid-state characterization of cocrystals involves powder X-ray diffraction (PXRD) which determines crystal structure and confirms phase purity. It enables the identification of alternative polymorphs and hydrates [14]. Single crystal X-ray diffraction (SCXRD) precisely defines molecular packing, conformation, and interactions through unit cell parameters. However, it requires high-quality single crystals [15]. The solid-state NMR spectroscopy can distinguish local environments when long-range order is absent in PXRD. It aids in ambiguous structures [16].

Spectroscopic characterization of cocrystals involves fourier transform infrared (FTIR) spectroscopy that identifies functional group interactions through peak shifts. It confirms hydrogen bonding and salt formation [17]. Raman spectroscopy fingerprint regions characterize molecular arrangements and complexation [18] and thermal methods like DSC (differential scanning calorimetry) and TGA (thermogravimetric analysis) quantify endothermic and exothermic solid-state transitions and degradation kinetics [19]. The morphology of cocrystals can be studied using microscopic techniques such as scanning electron microscopy (SEM) visualizes crystal habits, defects, and particle morphology at micrometer scales [20], and hot-stage microscopy monitors *in situ* phase changes and fusion temperatures.

10.3 Enhancement of Therapeutic Efficacy with Cocrystals

Cocrystals have shown great potential in enhancing the therapeutic efficacy of pharmaceutical drugs. One of the major ways in which cocrystals achieve this is by improving the solubility and dissolution rate of poorly water-soluble drug molecules. By incorporating such drugs into crystalline frameworks together with suitable coformers, it is possible to modify the intermolecular interactions and crystal packing of the drugs, which in turn improves their hydrophilicity and wettability [1]. This allows for better and faster absorption of the drugs from the gastrointestinal tract after oral administration. Increased solubility and dissolution rate lead to higher concentrations of drugs being absorbed systematically into the bloodstream [21]. This ensures that therapeutically effective doses are achieved more efficiently. In addition, cocrystals have been found to possess properties like enhanced stability and better flowability and compressibility during formulation. All these advantages of cocrystals ultimately help improve bioavailability and therapeutic outcomes for many drugs that otherwise face delivery challenges due to poor solubility and permeability [22, 23]. There are several ways in which cocrystals can enhance the therapeutic efficacy of drugs: Improving physicochemical properties, enhancing stability and shelf life, improving bioavailability and pharmacokinetics, combination therapy, and improving the targeted delivery of drugs [24]. The role of cocrystals in enhancing therapeutic efficacy of drugs is summarized in Table 10.1.

Table 10.1 Role of cocrystals in enhancing therapeutic efficacy of drugs.

Properties of drug improved	Drug	Coformer	Observed improvement	Reference
Improving physicochemical properties				
Melting point	Propofol	Isonicotinamide	Increased melting point (from 18°C to 68°C), enabling conversion to a stable solid form	[26]
	Lithium chloride	Glucose	Modestly improved hygroscopic stability under 40% RH compared to the drug substance alone	[29]
	Metoclopramide	Oxalic Acid	Effectively inhibited hydration, improving stability	[30]
	S-Oxiracetam	Gallic Acid	Less hygroscopic than S-oxiracetam alone or the racemic parent drug when stored at 25°C/60% RH for 12 months	[31]

(*Continued*)

Table 10.1 Role of cocrystals in enhancing therapeutic efficacy of drugs. (*Continued*)

Properties of drug improved	Drug	Coformer	Observed improvement	Reference
	Theophylline	Flavonoids	Greater resistance to hydration than theophylline alone when subjected to humidity cycling	[32]
Solubility and dissolution	Naproxen	Theophylline	Approximately twofold higher dissolution rates compared to naproxen alone	[35]
	Sulfadimidine	4-Aminosalicylic Acid	~80% dissolution within 5 min, compared to only ~30% for sulfadimidine alone	[36]
	Isoniazid	Curcumin	Provided a controlled release of isoniazid over 48 hours due to the incorporation of the insoluble curcumin	[34]

(*Continued*)

Table 10.1 Role of cocrystals in enhancing therapeutic efficacy of drugs. (*Continued*)

Properties of drug improved	Drug	Coformer	Observed improvement	Reference
Enhancing bioavailability and pharmacokinetics				
Permeability	Acyclovir	Fumaric acid and glutaric acid	Increased permeability across rat skin by 2.8- and 4-fold, respectively, compared to acyclovir alone	[56]
	Furosemide	Anthranilamide	Increased cumulative permeation and flux across a cellulose membrane compared to furosemide alone	[57]
	Indomethacin	Saccharine	Enhanced permeability across intestinal epithelial cells without cellular damage	[58]

(*Continued*)

Table 10.1 Role of cocrystals in enhancing therapeutic efficacy of drugs. (*Continued*)

Properties of drug improved	Drug	Coformer	Observed improvement	Reference
Absorption and distribution	Baicalein	Caffeine	4.1-fold higher oral bioavailability (AUC) compared to baicalein alone, attributed to inhibition of degradation.	[59]
	Celecoxib-tramadol hydrochloride		Modulated the release kinetics of both drugs, improving bioavailability and safety	[54, 55]
Increasing stability of drug molecules				
Improving stability	Lithium chloride	Glucose	Modestly improved hygroscopic stability under 40% RH compared to the drug substance alone	[29]
	Metoclopramide	Oxalic acid	Effectively inhibited hydration, improving stability	[30]

(*Continued*)

Table 10.1 Role of cocrystals in enhancing therapeutic efficacy of drugs. (*Continued*)

Properties of drug improved	Drug	Coformer	Observed improvement	Reference
	S-Oxiracetam	Gallic acid	Less hygroscopic than S-oxiracetam alone or the racemic parent drug when stored at 25°C/60% RH for 12 months	[31]
	Theophylline	Flavonoids	Greater resistance to hydration than theophylline alone when subjected to humidity cycling	[32]

10.3.1 Improving the Physicochemical Properties of Drugs

The addition of coformers into the APIs crystal lattice, the cocrystals allow for tuning various physicochemical properties such as solubility, dissolution rate, melting point, hygroscopicity, and mechanical behavior [25]. The cocrystallization process and different factors affecting the solubility and stability of cocrystals are summarized in Figure 10.2.

10.3.1.1 Melting Point

Cocrystallization can be utilized for increasing the melting point of APIs and converting liquid drugs into solid dosage forms [1]. McKellar *et al.* implemented a cocrystal approach for obtaining the solid form of propofol utilizing iso-nicotinamide as the coformer. The obtained cocrystal was found to exhibit an increased melting point (approximately 50°C higher than pure drug) and was a stable solid at room temperature. Experimental thermal analysis data revealed that propofol has a melting endotherm at 18°C, whereas the cocrystal showed a distinct melting endotherm at 68°C, representing over a 50°C increase in melting point. This cocrystal formation allowed the conversion of liquid propofol into a thermodynamically stable crystalline solid suitable for manufacturing as oral or other solid dosage forms. The increased melting point addresses issues with instability, pain during injection, and hyperlipidemia associated with the liquid propofol formulation [26].

10.3.1.2 *Hygroscopicity*

Cocrystallization can quantitatively improve the hygroscopic stability of APIs by inhibiting crystalline water formation through tighter hydrogen bonding networks and blocking hydration sites during pharmacy manufacturing and storage conditions [27]. Lithium chloride is used to treat neuropsychiatric disorders but is extremely hygroscopic. The lithium chloride-glucose cocrystal showed lower hygroscopicity under 40% RH compared to the drug substance alone, as assessed by equilibrium moisture content testing [28]. Metoclopramide is prone to crystallize water during manufacturing, triggering an ion-exchange reaction with excipients. The metoclopramide-oxalic acid cocrystal effectively inhibited hydration, improving stability [29]. Oxiracetam is a racemic drug containing the more pharmacologically active S-enantiomer which has higher hygroscopicity. Hygroscopic testing showed the S-oxiracetam-gallic acid cocrystal was less hygroscopic than S-oxiracetam alone or the racemic parent drug when stored at 25°C/60% RH for 12 months [30]. Theophylline-flavonoid cocrystals showed improved resistance to moisture than theophylline alone when subjected to humidity cycling between 25°C/30% RH and 40°C/75% RH for 7 days [31].

These case studies demonstrated how cocrystallization can be an effective strategy to address the challenge of drug hygroscopicity. By judiciously selecting coformers that can form robust hydrogen bonding networks and restrict water uptake within the crystal lattice, cocrystals of compounds like lithium chloride, metoclopramide, oxiracetam, and theophylline exhibited markedly improved resistance to hydration compared to the respective parent drugs. This enhanced hygroscopic stability is expected to translate into improved shelf life, processing characteristics, and ultimately, more consistent therapeutic outcomes for patients.

10.3.1.3 *Solubility and Dissolution*

Cocrystallization can enhance the dissolution behavior and rates of poorly soluble drugs through the formation of new supramolecular structures between APIs and coformers [32]. Controlled-release profiles can also be achieved by incorporating insoluble excipients into cocrystal lattices [33]. Naproxen-theophylline cocrystals produced by solvent-drop grinding exhibited around 2-fold higher dissolution rates than naproxen alone when evaluated using a USP Type II paddle apparatus at 50 rpm in 0.1 N HCl [34]. Cocrystals of sulfadimidine-4-aminosalicylic acid demonstrated enhanced dissolution behavior compared to the drug substance alone, with ~80% dissolution within 5 minutes for the cocrystal versus only ~30% for sulfadimidine on its own, as assessed by a USP Type I basket method [35].

In the case of the tyrosine kinase inhibitor, dasatinib, the formation of cocrystals with various carboxylic acid coformers, such as 4-hydroxybenzoic acid, 4-chlorobenzoic acid, and sorbic acid, has

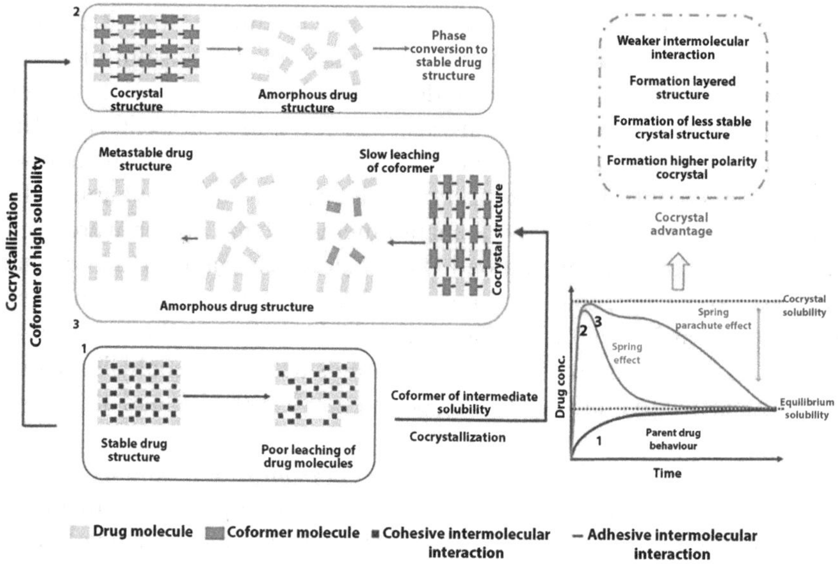

Figure 10.2 Schematic representation of the cocrystallization process and the different factors affecting the solubility and stability of cocrystals. The drug molecule in stable crystalline form has poor water solubility due to cohesive intermolecular interaction (1) Upon cocrystallization with coformers of higher solubility, a new crystal is formed. Upon addition of solvent, the conformer molecules get dissolved rapidly resulting in an amorphous state of drug molecule that may be converted into a stable form (2) or metastable form with better solubility (3) Spring parachute effect: The solubility of the drug in the amorphous cocrystal is higher than that of the pure drug, and the dissolution rate is faster. The drug molecules are released faster from the amorphous cocrystal into the solution, and the concentration of the drug in the solution increases rapidly. This is known as the "spring parachute effect." The figure is adapted from (Singh *et al.*, 2023; https://doi.org/10.3390/pharmaceutics15041161) published under a CC4 license.

demonstrated significant improvements in solubility and dissolution characteristics [36]. The cocrystallization process facilitated the development of new crystalline forms of dasatinib, which exhibited solubility enhancements up to 6-fold higher than the pure drug. This substantial increase in solubility is attributed to the formation of strong intermolecular interactions, such as hydrogen bonding and π-π stacking, between dasatinib and the coformers, as revealed by detailed spectroscopic and computational analyses. The improved solubility and dissolution properties of the dasatinib cocrystals have the potential to enhance the drug's bioavailability and therapeutic efficacy, ultimately leading to improved patient outcomes. Isoniazid-curcumin cocrystals were synthesized to achieve sustained release of isoniazid. Dissolution tests in pH 1.2 buffer showed only 28% of isoniazid was dissolved from the cocrystals after 15 minutes, whereas isoniazid from a physical mixture was fully dissolved. This demonstrated the cocrystal afforded controlled release over 48 hours due to the integration of the insoluble curcumin in the crystal lattice [33].

These enhancements in solubility and dissolution are crucial for enabling complete and consistent absorption of the active pharmaceutical ingredients, thereby ensuring that therapeutically effective concentrations are achieved at the site of action. Improved bioavailability translates to better therapeutic outcomes for patients, particularly for drugs that suffer from poor solubility and permeability issues. Furthermore, the ability to modulate drug release profiles through cocrystal design offers opportunities to optimize pharmacokinetics and tailor drug delivery to specific clinical needs.

10.3.1.4 Mechanical Behavior

Cocrystallization can modify the mechanical properties of drugs for improved solid dosage form manufacturing by tuning plasticity, slip plane activity, intermolecular forces, and packing arrangements [1]. The compaction behavior and mechanical properties of caffeine-glutaric acid cocrystal polymorphs were tested using a hardness tester and analysis of powder Brillouin light scattering spectra. The caffeine-3-nitrobenzoic acid cocrystal form I was found to exhibit better plastic deformation ability than form II due to the presence of more slip planes in its 2D layered crystal structure. This improved plasticity allows for better compaction properties during tablet manufacturing [37].

In comparison to pure celecoxib, the celecoxib–L-proline cocrystals presented lower punch sticking propensity when evaluated using a single punch tableting machine at compaction pressures 100 and 250 MPa. The cocrystal formation enhanced interlayer bonding energies, deactivated slip planes, and weakened API-API interactions to reduce punch sticking [38]. The mechanical properties of chlorzoxazone-picolinic acid cocrystals were assessed through compaction analysis. Tablets produced with the cocrystals achieved tensile strengths of about 1.6 MPa at 250 MPa compaction pressure without tablet defects (capping or lamination). The formation of cocrystal-based tablets of chlorzoxazone improved the mechanical properties which were just 0.8 MPa at compaction pressure 50 MPa, reflecting enhanced plastic deformation and bonding ability due to the cocrystal's denser packing structure [39]. Nanoindentation experiments revealed that caffeine-glutaric acid cocrystal polymorphs exhibited anisotropic mechanical responses depending on the crystalline face, with form I being softer and form II harder due to differences in intermolecular interactions and potential slip planes [37].

As shown in the case studies, cocrystals of caffeine-glutaric acid, celecoxib–L-proline, and chlorzoxazone-picolinic acid exhibited enhanced plastic deformation, reduced punch sticking propensity, and improved tablet tensile strength compared to the respective parent drugs. These improvements in mechanical behavior are crucial for the successful development and manufacturing of solid dosage forms, as they enable better compaction, reduced processing issues, and more robust final products. Ultimately, the ability to engineer cocrystals with optimized mechanical properties can lead to more consistent drug delivery and improved therapeutic outcomes for patients.

10.3.1.5 Optical Properties

Cocrystallization can modify the colors and fluorescence of pharmaceutical actives by tuning intermolecular pi-pi stacking, hydrogen bonding, and charge transfer interactions in the crystal lattice [1]. Cocrystal form I of furosemide and 4,4-bipyridine appeared pale yellow while form II appeared orange. It was proposed that the different p-stacking patterns and hydrogen bonding interactions in the two polymorphic structures led to their distinct colors [40]. Menadione is known to undergo [2+2] photodimerization upon exposure to sunlight. However, cocrystals of menadione with 1-hydroxy-2-naphthoic acid, 6-hydroxy-2-naphthoic acid, and sulfamerazine did not show any noticeable color change or degradation after illumination for 5 days, whereas menadione alone changed color noticeably. It was proposed that the enhanced pi-pi stacking between menadione and the coformers acted to block menadione molecules from approaching each other to undergo photocyclization [41]. Phloretin cocrystals with nicotinamide showed strong yellowish-green fluorescence under UV light, whereas the phloretin-isonicotinamide cocrystal did not fluoresce. The distinct luminescent properties were attributed to differences in packing arrangement and intermolecular interactions between the molecules in the two cocrystals [42].

As demonstrated, cocrystals exhibited distinct color variations and luminescent behavior compared to pure drug substances, attributed to altered intermolecular interactions and packing arrangements. Such modifications in optical characteristics may provide insights into the underlying structural changes that can impact the stability, solubility, and therapeutic efficacy of the drug.

10.3.2 Enhancing Stability and Shelf Life

Improving the stability of drugs during manufacturing, storage, and transportation is essential for pharmaceutical applications. Drug stability is impacted by various degradation pathways such as hydrolysis, oxidation, isomerization, and dimerization that can occur in the solid state [43]. Cocrystal engineering has been shown to effectively enhance the stability of drug molecules through the manipulation of crystal-packing interactions [27]. When a drug is incorporated into a crystalline lattice with a coformer, its molecular environment is altered in a way that can inhibit degradation reactions (Figure 10.3). For example, cocrystallization can increase the distance between reactive sites on drug molecules, prevent close contacts that catalyze degradation, or place the drug in a microenvironment with

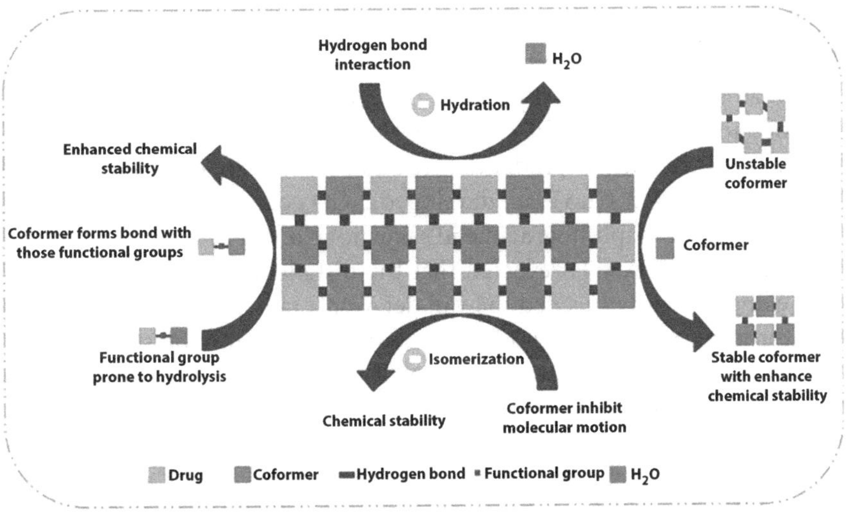

Figure 10.3 Coformer-drug interactions are the key to enhancing drug stability.

opposing properties that shield it from degradation [1]. Several stability-enhanced cocrystals have demonstrated improved long-term stability and resistance to degradation under stress conditions like heat, light, and humidity. By improving drug stability through cocrystallization, the shelf life of pharmaceutical products can be prolonged, allowing for extended expiration dates that facilitate manufacturing, distribution, and patient usage of medicines [27]. Major ways in which cocrystals can enhance the stability and shelf life of pharmaceutical products:

Coformers interact with drug molecules through various forces, such as hydrogen bonding, electrostatic interactions, and Van der Waals forces. This helps stabilize the drug molecule and prevent its degradation. Coformers can also protect the drug molecule from the effects of the environment, such as moisture and light. The overall effect of these interactions is to enhance the chemical stability of the drug molecule and improve its therapeutic efficacy. The figure is adapted from (Singh *et al.*, 2023; https://doi.org/10.3390/pharmaceutics15041161) published under a CC4 license.

10.3.2.1 Changing the Reaction Cavity/Environment

Cocrystallization can increase the distance between reactive sites on drug molecules that undergo degradation reactions like hydrolysis, preventing close contacts that can catalyze the reactions [1]. This enlarges the reaction cavity and sterically hinders degradation. This is demonstrated in the case of nicorandil (NCD). NCD is an anti-anginal drug that readily undergoes degradation under humid, heat, and mechanical stress conditions through a cyclic degradation pathway. In the crystal structure of NCD, the distance between the nitrogen atom (N1) of the pyridine ring and the carbon atom (C8) of the ethyl group is 3.367 Å, which is shorter than the sum of van der Waals radii of C and N. This close contact between N1 and C8 facilitates the degradation process by stabilizing the carbocation (C8+) intermediate. Guo *et al.* developed NCD cocrystals with coformers like 1-hydroxy-2-naphthoic acid and salicylic acid. The X-ray crystal structures of these cocrystals revealed that the distances between N1 and C8 were elongated and exceeded the van der Waals radii sum. As a result, the C8+ carbocation in the degradation intermediate cannot be stabilized by N1, inhibiting the catalyzed effect on NCD decomposition. Consequently, all NCD cocrystals exhibited significantly enhanced hydrothermal stability compared to NCD alone. This case study illustrates how cocrystallization with suitable coformers can enlarge reaction cavities and increase distances between reactive sites, sterically hindering degradation reactions from occurring. This imparts greater stability to labile drug molecules like NCD [44].

10.3.2.2 Alteration of Solid-State Properties

Properties like crystallinity, density, and polymorphic form influence molecular mobility and packing interactions in crystals. Cocrystals can modify these properties in a way that restricts molecular movement and interactions leading to degradation. Epalrestat (EPR) is susceptible to Photoisomerization upon exposure to light. It readily undergoes an E,Z to Z,Z photoisomerization that affects its pharmacological properties. Putra *et al.* reported that, when EPR was cocrystallized with betaine (BET), the resultant cocrystal showed no color change or degradation after illumination for 24 hours, unlike EPR alone which changed color. The X-ray crystal structures revealed that EPR-BET cocrystals possess strong intermolecular interactions and a reduced photoisomerization reaction cavity compared to the EPR form. This decreases molecular mobility and restricts the orientation that EPR molecules can adopt when exposed to light. As a result, the photoisomerization is suppressed [45].

Another example is ambrisentan (AMT), an antihypertensive drug prone to hydrolysis degradation. Its cocrystal with glycylglycine (GG) connected AMT molecules via carbonyl···OH hydrogen bonds and OH···π interactions. These strong intermolecular interactions within the lattice stabilize AMT, enhancing its hydrolytic stability compared to pure AMT [46]. In both examples, judicious cocrystallization with coformers optimized the solid-state packing and reduced molecular mobility through enhanced intermolecular interactions. As demonstrated, carefully designed cocrystals can restrict molecular mobility and interactions within the crystal lattice, thereby suppressing degradation pathways that would otherwise compromise the drug's potency and consistency of therapeutic effect. These improvements in chemical stability directly translate to better preservation of the active pharmaceutical ingredient's pharmacological activity during manufacturing, storage, and *in vivo* administration. By mitigating the risk of drug degradation, the cocrystal approach helps ensure the reliable delivery of the intended therapeutic dose to the patient, ultimately enhancing the overall efficacy and safety of the treatment.

10.3.2.3 Manipulation of Hydrogen Bonding Pattern

Strong and strategic hydrogen bonds formed between the drug and coformer disrupt the formation of hydrogen bonding motifs that can catalyze degradation reactions. This shields reactive functional groups on drug molecules. Adefovir dipivoxil is an oral prodrug of the antiviral drug adefovir [47], but it suffers from poor chemical stability due to two degradation pathways—hydrolysis of the pivaloyloxymethyl moiety and dimerization of the adenine ring. Gao *et al.* designed cocrystals of adefovir dipivoxil with the weakly acidic coformer saccharin and the basic coformer nicotinamide. They found that the hydrolytic stability of adefovir dipivoxil followed the order: Adefovir dipivoxil-saccharin cocrystal < Adefovir dipivoxil < Adefovir dipivoxil-nicotinamide cocrystal. After 30 days of chemical stability testing, the adefovir dipivoxil-saccharin cocrystal showed the lowest content of degradation product among the three forms. The researchers attributed this to saccharin's weakly acidic nature. Through hydrogen bonding interactions, saccharin generates a weakly acidic microenvironment around the adefovir dipivoxil molecule in the cocrystal. This weak acidity inhibits the hydrolysis degradation pathway of adefovir dipivoxil. In contrast, nicotinamide provides a basic environment that accelerates hydrolysis [48]. So, in this example, judiciously selecting saccharin as the coformer manipulated the hydrogen bonding network to place adefovir dipivoxil in a weakly acidic crystal environment. This hindered its hydrolytic degradation pathway and enhanced the chemical stability of the cocrystal.

Thus, the strategic formation of strong hydrogen bonds between drug molecules and coformers has emerged as an effective approach for improving the chemical stability of pharmaceuticals and, consequently, their therapeutic efficacy. By disrupting the hydrogen bonding motifs that can catalyze degradation pathways, cocrystallization creates a protective microenvironment around the drug substance. This enhanced stability translated to better preservation of the drug's pharmacological potency, enabling more consistent and reliable delivery of the intended therapeutic dose. Continued exploration of hydrogen bonding-driven cocrystal design will further expand the utility of this approach in developing robust, high-efficacy drug formulations.

10.3.2.4 Influencing Solid-State Microenvironment

Some coformers are selected such that they provide an acidic/basic/redox microenvironment unfavorable for degradation pathways like oxidation to occur. This chemically stabilizes the drug molecule. EPR is unstable to photoisomerization upon light exposure. The XRD structure of EPR showed a large reaction cavity, allowing high molecular mobility and susceptibility to isomerization. EPR was cocrystallized with betaine (BET).

The BET-EPR cocrystal XRD structure showed smaller and more crowded reaction cavities compared to EPR alone, due to hydrogen bonding interactions with BET restricting molecular mobility. When exposed to light, EPR powder changed color significantly, while the BET-EPR cocrystal remained unchanged. The cocrystal microenvironment suppressed isomerization by limiting molecular motion within the dense crystal structure [45].

Thus, by incorporating coformers that provide an unfavorable microenvironment for the degradation pathways of drugs, cocrystals can effectively shield the drug molecule. These stability improvements directly translate to better preservation of the drug's pharmacological potency, enabling more reliable and consistent delivery of the intended therapeutic dose.

10.3.2.5 Increased Thermal Stability

Stronger and denser packing of molecules in the lattice and higher lattice energies of cocrystals can impart greater thermal stability by elevating melting points and raising the energy requirement for degradation reactions. PFD-fumaric acid and PFD-trimesic acid cocrystals were found to have increased heat of fusion compared to PFD alone, as measured through solubility experiments. The heat of fusion is a thermal property, representing the amount of energy required to melt a substance. A higher heat of fusion indicates stronger interactions holding the solid structure together. The authors attributed the increased heat of fusion in the cocrystals to stronger intermolecular forces arising from hydrogen bonding interactions between PFD and the coformers in the crystalline lattice. These stronger interactions lead to higher thermal lattices and binding energies. While thermal stability/degradation experiments are not reported, the implication is that by requiring more energy to disrupt the crystalline structure through melting, the cocrystals may have improved stability against thermal decomposition compared to PFD alone. The stronger interactions make the cocrystal structure more robust [49].

Thus, cocrystallization has demonstrated the ability to enhance the thermal stability of drug substances, which is crucial for ensuring the therapeutic efficacy and safety of pharmaceutical products. The enhanced thermal stability directly translates to improved resistance against degradation, helping maintain the potency and consistency of the therapeutic effect. Continued research into the thermodynamic properties of cocrystals will further elucidate the mechanisms by which they can be leveraged to develop more stable and efficacious drug formulations.

10.3.3 Improving Bioavailability and Pharmacokinetics

Pharmacokinetics refers to the study of how drugs are absorbed, distributed, metabolized, and eliminated by the body. Cocrystals can influence these processes by altering the drug's physicochemical properties, such as its stability, permeability, and rate of metabolism [50]. The bioavailability of a drug refers to its ability to reach the systemic circulation and produce the desired therapeutic effect. However, many drugs face challenges in achieving sufficient bioavailability due to their poor solubility in biological fluids. Cocrystals provide a solution to this problem by forming new crystalline structures that enhance drug solubility [23].

Low aqueous solubility often translates to poor and erratic oral absorption, which can significantly impact the efficacy of some drugs. By modifying a drug's physicochemical properties through cocrystallization, it is possible to improve solubility and dissolution rates. Enhanced dissolution leads to higher drug concentration levels in the gastrointestinal fluid, allowing for more complete absorption. Several studies have demonstrated that cocrystals can remarkably increase a drug's exposure in the body (AUC) and maximum observed drug concentration (Cmax) after oral administration [46, 51]. The ability of cocrystals to modify release profiles and potentially bypass first-pass liver metabolism also affects absorption and distribution characteristics [23]. These pharmacokinetic advantages can translate to clinically meaningful outcomes such as higher efficacy, fewer required doses, and a better safety profile.

10.3.3.1 The Mechanism for Enhancing the Bioavailability and Improve Pharmacokinetics of Drugs

10.3.3.1.1 Increased Solubility and Dissolution Rate

By forming cocrystals, the aqueous solubility and dissolution rate of poorly water-soluble drugs can be significantly improved. This leads to higher absorption in the gastrointestinal tract and increased bioavailability [1]. Cheney *et al.* prepared 12 meloxicam cocrystals with aromatic/aliphatic carboxylic acid coformers. Non-sink dissolution studies showed the cocrystals underwent solvent-mediated phase transformation to meloxicam. To study the impact on oral absorption, a single-dose PK study in rats was conducted. Most cocrystals exhibited higher intestinal absorption than meloxicam, as evidenced by higher mean serum concentrations at matched time points. Linear regression analysis correlated mean *in vitro* dissolution profiles to mean serum concentration data, indicating cocrystals with higher dissolution rates likely yield better bioavailability through enhanced release [52].

Cocrystals often exhibit supersaturation behavior during dissolution due to higher energy crystalline forms. However, bioavailability may not be realized if precipitation occurs rapidly [1]. Dihydromyricetin (DMY) is a class IV drug with low solubility and permeability. DMY-caffeine and DMY-urea cocrystals were prepared. *In vitro* dissolution showed this maintained supersaturation initially but rapidly precipitated DMY dihydrate. Various polymers and surfactants were screened as crystallization inhibitors. Polyvinylpyrrolidone K30 (PVP K30) most effectively stabilized the supersaturated solutions, maintaining 6-fold higher concentrations than DMY for over 300 minutes. Rats were dosed with the cocrystals suspended in 0.5% sodium carboxymethylcellulose (NaCMC) or 2% PVP K30. DMY-urea with NaCMC failed to improve bioavailability versus DMY alone, attributed to rapid (<10 min) precipitation *in vivo*. However, DMY-urea with PVP K30 achieved a 4.9-fold higher AUC than DMY due to PVP K30 stabilizing supersaturation. DMY-caffeine alone provided 3.9-fold higher AUC than DMY through maintaining supersaturation longer (~30 min). With PVP K30, a further improvement to 5.3-fold was seen [53].

10.3.3.1.2 Modified Release Properties

Some cocrystals exhibit modified release behavior compared to their parent drugs. This includes sustained or delayed release, which can improve pharmacokinetic profiles [1]. Celecoxib has a low solubility of 7 μg/mL and belongs to BCS class II. However, rapid absorption of tramadol (BCS class I) is associated with a higher incidence of adverse effects likely due to peak concentrations. CTC was formulated to modify the PK profiles of both drugs [54]. Intrinsic dissolution rate (IDR) studies showed CTC decreased tramadol IDR 7-fold while increasing celecoxib IDR 3-fold compared to drugs alone. This suggested CTC could provide sustained release of tramadol for safety while improving celecoxib dissolution [54, 55].

Beagle dog PK studies were then performed comparing CTC tablets, commercial tramadol, and celecoxib products. CTC exhibited higher celecoxib Cmax and AUC than Celebrex, demonstrating it enhanced celecoxib bioavailability through modified dissolution. Meanwhile, CTC resulted in lower tramadol peak concentration and AUC relative to the commercial product. Human PK studies found CTC also reduced tramadol Cmax and AUC versus immediate-release tramadol, maintaining similar overall exposure. Thus, CTC was successful in modulating the release of each drug to enable improved bioavailability of celecoxib while potentially enhancing tramadol's safety profile [54, 55].

10.3.3.1.3 Enhanced Permeability

The permeation of drugs across biological membranes affects the rate and extent of absorption. Cocrystals may improve permeability through their interactions with membrane transporters or altered solid-state properties. Acyclovir is a BCS class III drug with good solubility but low permeability. Yan *et al.* synthesized acyclovir-fumaric acid and acyclovir-glutaric acid cocrystals. Permeability studies across rat skin showed the cocrystals increased the permeated amount of acyclovir by 2.8 and 4-fold respectively, compared to pure acyclovir [56]. This demonstrated cocrystals can improve the bioavailability of low-permeable drugs by enhancing permeability.

Furosemide is a BCS class IV drug with both low solubility and permeability. Banik *et al.* prepared different cocrystals and found that furosemide-anthranilamide cocrystals exhibited higher cumulative permeated amount/flux across a cellulose membrane compared to pure furosemide or other cocrystals [57]. This provided direct evidence that cocrystallization can simultaneously improve solubility and permeability for class IV drugs. In Caco-2 cellular studies, Ferretti *et al.* found that the indomethacin-saccharine cocrystal enhanced the apparent permeability of Indomethacin without affecting cellular integrity [58]. This suggests cocrystals can enhance permeability even at the level of the intestinal epithelium, possibly via drug-coformer interactions in solution-altering membrane diffusion.

These case studies elucidate different mechanisms like increased membrane flux, synergistic effects with permeability-enhancing coformers, and the absence of toxicity that underlie permeability enhancement of cocrystals—a key advantage for bioavailability improvement of BCS class III and IV drugs. Proper selection of coformers is important to manifest this potential of cocrystals. One case study analyzed was indomethacin cocrystals. Ferretti *et al.* studied the permeability of Indomethacin cocrystals across intestinal epithelial NCM460 cell monolayers. Cocrystal 1 containing 2-hydroxy-4-methyl-pyridine damaged cellular integrity and increased Indomethacin apparent permeability compared to pure Indomethacin. Conversely, physical mixture 1 did not affect cells. Cocrystal 3 containing saccharine improved Indomethacin permeability without cellular damage [58]. This demonstrated that cocrystals can modulate membrane permeability differently than physical mixtures.

These case studies signify how cocrystals can alter the intestinal/skin permeation of drugs in addition to solubility enhancement. By simultaneously improving solubility and membrane permeability, cocrystals represent a promising strategy for enhancing the bioavailability of BCS class III and IV drugs with poor solubility and/or permeability characteristics. The ability of cocrystals to modify diffusion/release rates through biological membranes expands their applications in improving oral drug delivery.

10.3.3.1.4 Modified Degradation

Drug stability influences bioavailability. Cocrystals can alter the degradation pathways of unstable drugs through interactions or modified microenvironments, increasing the fraction absorbed. One of the ways cocrystals can improve bioavailability is by protecting the drug from degradation in the gastrointestinal (GI) tract. Baicalein is a natural flavonoid with low solubility and permeability (BCS class IV). Zhu *et al.* synthesized a baicalein-caffeine cocrystal and conducted pharmacokinetic experiments in rats. They found that the cocrystal showed a 4.1-fold higher $AUC_{0\text{-}24h}$ compared to pure baicalein, which was attributed to the higher dissolution rate of the cocrystal. Interestingly, the physical mixture of baicalein and caffeine also improved $AUC_{0\text{-}24h}$ by 2-fold compared to pure baicalein. This improvement was explained by the synergic effect of caffeine on baicalein absorption by increasing intestinal blood flow [59].

Caffeine is known to inhibit the metabolism of some drugs like acetaminophen and ergotamine tartrate when co-administered, leading to higher plasma concentrations. It is proposed that caffeine may protect baicalein from degradation and first-pass metabolism in a similar way, allowing more baicalein to absorb into the systemic circulation. While the exact mechanism was not investigated, this case study demonstrates how a coformer like caffeine can potentially improve bioavailability by modifying drug degradation kinetics in the GI tract through metabolic inhibition effects [59]. Another example is celecoxib-tramadol hydrochloride cocrystal (CTC) discussed for the treatment of acute pain. Animal and human pharmacokinetic studies found that CTC exhibited a potential to simultaneously improve the bioavailability of celecoxib while improving the safety of tramadol hydrochloride by modifying its release kinetics. This suggests the cocrystal form protected both drugs from early degradation in the GI tract, allowing their plasma concentrations to increase in a controlled manner [54].

These case studies provide evidence that cocrystals can enhance bioavailability by modifying drug degradation pathways through coformer-drug interactions like metabolic inhibition, thereby protecting more drugs from degradation so they can successfully permeate and distribute in the body. Careful selection of appropriate coformers is important from a safety perspective as well.

10.3.4 Combination Therapy

Cocrystals have emerged as a promising strategy for combination drug therapy [60]. A cocrystal is a multicomponent crystal consisting of two or more molecules generally in a stoichiometric ratio [1]. Cocrystals differ from traditional pharmaceutical salts and solvates in that the component molecules are in their neutral/parent form. This property allows for combining two drug molecules without altering their intrinsic properties or pharmacology. Cocrystals offer several advantages for combination therapies. First, cocrystallization allows multiple drug components to be effectively delivered in a single dosage form at predefined stoichiometric ratios. This can help ensure consistent and predictable pharmacokinetics compared to physical mixtures or co-administration of separate dosage forms [21].

To date, over a dozen cocrystal-based drug products have received market approval from regulatory agencies. Clinical studies have demonstrated the efficacy and safety of these cocrystal combinations for various therapeutic indications. Entresto® is a pharmaceutical cocrystal approved for the treatment of chronic heart failure. It consists of a 1:1 cocrystal of valsartan and sacubitril, targeting both the renin-angiotensin-aldosterone system (RAAS) and neprilysin pathways. In clinical trials, Entresto® demonstrated reduced cardiovascular death and hospitalization compared to enalapril, an ACE inhibitor. This improved efficacy is attributed to the cocrystal's ability to deliver both actives simultaneously at their optimal 1:1 ratio [7].

Tramadol-celecoxib cocrystal (CTC) is a 1:1 cocrystal of tramadol hydrochloride and celecoxib that has been extensively studied by Almansa *et al.* The experimental studies have revealed several key findings about CTC. Dissolution studies have shown that the IDR of celecoxib is increased in CTC, while tramadol hydrochloride is released at a slower rate compared to when the drugs are administered individually. In terms of pharmacokinetics, clinical studies have found that tramadol is absorbed more slowly from CTC, resulting in a reduction in peak plasma concentration. On the other hand, celecoxib is absorbed faster from CTC compared to its single-entity form. Moreover, phase II clinical trials have demonstrated that CTC has a superior analgesic effect for moderate-severe acute post-surgical pain management when compared to tramadol, celecoxib alone, and their combination. The mechanism of action underlying these findings is believed to be the optimized dissolution properties of CTC, which positively modify the pharmacokinetic profile of both drugs. Tramadol's slower release reduces its peak plasma concentration, while celecoxib's faster release increases its bioavailability. This synchronized plasma concentration-time profile is hypothesized to produce the enhanced analgesic synergy observed clinically. In conclusion, the experimental evidence suggests that CTC, with its modulation of the dissolution and absorption properties of tramadol and celecoxib, offers clinically relevant advantages in terms of pharmacokinetics and efficacy over conventional combination or single-drug formulations [55].

Meloxicam and aspirin both have anti-inflammatory effects, but they differ in their selectivity for COX enzymes. Meloxicam primarily targets COX-2, whereas aspirin inhibits both COX-1 and COX-2. Cheney *et al.* have reported a 1:1 cocrystal of meloxicam and aspirin (MA) in their study. The experimental studies shed light on the combined therapeutic effect of MA. Solubility experiments using phosphate buffer have shown that MA exhibits a 44-fold increase in solubility at pH 7.4 compared to meloxicam alone. This increased solubility is expected to enhance the dissolution rate as well, potentially leading to the complete absorption of drugs. Pharmacokinetic studies in rats have demonstrated that MA has approximately 4 times higher oral bioavailability compared to meloxicam alone, which correlates with its improved solubility. This enhanced bioavailability suggests that the cocrystallization of MA allows for a greater concentration of the drugs to be absorbed and available systemically, potentially enhancing their therapeutic effectiveness [52].

While both andrographolide and salicylic acid exhibit anti-inflammatory properties, they exert their effects through different molecular mechanisms. Andrographolide targets multiple signaling pathways involved in inflammation, immunomodulation, and oxidative stress, while salicylic acid primarily acts as a COX inhibitor to reduce inflammation and pain. In their study, Suresh *et al.* reported the successful development of a 1:1 cocrystal of andrographolide and salicylic acid (AS). Notably, AS demonstrated a remarkable 12-fold increase in solubility compared to andrographolide alone, as evidenced by saturation solubility experiments. Furthermore, *in vitro* dissolution studies revealed that the dissolution rate of AS was approximately 3 times faster than that of andrographolide alone. The relative drug release of AS was estimated to be approximately 2 times higher than andrographolide, as observed in dynamic dissolution studies. Importantly, AS exhibited complete inhibition of the chemical transformation of andrographolide to its inactive sulfate metabolite, as confirmed analytically. These experimental findings collectively demonstrate that cocrystallization between andrographolide and salicylic acid effectively modulated the physiochemical properties of andrographolide. This modulation was achieved through molecular-level interactions between the two components, without any chemical modification to andrographolide. The significant enhancement in solubility, dissolution rate, and drug release observed in the AS cocrystal provides valuable insights into the potential of this combination therapy approach [61].

While isoniazid primarily targets cell wall synthesis, pyrazinamide acts on the intracellular environment of *Mycobacterium tuberculosis*. The combination of these two drugs in tuberculosis treatment is synergistic, as they target different aspects of bacterial metabolism and provide a more comprehensive approach to eradicating the infection. In their study, Yan *et al.* reported the synthesis of a drug-bridge-drug ternary cocrystal composed of isoniazid, pyrazinamide, and fumaric acid. The physicochemical characterizations revealed that the ternary cocrystal exhibited an optimized formulation capacity, displaying favorable compressibility and tableting properties. Dissolution studies demonstrated significantly enhanced dissolution rates of isoniazid and pyrazinamide from the ternary cocrystal compared to physical mixtures. Furthermore, pharmacokinetic experiments conducted in rats revealed synergistic effects between the three components of the ternary cocrystal, leading to improved bioavailability when compared to physical mixtures. The mechanism of action involves the formation of a cocrystalline network through multiple molecular interactions between isoniazid, pyrazinamide, and the bridging fumaric acid moiety. This cocrystalline network effectively modulates the formulation capacity and optimizes dissolution behavior *in vitro*, resulting in enhanced bioavailabilities of isoniazid and pyrazinamide *in vivo*. These findings provide a new rationale for the design of effective anti-tuberculosis combination therapies, highlighting the potential of the isoniazid-pyrazinamide-fumaric acid cocrystal in achieving synergistic anti-tuberculosis effects compared to simple physical mixtures [62].

10.3.5 Improving the Targeted Delivery of Drugs

Nano-cocrystals are a novel class of drug delivery system that combines the advantages of nano-crystals and cocrystals. By reducing the particle size of cocrystals into the nanoscale range, nano-cocrystals are able to exploit the benefits associated with both technologies for targeted drug therapy. As successors to nano-crystals, nan-ococrystals have the potential for superior targeting capabilities over either nano-crystals or cocrystals alone. Being nanoparticles, nano-cocrystals are able to utilize mechanisms like the enhanced permeability and retention effect to passively accumulate in pathological sites like tumor tissues for targeted action. In addition, the large surface area of nano-cocrystals allows for opportunities to selectively functionalize their surface with targeting ligands. This enables active targeting abilities for improved localization and intracellular delivery of therapeutic payloads. Moreover, the nanoscale size range of nano-cocrystals serves to extend their plasma circulation time, improving the proportion of drug accumulated at target sites. The combined properties of nano-cocrystals thus present promising prospects for achieving the efficient, specific release of pharmaceutical actives directly at the site of intended therapy [63].

Witika *et al.* produced lamivudine-zidovudine nano-cocrystals using a sonochemical synthesis method in the presence of surfactants and polymers. Their goal was to stabilize the nano-sized cocrystals and control their critical quality attributes such as particle size, polydispersity index, and zeta potential. They characterized the nano-cocrystals using techniques such as X-ray powder diffraction, Raman spectroscopy, FTIR spectroscopy, and differential scanning calorimetry. This confirmed the cocrystalline nature of the lamivudine-zidovudine nanoparticles. They then tested the cytotoxicity of the nano-cocrystals versus a physical mixture of the APIs in HeLa cells. Interestingly, the experimental results showed significantly lower cytotoxicity for the nano-cocrystals compared to the mixture, with cell viability remaining over 80% even at high concentrations of 100 μM. This demonstrates the ability of the nano-cocrystals to safely release the drugs after cellular absorption. In a follow-up study, Witika *et al.* further optimized the sonochemical synthesis using a quality-by-design approach to producing nano-cocrystals with predefined properties for drug delivery applications [64].

10.4 Regulatory Considerations for Cocrystal Approval

The cocrystal drug product development workflow is shown in Figure 10.4. It requires addressing certain regulatory aspects. Both the US Food and Drug Administration (FDA) and the European Medicines Agency (EMA) have provided guidance on the regulatory classification and development of cocrystals [65]. In 2013, the FDA issued a guidance document titled "Regulatory Classification of Pharmaceutical Cocrystals" which defined cocrystals as "crystalline materials composed of two or more molecules within the same crystal lattice." The guidance explains that cocrystals constitute drug product intermediates and not new active pharmaceutical ingredients (APIs). FDA classifies cocrystals as a special case of solvates or

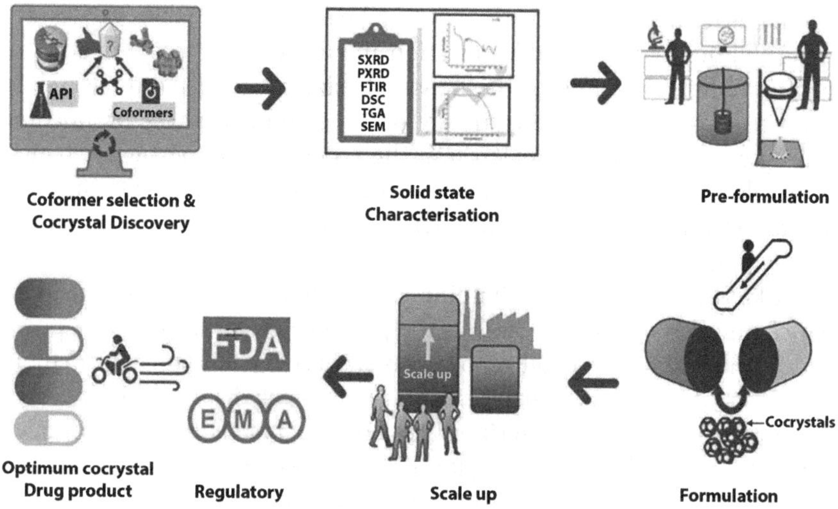

Figure 10.4 From discovery to delivery: The workflow of coformer selection and cocrystal drug product development. The cocrystallization process starts with the selection of the API and the coformer. The cocrystal discovery is based on screening methods and solubility studies to identify the most promising cocrystal candidates. The solid-state characterization of the cocrystals is performed using various analytical techniques. The preformulation studies are carried out to optimize the cocrystal formulation and to assess its stability. The scale-up process involves optimizing the manufacturing process and scale-up of the cocrystal. The regulatory approval is required before the cocrystal product can be commercialized. The figure is adapted from (Singh *et al.*, 2023; https://doi.org/10.3390/pharmaceutics15041161) published under a CC4 license.

hydrates, recognizing their potential to modify different properties associated with therapeutic efficacy without altering the active moiety [66]. The EMA reflection paper from 2014 defines cocrystals as "homogeneous (single phase) crystalline structures made up of two or more components in a specific stoichiometric ratio where the arrangement in the crystal lattice is not based on ionic bonds." Thus, EMA views cocrystals similarly to salts, considering them eligible for regulatory approval as new substances if shown to impact efficacy, safety, or pharmacokinetics compared to the reference drug [67]. Some key regulatory considerations for developing cocrystals include:

- Demonstrating equivalence to the reference listed drug in terms of pharmacokinetic parameters like rate and extent of

absorption. Studies may be needed to show the dissociation of cocrystal components.
- Establishing stability of the cocrystal form under intended storage and processing conditions as per ICH guidelines.
- Characterizing the cocrystal components and ensuring quality, stability, and cGMP compliance.
- Demonstrating process validation and manufacturing reproducibility for the cocrystal product.
- Providing scientific rationale for the selection of pharmaceutically acceptable and non-toxic excipients and coformers.
- Addressing any potential patent issues related to the cocrystal compositions.

Addressing these regulatory aspects helps establish a cocrystal product as a therapeutic equivalent to the reference drug, facilitating its approval and market access. The story of sacubitril-valsartan's evolution from a scientific concept into the approved heart failure treatment Entresto® was a long yet fruitful journey guided by rigorous clinical research [7]. Its beginnings trace back to discoveries regarding neprilysin's role in degrading peptides vital to cardiovascular homeostasis. Researchers recognized inhibiting this enzyme's activity could help manage heart failure symptoms by allowing therapeutic peptides' increased circulation. This realization spurred sacubitril's early development as a neprilysin inhibitor. Initial studies explored its blood pressure-lowering and cardiac function-improving capabilities, yielding promising results. They set the stage for sacubitril's pairing with valsartan, an angiotensin receptor blocker, into a novel combination therapy. Its first clinical trial testing sacubitril/valsartan's safety and efficacy marked the beginning of a trial-guided odyssey [68].

The landmark PARADIGM-HF trial would prove most pivotal, comparing the combination to enalapril, the standard treatment. Its overwhelming success in preventing cardiovascular deaths and hospitalizations stopped the trial early. Presented to regulators, the FDA granted priority review and conditional approval on July 7, 2015, recognizing sacubitril/valsartan's life-saving potential as Entresto [69].

Post-approval commitments further reinforced Entresto's benefits, expanding indications based on thorough investigations. Today, it exemplifies cocrystals' potential for progressing medicine by refining treatments at the molecular level. Sacubitril/valsartan's pioneering approval sets the stage for future innovation in applying multifaceted drug research. Most importantly, it underscores cocrystals' life-saving power through translational efforts linking discovery to real-world care.

10.5 Conclusion and Future Aspects

Cocrystallization is emerging as an important approach in pharmaceutical sciences for modulating the physiochemical and performance properties of drugs. Significant advances have been made in the past two decades towards understanding cocrystal formation and characterization. The fundamental principles of supramolecular synthon-mediated interactions between API and coformer provide a scientific basis for the rational design of cocrystals. Considerable research efforts have demonstrated the ability of cocrystals to improve the solubility, dissolution, permeability, bioavailability, and stability profiles of numerous drugs. This in turn enhances their therapeutic efficacy by enabling complete and consistent absorption of intended drug concentrations *in vivo*. Regulatory guidelines have accepted cocrystals as eligible for approval provided their clinical equivalence to reference products is demonstrated. Several real-world examples of marketed cocrystal products validated the technology. However, further exploration of multicomponent systems, advanced formulations, targeted delivery, and combination therapies hold promise to fully realize cocrystals' potential. Addressing challenges like scalable manufacturing and understanding long-term behavior also requires ongoing investigation. Looking ahead, cocrystallization represents a versatile means to refine early and clinical-stage drug candidates. Strategic application of computational and high-throughput methodologies could aid the discovery of optimized crystal forms. Continued evaluation in well-designed clinical trials will strengthen understanding of structure-activity relationships and true patient impact. With diligent research linking discovery to practice, cocrystals are positioned to advance medicinal care by enabling the delivery of novel and improved medicines to benefit public health worldwide.

The future of cocrystal research is promising with many opportunities yet to be explored. Areas of focus include developing multicomponent frameworks incorporating two or more active pharmaceutical ingredients to generate new combination therapies. Long-acting controlled-release cocrystal platforms integrated with biocompatible polymers could improve adherence and pharmacokinetics. Engineering targeted delivery through functionalizing porous cocrystal surfaces with antibodies or ligands may enhance efficacy. Exploring non-oral routes such as pulmonary cocrystal designs may unlock new administration pathways. Application of Quality by Design and process analytical technologies can facilitate scalable and reproducible manufacturing. Uncovering excipient cocrystals may confer benefits like improved stability and compression properties. Innovative crystallization processes leveraging mechanochemistry, 3D printing, or emerging technologies present opportunities. Further elucidating structure-property relationships through computational modeling may provide valuable insights. Additional clinical evaluations of existing and new cocrystal products hold promise to expand therapeutic applications into new disease areas. With consistent progress in these areas, cocrystals have significant potential to transform drug development strategies and enhance outcomes for patients worldwide.

Acknowledgment

The authors are very thankful to Guru Nanak Institute of Pharmaceutical Science and Technology for providing library and infrastructure facilities.

References

1. Guo, M., Sun, X., Chen, J., Cai, T., Pharmaceutical cocrystals: A review of preparations, physicochemical properties and applications. *Acta Pharm. Sin. B*, 11, 8, 2537–2564, 2021.
2. Cerreia Vioglio, P., Chierotti, M.R., Gobetto, R., Pharmaceutical aspects of salt and cocrystal forms of APIs and characterization challenges. *Adv. Drug Deliv. Rev.*, 117, 86–110, 2017.
3. Bhalani, D.V., Nutan, B., Kumar, A., Singh Chandel, A.K., Bioavailability Enhancement Techniques for Poorly Aqueous Soluble Drugs and Therapeutics. *Biomedicines*, 10, 2055, 2022.
4. Rathi, R. and Singh, I., Multicomponent crystal compromising dasatinib and selected co-crystals formers: a patent evaluation of EP2861589B1. *Pharm. Pat. Anal.*, 11, 1, 15–21, 2021.
5. Garnero, C., Zoppi, A., Aloisio, C., Longhi, M. R., Technological delivery systems to improve biopharmaceutical properties. In *Nanoscale Fabrication, Optimization, Scale-Up and Biological Aspects of Pharmaceutical Nanotechnology*, pp. 253–299, William Andrew Publishing, UK, 2018.
6. Kumar Bandaru, R., *et al.*, Recent Advances in Pharmaceutical Cocrystals: From Bench to Market. *Front. Pharmacol.*, 12, 780582, 2021.
7. Singh, M., *et al.*, Cocrystals by Design: A Rational Coformer Selection Approach for Tackling the API Problems. *Pharmaceutics*, 15, 4, 1161, 2023.
8. Rathi, R., Kushwaha, R., Goyal, A., Singh, I., Oxaliplatin-flavone pharmaceutical co-crystal-CN111205332A: patent spotlight. *Pharm. Pat. Anal.*, 11, 5, 147–154, 2022.

9. Jain, S., Kaur, S., Rathi, R., Nagaich, U., Singh, I., Application of co-processed excipients for developing fast disintegrating tablets: A review. *Polym. Med.*, 53, 1, 59–68, 2023.
10. Jug, M. and Mura, P.A., Grinding as Solvent-Free Green Chemistry Approach for Cyclodextrin Inclusion Complex Preparation in the Solid State. *Pharmaceutics*, 10, 4, 189, 2018.
11. Rathi, R., Sanshita, Singh, I., Hot Melt Extrusion Technique for Developing Pharmaceutical Co-crystals: A Review. *Drug Deliv. Lett.*, 14, 2, 97–108, 2024.
12. Solares-Briones, M., *et al.*, Mechanochemistry: A Green Approach in the Preparation of Pharmaceutical Cocrystals. *Pharmaceutics*, 13, 12, 790, 2021.
13. Thakur, R. and Gupta, R.B., Rapid expansion of supercritical solution with solid cosolvent (RESS-SC) process: Formation of 2-aminobenzoic acid nanoparticle. *J. Supercrit. Fluids*, 37, 307–315, 2006.
14. Yeh, K.L. and Lee, T., Intensified crystallization processes for 1: 1 drug–drug cocrystals of sulfathiazole–theophylline, and sulfathiazole–sulfanilamide. *Cryst. Growth Des.*, 18, 3, 1339–49, 2018.
15. Rodríguez, I., Gautam, R., Tinoco, A.D., Using X-ray Diffraction Techniques for Biomimetic Drug Development, Formulation, and Polymorphic Characterization. *Biomimetics (Basel Switzerland)*, 6, 1, 1, 2020.
16. Lecoq, L., Fogeron, M.-L., Meier, B.H., Nassal, M., Böckmann, A., Solid-State NMR for Studying the Structure and Dynamics of Viral Assemblies. *Viruses*, 12, 10, 1069, 2020.
17. Mohamed, M.A., Jaafar, J., Ismail, A.F., Othman, M.H.D., Rahman, M.A., Chapter 1 - Fourier Transform Infrared (FTIR) Spectroscopy, in: Hilal N, Ismail AF, Matsuura T, Oatley-Radcliffe DBTMC, editors. *Membrane Characterization*, UK: Elsevier, p. 329, 2017.
18. Cialla-May, D., *et al.*, Raman Spectroscopy and Imaging in Bioanalytics. *Anal. Chem.*, 94, 86–119, 2022.
19. Byrn, Stephen R.., Zografi, George., Chen, Xiaoming (Sean)., Chen, Sean. *Solid-State Properties of Pharmaceutical Materials*. Germany: Wiley, 2017.
20. Liu, F., Wu, J., Chen, K., Xue, D., Morphology Study by Using Scanning Electron Microscopy, in: *Microsc. Sci. Technol. Appl. Educ.*, 2010.
21. Rathi, R., Kaur, S., Singh, I., A Review on Co-crystals of Herbal Bioactives for Solubility Enhancement: Preparation Methods and Characterization Techniques. *Cryst. Growth Des.*, 22, 3, 2023–42, 2022.
22. Bashimam, M. and El-Zein, H., Pharmaceutical cocrystal of antibiotic drugs: A comprehensive review. *Heliyon*, 8, e11872, 2022.
23. Emami, S., Siahi-Shadbad, M., Adibkia, K., Barzegar-Jalali, M., Recent advances in improving oral drug bioavailability by cocrystals. *Bioimpacts*, 8, 305–320, 2018.
24. Kale, D.P., Zode, S.S., Bansal, A.K., Challenges in translational development of pharmaceutical cocrystals. *J. Pharm. Sci.*, 106, 2, 457–70, 2017.
25. Kumar, S. and Nanda, A., Pharmaceutical Cocrystals: An Overview. *Indian J. Pharm. Sci.*, 79, 6, 2017.

26. McKellar, S.C., Kennedy, A.R., McCloy, N.C., McBride, E., Florence, A.J., Formulation of Liquid Propofol as a Cocrystalline Solid. *Cryst. Growth Des.*, 14, 2422–2430, 2014.
27. Dhondale, M. R., Thakor, P., Nambiar, A. G., Singh, M., Agrawal, A. K., Shastri, N. R., Kumar, D. Co-crystallization approach to enhance the stability of moisture-sensitive drugs. *Pharmaceutics*, 15, 1, 189, 2023.
28. Duggirala, N.K., Smith, A.J., Wojtas, L., Shytle, R.D., Zaworotko, M.J., Physical Stability Enhancement and Pharmacokinetics of a Lithium Ionic Cocrystal with Glucose. *Cryst. Growth Des.*, 14, 6135–6142, 2014.
29. Nugraha, Y.P. and Uekusa, H., Suppressed hydration in metoclopramide hydrochloride by salt cocrystallisation. *CrystEngComm*, 20, 2653–2662, 2018.
30. Wang, Z.-Z., Chen, J.-M., Lu, T.-B., Enhancing the Hygroscopic Stability of S-Oxiracetam via Pharmaceutical Cocrystals. *Cryst. Growth Des.*, 12, 4562–4566, 2012.
31. Huang, S., Xue, Q., Xu, J., Ruan, S., Cai, T., Simultaneously Improving the Physicochemical Properties, Dissolution Performance, and Bioavailability of Apigenin and Daidzein by Co-Crystallization With Theophylline. *J. Pharm. Sci.*, 108, 2982–2993, 2019.
32. Almarsson, O. and Zaworotko, M.J., Crystal engineering of the composition of pharmaceutical phases. Do pharmaceutical co-crystals represent a new path to improved medicines? *Chem. Commun.*, 1889–1896, 2004.
33. Xuan, B., Wong, S. N., Zhang, Y., Weng, J., Tong, H. H., Wang, C., Chow, S. F., Extended release of highly water-soluble isoniazid attained through cocrystallization with curcumin. *Cryst. Growth Des.*, 20, 3, 1951–1960, 2020.
34. Latif, S., Ijaz, Q. A., Hameed, M., Fatima, K., Hussain, A., Arshad, M. S., Abbas, N., Improvement of physico-mechanical and pharmacokinetic attributes of naproxen by cocrystallization with L-alanine. *J. Drug Deliv. Sci. Technol.*, 61, 102236, 2021.
35. Grossjohann, C., *et al.*, Polymorphism in sulfadimidine/4-aminosalicylic acid cocrystals: solid-state characterization and physicochemical properties. *J. Pharm. Sci.*, 104, 1385–1398, 2015.
36. Abdul Azeeze, M.S.T., Kandasamy, S., Palanisamy, P., Palanichamy, P., Subramania Nainar, M., Cocrystallization of Dasatinib with different acids improves the solubility and physicochemical properties: A combined experimental and theoretical studies. *Chem. Phys. Impact*, 7, 100247, 2023.
37. Mishra, M.K., Mishra, K., Narayan, A., Reddy, C.M., Vangala, V.R., Structural Basis for Mechanical Anisotropy in Polymorphs of a Caffeine–Glutaric Acid Cocrystal. *Cryst. Growth Des.*, 20, 6306–6315, 2020.
38. Wang, C., Paul, S., Sun, D.J., Nilsson Lill, S.O., Sun, C.C., Mitigating Punch Sticking Propensity of Celecoxib by Cocrystallization: An Integrated Computational and Experimental Approach. *Cryst. Growth Des.*, 20, 4217–4223, 2020.

39. Roy, P. and Ghosh, A., Mechanochemical cocrystallization to improve the physicochemical properties of chlorzoxazone. *CrystEngComm*, 22, 4611-4620, 2020.
40. Sangtani, E., *et al.*, Furosemide Cocrystals with Pyridines: An Interesting Case of Color Cocrystal Polymorphism. *Cryst. Growth Des.*, 15, 5858–5872, 2015.
41. Zhu, B., Wang, J.-R., Zhang, Q., Mei, X., Improving Dissolution and Photostability of Vitamin K3 via Cocrystallization with Naphthoic Acids and Sulfamerazine. *Cryst. Growth Des.*, 16, 483–492, 2016.
42. Huang, S., Xu, J., Peng, Y., Guo, M., Cai, T., Facile Tuning of the Photoluminescence and Dissolution Properties of Phloretin through Cocrystallization. *Cryst. Growth Des.*, 19, 6837–6844, 2019.
43. Rathi, R., Guggal, Y., Twinkle, Sandhu, V., Singh, I., Exploring the Patent Landscape and Regulatory Prospective on Pharmaceutical Cocrystals. *J. Pharm. Technol. Res. Manag.*, 11, 49–60, 2023.
44. Guo, C., *et al.*, Pharmaceutical Cocrystals of Nicorandil with Enhanced Chemical Stability and Sustained Release. *Cryst. Growth Des.*, 20, 6995–7005, 2020.
45. Putra, O.D., Umeda, D., Nugraha, Y.P., Nango, K., Yonemochi, E., Uekusa, H., Simultaneous Improvement of Epalrestat Photostability and Solubility via Cocrystallization: A Case Study. *Cryst. Growth Des.*, 18, 373–379, 2018.
46. Haneef, J., Markad, D., Chadha, R., Interaction Map Driven Cocrystallization of Ambrisentan: Structural and Biopharmaceutical Evaluation. *Cryst. Growth Des.*, 20, 4612–4620, 2020.
47. Yuan, L.-C., Dahl, T.C., Oliyai, R., Effect of Carbonate Salts on the Kinetics of Acid-Catalyzed Dimerization of Adefovir Dipivoxil. *Pharm. Res.*, 17, 1098–1103, 2000.
48. Gao, Y., *et al.*, Coformer selection based on degradation pathway of drugs: A case study of adefovir dipivoxil–saccharin and adefovir dipivoxil–nicotinamide cocrystals. *Int. J. Pharm.*, 438, 327–335, 2012.
49. Kumari, N., Bhattacharya, B., Roy, P., Michalchuk, A.A.L., Emmerling, F., Ghosh, A., Enhancing the Pharmaceutical Properties of Pirfenidone by Mechanochemical Cocrystallization. *Cryst. Growth Des.*, 19, 6482–6492, 2019.
50. Shan, N., Perry, M.L., Weyna, D.R., Zaworotko, M.J., Impact of pharmaceutical cocrystals: the effects on drug pharmacokinetics. *Expert Opin. Drug Metab. Toxicol.*, 10, 1255–1271, 2014.
51. Alhalaweh, A., Ali, H.R.H., Velaga, S.P., Effects of Polymer and Surfactant on the Dissolution and Transformation Profiles of Cocrystals in Aqueous Media. *Cryst. Growth Des.*, 14, 643–648, 2014.
52. Cheney, M.L., Weyna, D.R., Shan, N., Hanna, M., Wojtas, L., Zaworotko, M.J., Coformer selection in pharmaceutical cocrystal development: a case study of a meloxicam aspirin cocrystal that exhibits enhanced solubility and pharmacokinetics. *J. Pharm. Sci.*, 100, 2172–2181, 2011.

53. Wang, C., Tong, Q., Hou, X., Hu, S., Fang, J., Sun, C.C., Enhancing Bioavailability of Dihydromyricetin through Inhibiting Precipitation of Soluble Cocrystals by a Crystallization Inhibitor. *Cryst. Growth Des.*, 16, 5030–5039, 2016.
54. Videla, S., *et al.*, Single-dose pharmacokinetics of co-crystal of tramadol-celecoxib: Results of a four-way randomized open-label phase I clinical trial in healthy subjects. *Br. J. Clin. Pharmacol.*, 83, 2718–2728, 2017.
55. Almansa, C., Mercè, R., Tesson, N., Farran, J., Tomàs, J., Plata-Salamán, C.R., Co-crystal of Tramadol Hydrochloride–Celecoxib (ctc): A Novel API–API Co-crystal for the Treatment of Pain. *Cryst. Growth Des.*, 17, 1884–1892, 2017.
56. Yan, Y., Chen, J.-M., Lu, T.-B., Simultaneously enhancing the solubility and permeability of acyclovir by crystal engineering approach. *CrystEngComm*, 15, 6457–6460, 2013.
57. Banik, M., Gopi, S.P., Ganguly, S., Desiraju, G.R., Cocrystal and Salt Forms of Furosemide: Solubility and Diffusion Variations. *Cryst. Growth Des.*, 16, 5418–5428, 2016.
58. Ferretti, V., *et al.*, Indomethacin Co-Crystals and Their Parent Mixtures: Does the Intestinal Barrier Recognize Them Differently? *Mol. Pharmaceutics*, 12, 1501–1511, 2015.
59. Zhu, B., Zhang, Q., Wang, J.-R., Mei, X., Cocrystals of Baicalein with Higher Solubility and Enhanced Bioavailability. *Cryst. Growth Des.*, 17, 1893–1901, 2017.
60. Thipparaboina, R., Kumar, D., Chavan, R.B., Shastri, N.R., Multidrug co-crystals: towards the development of effective therapeutic hybrids. *Drug Discovery Today*, 21, 481–490, 2016.
61. Suresh, K., Goud, N.R., Nangia, A., Andrographolide: Solving Chemical Instability and Poor Solubility by Means of Cocrystals. *Chem. Asian J.*, 8, 12, 3032–3041, 2013.
62. Liu, F., Song, Y., Liu, Y.-N., Li, Y.-T., Wu, Z.-Y., Yan, C.-W., Drug-Bridge-Drug Ternary Cocrystallization Strategy for Antituberculosis Drugs Combination. *Cryst. Growth Des.*, 18, 1283–1286, 2018.
63. Rathi, R., Nagpal, M., Kaur, M., Ballouria, P., Dutta, R., Aggarwal, G., Arora, S., Treatment of tuberculosis in nano era: recent avenues. *Curr. Bioact. Comp.*, 18, 7, 13–26, 2022.
64. Witika, B.A., Smith, V.J., Walker, R.B., Top-Down Synthesis of a Lamivudine-Zidovudine Nano Co-Crystal. *Crystals*, 11, 1, 33, 2020.
65. Schultheiss, N. and Newman, A., Pharmaceutical Cocrystals and Their Physicochemical Properties. *Cryst. Growth Des.*, 9, 2950–2967, 2009.
66. Yadav, A.V., Shete, A.S., Dabke, A.P., Kulkarni, P.V., Sakhare, S.S., Co-crystals: a novel approach to modify physicochemical properties of active pharmaceutical ingredients. *Indian J. Pharm. Sci.*, 71, 359–370, 2009.
67. Nugrahani, I. and Jessica, M.A., Amino Acids as the Potential Co-Former for Co-Crystal Development: A Review. *Molecules*, 26, 11, 3279, 2021.

68. Haiduc, I., Tiekink, E., Schpector, J., The Importance of Pi-Interactions in Crystal Engineering: Frontiers in Crystal Engineering, in: *The Importance of Pi-Interactions in Crystal Engineering: Frontiers in Crystal Engineering*, pp. 301–322, 2012.
69. Sugden, I.J., Braun, D.E., Bowskill, D.H., Adjiman, C.S., Pantelides, C.C., Efficient Screening of Coformers for Active Pharmaceutical Ingredient Cocrystallization. *Cryst. Growth Des.*, 22, 4513–4527, 2022.

11

Application of Pharmaceutical Cocrystals in Improving Mechanical Properties and Stability

Sushant Sharma, Ritu Rathi, Ayushi Taneja, Poonam Arora and Inderbir Singh*

Chitkara College of Pharmacy, Chitkara University, Punjab, India

Abstract

Cocrystals, defined as crystalline complexes formed by two or more constituents, usually involving a drug and a coformer, present a promising strategy to improve the physicochemical characteristics of pharmaceuticals. These solid-state materials can enhance stability, bioavailability, and solubility, addressing common challenges in drug formulation. One of the significant advantages of cocrystals lies in their ability to change the mechanical properties of active pharmaceutical ingredients (APIs). Many drugs face issues such as poor compressibility and low stability, which can hamper their efficacy and manufacturability. By integrating cocrystals, these limitations can be mitigated. For instance, the formation of a cocrystal can stabilize an API that is otherwise prone to degradation under environmental conditions, thereby prolonging its shelf life. Furthermore, cocrystals often exhibit improved mechanical properties, which are critical for tablet formulation. The incorporation of coformers can lead to a more favorable solid-state structure, enhancing flowability and compressibility, ultimately resulting in a lower risk of brittleness and better tablet hardness. Additionally, cocrystals can enable better dissolution rates, enhancing bioavailability and therapeutic outcomes. By optimizing the physicochemical characteristics of drugs, cocrystals serve as a vital tool in pharmaceutical development, providing a strategic approach to overcoming stability issues and improving mechanical properties. This innovation not only fosters more efficient drug formulations but also opens pathways for re-evaluating existing medications, ensuring that they can meet contemporary therapeutic demands with greater efficacy. Here we review the cocrystals and their mechanism that led to enhanced stability and mechanical properties of various drugs.

**Corresponding author*: inderbir.singh@chitkara.edu.in

Inderbir Singh, Subrata Mallick and Ritu Rathi (eds.) Cocrystals in Pharmaceutical Sciences: Design to Applications, (329–356) © 2025 Scrivener Publishing LLC

Keywords: Tablets, hygroscopicity, compressibility, plasticity, cocrystals, tabletability

11.1 Introduction

Pharmaceutical cocrystals are multicomponent systems comprising at least one API and other components (called coformer) in a stoichiometric ratio that are held together through noncovalent interactions (e.g., hydrogen bonding or van der Waals interactions) [1]. Cocrystals offer improved drug performance by combining the pharmacological qualities of the drug molecule with a hydrogen-bonded coformer in the crystal lattice [2]. The ionization of both the components inside the cocrystal can be neutral, ionic, or zwitterion and can have its own set of polymorphs and solvates. With such structural diversity, cocrystals are beneficial in improving the physicochemical properties of drugs. Cocrystallization is a promising technique for enhancing the physicochemical characteristics of APIs without changing their chemical structure [3]. Drugs physicochemical characteristics, including their melting point, tabletability, solubility, stability, bioavailability, and permeability, can be improved by pharmaceutical cocrystals [4]. At present, the BCS Class II and IV comprise 40% of already marketed medications and about 90% of new chemical entities. These classes suffer from issues of low bioavailability and poor water solubility [5]. This limits the drug absorption in the gastrointestinal tract, which in turn limits the therapeutic efficacy of the drug. The physicochemical characteristics of pharmaceutical solids have a significant impact on how well medicinal products work [6]. Poor mechanical and physicochemical qualities, such as the solubility, permeability, bioavailability, dissolving rate, hygroscopicity, tabletability, compressibility, and stability of APIs and excipients, are a common cause of medication failure in the clinical development stages [7, 8].

The study of the mechanical properties of pharmaceutical APIs is critical for understanding the nature of the material and developing tablets with suitable tensile strength. Poor mechanical properties can lead to several tablet defects such as capping, fragmentation, and cracking, which would be difficult for scale-up as well. Therefore a powder should have sufficient mechanical properties, which include elasticity, plasticity, and hardness. Cocrystallization is one of the leading approaches that can improve the mechanical properties of powders which also results in improved tabletability [9]. Cocrystal-based tablets can display variable tensile strength, which can either increase or decrease. This variability is often due to changes in elasticity and plasticity introduced by the cocrystal formation. Cocrystals can improve powder tableting performance through appreciation of their unique structural features, including the presence of salts, strong and weak hydrogen bonds, and flat layers. These structural advantages lead to enhanced mechanical qualities in cocrystals, highlighting the potential of crystal engineering to optimize pharmaceutical solids. The formation of cocrystals is therefore essential for improving the mechanical properties and processing characteristics of xenobiotics, surpassing the capabilities of drugs that do not form cocrystals [10]. Various research studies have already been conducted to examine the effect of cocrystals on mechanical properties. Research conducted by Chow *et al.* for exploring the mechanical properties, dissolution performance, and hygroscopicity of flurbiprofen. The flurbiprofen was cocrystallized with nicotinamide using a rapid-solvent evaporation technique. The resulting cocrystals showed improved tableting performance, dissolution behavior, and reduced hygroscopicity [11]. This research indicated the potential of cocrystals to improve the mechanical properties of pharmaceutical APIs.

In addition to mechanical properties, cocrystals have applications in altering the stability of pharmaceutical APIs. Crystallization can improve chemical stability, which includes resistance to degradation and preserves the drug's efficacy over time; physical stability, which maintains the drug's structural integrity and prevents undesirable transformations; and thermal stability, which ensures the cocrystal can withstand heat without decomposing. Traditional drugs often face challenges like polymorphic transformations, hygroscopicity, and poor compaction, leading to instability and reduced efficacy. Cocrystals address these issues by enhancing chemical stability through protective interactions that shield the active pharmaceutical ingredient (API) from environmental degradation. They also improve physical stability by maintaining a consistent crystal structure, which helps prevent unwanted phase changes. Furthermore, the robust bonding in cocrystals increases thermal stability, allowing drugs to endure higher temperatures without losing efficacy. Overall, cocrystals offer a more reliable and effective solution for drug tablets by overcoming many of the stability challenges associated with simple drugs. Dapagliflozin, an anti-hyperglycemic drug is unstable due to its highly hygroscopic nature and exposure to high temperatures. The preparation of 1:1 cocrystals of dapagliflozin and citric acid was capable of solving the issue. The cocrystals showed improved stability over high temperatures and high relative humidity, while compromising the dissolution performance and tableting performance [12]. This indicated the significance of cocrystals in enhancing the stability of pharmaceutical APIs.

11.2 Pharmaceutical Cocrystals Improve Mechanical Properties

Mechanical properties show a significant role in the manufacture of solid dosage forms that have all of the features required for a pharmaceutical API to be compressed in tablet form, such as compaction, tabletability, and tensile strength. The mechanical characteristics of crystalline constituents are essential as they are important in various manufacturing processes, including milling, granulation, tableting, coating, and blending. Pharmaceutical APIs exhibiting low elastic recovery and increased plastic deformation are predicted to offer good tableting behavior. In the case of solid materials, mechanical deformation includes fragmentation, plastic, elastic, and plastic. Elastic deformation allows a material to gain to its original shape after the deforming force is removed, due to the alterable stretching of intermolecular bonds, though this is typically restricted to minor strains. Plastic deformation, on the other hand, results in irreversible changes to the shape and size of the solid due to stress, caused by irreversible dislocation movement within the crystal structure [13]. Viscoelastic deformation is partially reversible and depends on the amount and rate of applied stress. Fragmentation occurs when the material fractures due to external stress if elastic, viscoelastic, or plastic deformations are insufficient to accommodate the strain, forming fragments.

Furthermore, many organic substances exhibit poor mechanical characteristics, which is an obstacle to manufacturing tablet formulations. It was found that material with improved plasticity properties can display superior compressibility, which remains irreversible even after stress elimination. Cocrystallization is the method that has been shown to successfully improve the mechanical characteristics of xenobiotics by changing the crystal packing and providing the necessary stability and mechanical strength. For organic constituents, low elastic recovery and increased plastic deformation are predicted by good tableting behavior. The behavior of bulk compaction would eventually be improved by crystal formations having slip planes, which would facilitate plastic deformation. Roy *et al.* reported one of the examples explained the mechanical behaviors of chlorzoxazone. Chlorzoxazone shows poor compressibility, so chlorzoxazone and picolinic acid cocrystals were developed, followed by tablet production. The tensile strength of tablets comprising chlorzoxazone only was less than 0.8 MPa at a compaction force of fifty MPa. Chlorzoxazone-picolinic acid cocrystals' tensile strength was on the other much higher reaching approx. 1 6 MPa at even with no cap or lamination (250 MPa). The higher adhesion strength is related to the more closely packed cocrystals compared with pure chlorzoxazone. Furthermore, the cocrystal structure's slip planes promote better plastic deformation in the material, which enhances compressibility [14]. One of the reasons that cocrystals advance the mechanical properties of drugs is due to the presence of slip planes which are crystallographic planes that have the weakest interplanar interaction in a given crystal and have been related with the highest molecular density and the major separation amid adjacent planes [15]. The existence of slip planes in crystal assembly enables sliding motion, hence providing better plasticity to the crystal which results in an enhanced inter-particulate bonding area of the compact. Kale *et al.* reported improved tableting performance of rivaroxaban-malonic acid (2:1) cocrystal than that of rivaroxaban due to the occurrence of a flat slip plane that enhanced the plasticity of rivaroxaban-malonic acid cocrystal [16].

A drug's "tabletability" refers to its ability to be transformed from bulk powder at a specific compaction pressure into tablets with a specific tensile strength [17]. High-speed tableting can be achieved by adequate blend flowability, mechanical strength, friability, and appropriate dissolving performance [18]. Cocrystal formation can be used to modify compaction characteristics and, consequently, the manufacturability of solid-state pharmaceuticals [19]. It has been found that crystallization improves tabletability, compressibility, packability, and powder flowability of drugs. The salt and cocrystals' compressibility and flowability are crucial to assessing tabletability more precisely. In the field of pharmaceutical research, the optimal formulation should be engineered to be stable, compact, and flowable at low punch force, and it should require as little excipient addition as possible [20].

Flowability in the context of cocrystals refers to the ease with which cocrystal powders can move and be handled during processes like mixing, granulation, and tablet compression. Good flowability ensures uniformity in dosage forms and prevents issues like segregation during processing. Favipiravir exhibited poor flowability at higher angles of repose, as demonstrated by Xinyuan *et al.* In comparison, its cocrystals displayed superior flow characteristics at lower angles of repose and outstanding compressibility of these samples [21]. The results of powder tableting supported the poor fluidity and compressibility of favipiravir. Favipiravir cocrystals have dramatically improved permeability and tabletability as a result of a characteristic layered structure and a decrease in molecular polarity [22]. The molecular configurations and packing techniques of the crystals also affect how tabletable crystalline powders are. Jie *et al.* showed that hesperetin cocrystals of temozolomide, have superior tableting behavior, and have better compaction properties than temozolomide powder. Over the whole compaction pressure range, they can be formed into intact tablets without the need for any excipients. Molecular configurations and packing techniques of crystals affect how tabletable crystalline powders are. Temozolomide's crystal structure is packed in a reticular pattern without any possible slip planes, which hinders its capacity to be plastically deformed and results in poor tabletability. Hesperetin exhibits significantly greater tabletability than temozolomide because its crystal structure is filled with two-dimensional layered structures that can generate slip planes. Consequently, the tabletability of the cocrystal is in between that of hesperetin and pure temozolomide [23]. When stress is detached, constituents with higher plasticity properties classically display greater compressibility, which is permanent. Nevertheless, a task in producing tablet preparations is the weak mechanical characteristics of many organic substances. It has been shown that cocrystallization can significantly enhance a drug's mechanical qualities by changing the crystal packing [24, 25] example here. For organic materials, better tableting behavior predicts greater plastic deformation and less elastic regaining. The mechanical characteristics and compaction performance of caffeine cocrystal polymorphs were assessed by Singaraju *et al.* Caffeine-3-nitrobenzoic acid cocrystal form I exhibits superior plastic deformation (compared to form II) which may be explained by its two-dimensional layered crystal structure, according to data from the mechanical and energy framework [26]. Mishra *et al.* used nanoindentation to examine the mechanical characteristics of caffeine-glutaric acid cocrystals on various crystalline faces. It was discovered that the anisotropic mechanical responses of polymorphs were contingent on the number of potential slip planes and the potency of intermolecular interactions within the crystal structure concerning the indentation path [27]. Its maximum plasticity and tabletability are congruent. Since the slip planes with the corrugated sheet structure only interact extremely weakly, hardness and plasticity are inversely related. This is also the fundamental cause of its distortion due to bending. The maximum tabletability indicates that bending crystals are significantly more appropriate for tableting than shearing crystals [28]. In a study, Malonic acid had a fairly excellent bonding strength, but due to its lesser flexibility, it showed "lower" tabletability. This behavior was influenced by the "intermediate" bonding stability and the lack of a slip plane. The primary reason for Rivaroxaban's "intermediate" tabletability was identified as differential surface topologies of slip planes. Because of its greater flexibility and bonding strength, rivaroxaban-malonic acid cocrystal solids had a noticeably "higher" tabletability. The enhanced mechanical behavior of rivaroxaban-malonic acid cocrystals was attributed to flat-layered topological slip through the plane, a higher degree of intermolecular contact, and a bigger spacing among neighboring crystallographic layers [29]. Deformation of organic molecule solids that is irreversible and pliable leads to increased BA; this type of deformation behavior is referred to as plasticity [30]. Slip systems and other crystallographic characteristics are what give single- and multicomponent cocrystals their plasticity (a rise in BA). Two factors that affect the BS of the crystal form are the degree of intermolecular contacts besides actual density [31]. It is important to note that "plasticity" only affects BA; hence, while assessing bulk deformation (compaction) behavior, one must also take into account the impact of BS. "Compressibility" is a term used to describe a powder material's capacity to experience a volume reduction upon the application of compaction pressure. Porosity versus compaction pressure is plotted to show compressibility, and the plot indicates the degree of the interparticulate bonding area (BA) increase. The capacity of powder material to form tablets with adequate tensile strength when densification occurs is known as compactibility [37]. It is shown as a plot of tablet porosity versus tensile strength, where the plot represents the bonding strength (BS) of a particular material. Plots showing compactibility reveal that the cocrystal (RIV-MAL Co) has a stronger bonding strength than both RIV and MAL at all compaction pressures. Higher BA and BS combined to give RIV-MAL Co a higher tabletability [31]. Significantly higher tabletability was shown by the cocrystal with the highest bond angle and bond strength. The porosity is another factor that influences tabletability. A study showed that porosity as a function of applied pressure makes up the powder compressibility profile. Beneath a certain stress, more compressible materials have a greater solid fraction besides a lower final porosity. The tensile strength of tablets is shown as a function of porosity in the compactibility profile. Therefore, powdered materials are said to be more compactible if they produce tougher compacts at a given porosity. Plotting the tablet's tensile strength against the applied compression pressure results in the creation of a tabletability profile. All materials experience initial fragmentation and particle rearrangement at mild stress (< 20 MPa). All materials first show elastic deformation as pressure rises beyond 20 MPa, with the observed strain being linearly proportional to the stress. The material will give and start to plastically flow or irrevocably deform if a critical pressure is reached [32]. Therefore, it is crucial to note in the conclusion that cocrystallization is an effective method for improving the API's tabletability. More examples of different drugs whose stability is enhanced by the formation of cocrystals are enumerated below in Table 11.1.

Table 11.1 Reported cocrystals for improving tabletability.

S. no.	Drug	Coformer	Technique	Inference	Ref.
1	Resveratrol	4-aminobenzamide	Liquid-assisted grinding (LAG)	The tabletability of the formulation is significantly improved as a result of cocrystal formation, which leads to an enhancement in bonding strength among the particles, which concludes in improved compression and cohesion during the tablet manufacturing process.	[65]
2	Isoniazid	4-aminobenzamide	Rapid solvent removal methods	Improved tensile strength as compared to pure drug.	[66]
3	Nebivolol hydrochloride	4-hydroxybenzoic acid	Liquid-assisted grinding and solvent evaporation method.	The development of cocrystals of nebivolol hydrochloride with 4-hydroxybenzoic acid elevated tabletability.	[55]

(Continued)

Table 11.1 Reported cocrystals for improving tabletability. (*Continued*)

S. no.	Drug	Coformer	Technique	Inference	Ref.
4	Flufenamic acid	Nicotinamide	Solvent Evaporation	It has been demonstrated that using cocrystals to increase tabletability works well.	[67]
5	Ibuprofen	Nicotinamide	Rapid solvent removal using rotary evaporation	Mechanical characteristics are enhanced by cocrystallization of nicotinamide and ibuprofen.	[11]
6	Flurbiprofen	Nicotinamide	Rapid solvent removal using rotary evaporation	The formation of cocrystal/cocrystallization of flurbiprofen with nicotinamide improved tabletability.	[11]
7	Nicorandil (NCR)	Fumaric acid	Slow evaporation	The tabletability of the drug was improved by the cocrystallization approach.	[24]
		Suberic acid			
8	Rivaroxaban	Malonic acid	Slow solvent evaporation	Cocrystallization managed to modify crystallographic features as compared to the parent constituents leading to the formation of a new multicomponent crystal phase, which results in enhanced tabletability.	[29]

(*Continued*)

Table 11.1 Reported cocrystals for improving tabletability. (*Continued*)

S. no.	Drug	Coformer	Technique	Inference	Ref.
9	Chlorzoxazone	Picolinic acid	Supercritical fluid method	Tabletability test results showed improved powder compaction behavior.	[14]
10	Ligustrazine	Salicylic acid (SA)	Solution cocrystallization	Ligustrazine suffers from over-compression at high pressures and SA has low plasticity which provides additional mechanical strength to the drug.	[68]
11	Palbociclib	Kakaempferol	Antisolvent cocrystallization	Cocrystals proved to be an effective approach for increasing tabletability.	[69]
12	Telmisartan	Oxalic acid	Ultrasound-assisted cocrystallization from solution (USSC) methods	Results showed enhanced tensile strength; hence, tablets can be easily formed at any compaction force.	[70]
13	Olaparib	Fumaric acid	Solvent evaporation	Improved tabletability as compared to pure drug and the most increased AUC0-24h and Cmax values that were 6.1 and 11.6 times of free OLA, respectively, have great potential to be advanced into the enhanced solid preparations of OLA.	[71]

(*Continued*)

Table 11.1 Reported cocrystals for improving tabletability. (*Continued*)

S. no.	Drug	Coformer	Technique	Inference	Ref.
14	Vanillin	6-chloro-2,4-dinitroaniline	Slow evaporation method	The cocrystal approach successfully modifies brittle crystals hence strengthening its mechanical properties.	[28]
	Ethylvanillin				
	Iso-vanillin				
	Ortho-vanillin				
	Ethylene diamine				
15	Caffeine	Methyl gallate	Slurry method	Improved tensile strength and powder compaction as the drug faces over compaction with severe lamination.	[72]
16	Paracetamol form II	Oxalic acid	Cocrystallization from solution	The layered structure of cocrystals improves their ability to form tablets more effectively.	[75]
		Theophylline			
		Naphthalene			
		Phenazine			
17	Theophylline	Methyl gallate	Solvent evaporation method	Enhanced elastic modulus, indentation values, and crystal slip planes which led to providing the required tabletability performance.	[73]

(*Continued*)

Table 11.1 Reported cocrystals for improving tabletability. *(Continued)*

S. no.	Drug	Coformer	Technique	Inference	Ref.
18	Paracetamol	Trimethyl glycine	Slurry method	Enhanced compression characteristics due to structural stability and improved crystal face	[74]
19	Sulfadimidine	4-aminosalicylic acid	Cocrystallization from solution	Crystal habit engineering of cocrystals leads to enhanced flow characteristics.	[75]
20	Adefovir dipivoxil	Stearic acid	Hot-melt extrusion method	The tabletability is improved due to the changed crystal habit introduced *via* the coformer.	[76]
21	Paracetamol	5-nitroisophthalic acid	Solvent evaporation method	Superior tabletability of cocrystal than the xenobiotics.	[77]
22	Voriconazole	Fumaric acid	Slurry method	Hardness is enhanced by the development of cocrystals as compared to its original form.	[78]
		Oxalate salt			
		4-aminobenzoic acid			
		4-hydroxybenzoic acid			

11.3 Pharmaceutical Cocrystals Improving Stability

Cocrystals (CCs) are defined as crystalline materials consisting of two or more diverse compounds that form a definite crystal lattice, often enhancing the stability, dissolution, hygroscopicity, tensile strength, solubility, tabletability, mechanical strength, and performance of xenobiotics [33]. One of the advantages of cocrystals is that they can upgrade stability against moisture, which is the utmost requirement for preventing degradation and maintaining the efficacy of hygroscopic drugs. Moreover, it can cocrystal improve also physical and chemical studies, and results have shown that cocrystals can exhibit reduced hygroscopicity compared to pure APIs, leading to enhanced resistance to moisture uptake and, consequently, improved formulation stability. One example is phenazopyridine which has a highly hygroscopic nature, due to which it is very difficult to form dosage form and also has a low therapeutic effect due to which manufacturing and efficacy are poor so to overcome this issue the cocrystal of phenazopyridine was synthesized using different coformers which are Benzoic acid, 4-hydroxyphenyl acetic acid, and saccharin. The CC was prepared *via* solvent evaporation technique. The CC showed improved solubility, but cocrystal with 4-hydroxyphenyl acetic acid showed higher solubility among all 3 CCs. Moreover, the hygroscopicity of phenazopyridine was also reduced and CC with saccharin was found to be least hygroscopic at 85% and 98% RH [33]. Another example is palmatine chloride which is hygroscopic. For improving hygroscopicity palmatine chloride (PMTCl) use of an unreported salt cocrystal of palmatine chloride combined with gallic acid (GA) has been designed and synthesized through the solvent evaporation method. Data of structure analysis demonstrates that the asymmetric unit of PMTCl-GA includes two PMT cations, two chloride anions, a GA-neutral molecule, and two methanol molecules. In the process of self-assembly of salt cocrystal GA neutral molecules and methanol participate which inhabit binding sites among chloride anions through strong hydration capability and external water molecules which reduces the binding ability of PMTCl-GA with external water molecules. The hygroscopic strength of PMTCl-GA showed significant enhancement when compared to that of PMTCl [34]. Additionally, cocrystals can improve strength, but they may also introduce new challenges associated with the solubility and bioavailability of the xenobiotic, necessitating careful evaluation of their physicochemical properties. While cocrystals present a promising approach to enhancing drug formulations, ongoing research is essential to address these challenges and optimize their use in pharmaceutical applications.

When a solid is exposed to a certain amount of water vapor at a particular temperature and relative humidity, a phenomenon known as hygroscopicity takes place. Through noncovalent interactions, primarily hydrogen bonding, the solid's surface and bulk properties are affected" [35, 36]. Hydroscopic active pharmaceutical ingredients (APIs) may absorb water at any point during processing (lyophilization, compaction, wet granulation, spray drying), storage, and transportation [36]. Highly hygroscopic solids can be described as "water-loving" because they have polar functional groups or water-attracting binding sites that allow them to form hydrogen bonds with water [37]. Moisture absorption is characteristically ruled by the following five factors: the solid's reactivity to water; the temperature of the surrounding air; the solid's exposed surface area; the velocity of the moist air; and the difference between the solid's equilibrium moisture concentration and the partial vapor pressure of the surrounding water [38]. In addition, the presence of polar groups, such as hydrogen bonding sites could also contribute to hygroscopicity which depends upon the drug's surface area, chemical composition, crystal structure orientation, and presence of any disorganized structure [35]. Physicochemical characteristics of drugs such as stability, crystallinity (both chemical and physical), solubility, dissolving behavior, wettability, bioavailability, powder flow property, powder compatibility, and hardness, can be hampered by hygroscopic behavior. The physicochemical properties of highly hygroscopic nutraceutical and pharmaceutical solids can be significantly altered by chemical degradation and/or solid-state that in turn affect their therapeutic efficacy and shelf life [39]. Moisture could have a major impact on the pharmacological properties of solid dosage forms as it favors chemical processes like hydrolysis, decreasing glass transition temperatures [40], and handling and formulation development [41]. Preventive measures can be used during formulation to prevent and/or minimize the hygroscopicity of active components, rather than depending on the modification of storage and manufacturing conditions to ensure the stability of hygroscopic compounds. Cocrystal appears to be one of the most guaranteed approaches for enhancing the stability of APIs [42]. Utilizing hydrophobic coformers, encapsulating drug molecules, adjusting crystal packing arrangement, encouraging hydrogen bonding interactions, and reducing solvent content through desolvation effects, cocrystals present a viable approach to decreasing the hygroscopicity of drugs [12, 43]. Yokoyama and colleagues reported the progress of a series of multiple-component crystals of 3-aminopyridine, a hygroscopic compound [44]. This is one of the initial case studies on physical stability enhancement concerning atmospheric humidity. Researchers found out that for up to 45 days, moisture held at 100% RH is not absorbed by the fatty acid cocrystals.

Hydrogen bonds between water molecules and exposed functional groups on the crystal surface are the primary means of interaction [45]. The majority of research has suggested that the mechanism underlying the cocrystal formation-induced modification of hygroscopicity is hydrogen bond creation between the coformer and the API. The hydrophilic functional groups of the API appear to be interacting with the coformer molecules, preventing these groups from interacting with water molecules. Metoclopramide hydrochloride and berberine chloride have shown success with this approach. According to Lue and colleagues, the cocrystallization of citric acid with berberine chloride, a highly hygroscopic chemical, results in increased stability against moisture. According to DVS isotherms, berberine chloride can absorb water at a very low RH of 10%. At 70% RH, it changes into berberine chloride dihydrate, resulting in a weight gain of around 10%. However, with the cocrystal approach, less than 2% of moisture got absorbed at up to 70% RH [46]. According to a different study, oxymatrine has an excellent capacity to create hydrogen bonds with molecules of water, which makes it extremely hygroscopic. In cocrystals, urea and oxymatrine form hydrogen bonds that occupy oxymatrine's hydrogen bond site, lowering hygroscopicity [47]. Using acesulfame (AceH) as a coformer, Aitipamula *et al.* employed the cocrystallization method to enhance the physicochemical characteristics of griseofulvin (GF). While amorphous GF transformed to crystalline form under high humidity conditions in less than a day, the synthesized cocrystal (GF)2·Ace-H monohydrate exhibits higher physical strength [48]. Another study showed that lithium chloride (LIC), which is used to treat neuropsychiatric disorders, has a high hygroscopicity and rapidly deliquefies, even at 11.30% relative humidity which constrains the drug's potential development. The development of LIC and glucose cocrystals showed improvement in hygroscopic stability under 40% relative humidity [49].

Cocrystals of nitrofurantoin with, p-hydroxybenzoic acid (PHBA), p-aminobenzoic acid (PABA), urea, and L-arginine were described by Vangala *et al.* and Cherukuvada *et al.* Among the cocrystals prepared by Cherukuvada *et al.*, only the nitrofurantoin–PABA cocrystal showed enhanced stability toward hydration [50, 51]. In contrast, anhydrous nitrofurantoin changes to its monohydrate within 7 weeks under the same circumstances. Nitrofurantoin–PHBA cocrystals exhibited high physical stability under 24 °C and 97% RH on prolonged storage (up to 13 weeks). Metoclopramide (MCP), a common antiemetic medication sold as metoclopramide hydrochloride (MCP HCl) salt98, is another example of a cocrystal that is stable in water. During the production process, crystallized water in the monohydrate form may accelerate the Maillard reaction with lactose and cause an ion exchange reaction with KBr98. The use of an oxalic acid cocrystal successfully reduced the hydration of MCP HCl. The enhanced stability against moisture can be attributed to the participation of all hydrogen bond donor and acceptor sites of MCP in the cocrystal formation [52]. An inhibitor of 5-lipoxygenase called zileuton (ZIL) is used to treat asthma. ZIL is available as a racemic combination in its anhydrous form. When exposed to high humidity, the moderately hygroscopic anhydrous form of ZIL (form A) changes into a monohydrate (form H). The cocrystallization method was used by Chen *et al.* to solve the hydration stability problem. Nicotinamide (ZIL–NIC) and isonicotinamide (ZIL–ISO) were used to create its cocrystals. When examined under controlled circumstances, it was discovered that both cocrystals were comparatively stable and did not experience phase transformation or disintegration for up to 4 weeks [53]. So, Cocrystal formation has shown that the decrease in hygroscopicity was caused by the development of hydrogen bonds amid the compounds that make up crystals, which decreased the amount of hydrogen bonding spots that could interact with water. More examples of different drugs whose stability is enhanced by the formation of cocrystals are enumerated below in Table 11.2.

11.4 Conclusion and Future Prospective

The future of cocrystals in enhancing the stability and mechanical properties of drugs holds significant promise for the pharmaceutical industry. As researchers continue to explore the potential of cocrystal technology, innovative strategies are emerging to address the persistent challenges of drug formulation. One key area of focus is the selection of novel coformers that can not only stabilize APIs but also tailor their mechanical properties to meet specific formulation requirements. Advances in computational chemistry and molecular modeling are enabling the design of bespoke cocrystals with optimized solubility and stability profiles, allowing for more effective drug delivery systems.

Moreover, the increasing interest in personalized medicine is likely to drive the development of cocrystals tailored to individual patient needs, enhancing therapeutic efficacy. Additionally, regulatory frameworks are gradually adapting to accommodate cocrystal formulations, which could streamline the approval process for these innovative products. As analytical techniques improve, the characterization of cocrystals will become more sophisticated, allowing for better prediction of performance in varied conditions. Furthermore, the integration of cocrystal technology with

Table 11.2 Reported cocrystals for improving stability.

S. no.	Drug	Coformer	Technique	Inference	Reference
1	Palmatine chloride	Gallic acid	Solvent evaporation method	Since Palmatine chloride is a hygroscopic drug, the cocrystal of palmatine chloride was formulated with gallic acid. The cocrystal showed increased hygroscopicity of palmatine chloride.	[34]
2	L-carnitine(LC)	Myricetin (MYR)	Slurry method	The stability of LC was improved *via* the cocrystal formation approach as 16-fold lower hygroscopicity was observed.	[54]
3	Ibuprofen	Nicotinamide	Rotary evaporation	The formation of cocrystal/cocrystallization of ibuprofen with nicotinamide improved hygroscopicity and tabletability.	[11]
4	Flurbiprofen	Nicotinamide	Rapid solvent evaporation using rotary evaporation	The formation of cocrystal/cocrystallization of flurbiprofen with nicotinamide improved hygroscopicity and tabletability.	[11]

(Continued)

Table 11.2 Reported cocrystals for improving stability. (*Continued*)

S. no.	Drug	Coformer	Technique	Inference	Reference
5	Nebivolol hydrochloride	4-hydroxybenzoic acid	Solvent evaporation method	The development of cocrystals of nebivolol hydrochloride with 4-hydroxybenzoic acid elevated stability.	[55]
6	Paracetamol:4,4-bipyridine	4,4-bipyridine	Solution-based methods	The cocrystal is thermally stable due to 4,4-bipyridine's extraordinary melting point.	[56]
7	Theophylline	Oxalic acid	Cooling crystallization	Oxalic acid resists the hydrate formation of cocrystals, which results in chemical stability.	[57]
8	Phenazopyridine	Benzoic acid 4-hydroxyphenylacetic acid Saccharin	Slow solvent evaporation	The stability of phenazopyridine is poor due to its hygroscopic nature. To overcome this, the CC of phenazopyridine was synthesized using different coformers that enhanced hygroscopic stability.	[33]
9	Temozolomide (TMZ)	(1) Succinic acid (2) Oxalic acid	Solvent drop grinding	TMZ mainly decomposes in an aqueous medium so to overcome these cocrystals of TMZ were formed that exhibited longer half-life and enhanced stability at 40°C and 75 % relative humidity.	[58]

(*Continued*)

Table 11.2 Reported cocrystals for improving stability. (*Continued*)

S. no.	Drug	Coformer	Technique	Inference	Reference
10	Ligustrazine	(1) P-aminobenzoic acid(PABA)	Slurry crystallization	The stability of ligustrazine was improved by significantly enhancing the hygroscopicity stability and the decreasing mass change of the following cocrystals at 90% relative stability.	[59]
		(2) Aminobenzoic acid (MABA)			
		(3) 3,5-Dinitrobenzoic acid (DNBA)			
11	Berberine chloride (BCl)	Citric acid	Wet granulation method	Cocrystals proved to be an effective approach for enhancing stability.	[46]
12	Berberine hydrochloride (BCl)	L(+)-lactic acid	Crystallization	The cocrystals of berberine hydrochloride and lactic acid showed improved hygroscopicity.	[60]
13	Oxymatrine (OMT)	Sulfanilamide	Cocrystallization	It was reported that OMT has high hygroscopicity which was stabilized by the formation of cocrystals.	[47]
		2-Ketoglutaric acid			
		Hydroxy-2-naphthoic acid			

(*Continued*)

Table 11.2 Reported cocrystals for improving stability. *(Continued)*

S. no.	Drug	Coformer	Technique	Inference	Reference
14	Indomethacin	Saccharin	Slow evaporation	The result showed that moisture uptake by cocrystals was almost negligible at 98% RH than Indomethacin.	[61]
15	Theophylline	Flufenamic acid	Cocrystallization	The formation of cocrystals ensures the reduction of the hygroscopicity of theophylline, indicating that this process effectively decreases the compound's ability to absorb moisture from the surroundings.	[62]
16	Flucytosine	Gallic acid	Solvent evaporation	The cocrystals were found to be stable at 70% to 95% RH conditions.	[63]
		Glutaric acid			
		2,3-dihydroxy benzoic acid			
17	Ipragliflozin	L-Proline	Solvent evaporation method	By the formation of cocrystals, drugs achieve stability against hydrate formation.	[64]
18	Ertugliflozin	l-pyroglutamic acid	Grinding method, slurry method	Stability enhanced.	[65]

(Continued)

Table 11.2 Reported cocrystals for improving stability. (*Continued*)

S. no.	Drug	Coformer	Technique	Inference	Reference
19	Valproic acid	Sodium valproate	Solvent evaporation method, hot-melt extrusion method	Achieves solid-phase stability and hygroscopicity.	[66]
20	Escitalopram oxalate	Oxalate	Cocrystallization from solution	Improved stability of drug while compared with the pure drug which does not form cocrystal.	[64]
21	Chloral hydrate	Betaine	Grinding method	Thermal stability was enhanced.	[64]

other advanced drug delivery platforms, such as nanocarriers and polymeric systems, may further enhance stability and bioavailability. Overall, the continued exploration of cocrystals is set to transform drug formulation strategies, resulting in more robust, effective, and patient-friendly therapies that can overcome current limitations in stability and mechanical properties, paving the way for a new era in pharmaceutical development.

Acknowledgments

The authors are thankful to Chitkara College of Pharmacy, Chitkara University, Punjab, India for their support, infrastructure, and library facilities.

References

1. Rathi, R. and Singh, I., Multicomponent crystal compromising dasatinib and selected co-crystals formers: a patent evaluation of EP2861589B1. *Pharm. Pat. Anal.*, 11, 1, 15–21, 2021.
2. Bolla, G. and Nangia, A., Pharmaceutical cocrystals: walking the talk. *Chem. Commun.*, 52, 54, 8342–8360, 2016.
3. Shaikh, R., Singh, R., Walker, G.M., Croker, D.M., Pharmaceutical Cocrystal Drug Products: An Outlook on Product Development. *Trends Pharmacol. Sci.*, 39, 1033–1048, 2018.
4. Morissette, S.L., Almarsson, O., Peterson, M.L., Remenar, J.F., Read, M.J., Lemmo., A.V., *et al.*, High-throughput crystallization: polymorphs, salts, cocrystals, and solvates of pharmaceutical solids. *Adv. Drug Deliv. Rev.*, 56, 275–300, 2004.
5. Ross, S.A., Lamprou, D.A., Douroumis, D., Engineering and manufacturing of pharmaceuticals cocrystals: a review of solvent free manufacturing technologies. *Chem. Commun.*, 52, 8772–8, 2016.
6. Qiao, N., Li, M., Schlindwein, W., Malek, N., Davies, A., Trappitt, G., Pharmaceutical cocrystals: An overview. *Int. J. Pharm.*, 419, 1–11, 2011.
4. Kumar, S. and Nanda, A., Pharmaceutical Cocrystals: An Overview. *Indian J. Pharm. Sci.*, 79, 6, 2017.
5. Kalepu, S. and Nekkanti, V., Insoluble drug delivery strategies: review of recent advances and business prospects. *Acta Pharm. Sin. B*, 5, 5, 442–4539, 2015.
6. Guo, M., Sun, X., Chen, J., Cai, T., Pharmaceutical cocrystals: A review of preparations, physicochemical properties, and applications. *Acta Pharm. Sin. B*, 11, 8, 2537–2564, 2021.

7. Leuenberger, H. and Rohera, B.D., Fundamentals of powder compression. I. The compactibility and compressibility of pharmaceutical powders. *Pharm. Res.*, 3, 12–22, 1986.
8. Devadasu, V.R., Deb, P.K., Maheshwari, R., Sharma, P., Tekade, R.K., *Dosage Form Design Considerations*, pp. 149–178, Elsevier, Netherlands, 2018.
9. Rathi, R., Kaur, S., Singh, I., A Review on Co-crystals of Herbal Bioactives for Solubility Enhancement: Preparation Methods and Characterization Techniques. *Cryst. Growth Des.*, 22, 3, 2023–42, 2022.
10. Juneja, P., Kaur, B., Odeku, O.A., Singh, I., Development of Corn Starch-Neusilin UFL2 Conjugate as Tablet Superdisintegrant: Formulation and Evaluation of Fast Disintegrating Tablets. *J. Drug Deliv.*, 2014, 1, 827035, 2014.
11. Chow, S.F., Chen, M., Shi, L., Chow, A.H., Sun, C.C., Simultaneously improving the mechanical properties, dissolution performance, and hygroscopicity of ibuprofen and flurbiprofen by cocrystallization with nicotinamide. *Pharm. Res.*, 29, 1854–1865, 2012.
12. Deng, J.H., Lu, T.B., Sun, C.C., Chen, J.M., Dapagliflozin-citric acid cocrystal showing better solid state properties than dapagliflozin. *Eur. J. Pharm. Sci.*, 104, 255–261, 2017.
13. Rathi, R., Sanshita, Singh, I., Hot Melt Extrusion Technique for Developing Pharmaceutical Co-crystals: A Review. *Drug Deliv. Lett.*, 14, 2, 97–108, 2024.
14. Roy, P. and Ghosh, A., Mechanochemical cocrystallization to improve the physicochemical properties of chlorzoxazone. *CrystEngComm*, 22, 27, 4611–4620, 2020.
15. Sun, C.C. and Kiang, Y.H., On the identification of slip planes in organic crystals based on attachment energy calculation. *J. Pharm. Sci.*, 97, 8, 3456–3461, 2008.
16. Kale, D.P., Puri, V., Kumar, A., Bansal, A.K., The role of crystallization-mediated altered crystallographic properties on the tableability of rivaroxaban and malonic acid. *Pharmaceutics*, 12, 6, 546–568, 2020.
17. Joiris, E., Martino, P.D., Berneron, C., Guyot-Hermann, A.M., Guyot, J.C., Compression behavior of orthorhombic paracetamol. *Pharm. Res.*, 15, 1122–1130, 1998.
18. Chen, H., Guo, Y., Wang, C., Dun, J., Sun, C.C., Spherical cocrystallization—an enabling technology for the development of high dose direct compression tablets of poorly soluble drugs. *Cryst. Growth Des.*, 19, 4, 2503–2510, 2019.
19. Chattoraj, S., Shi, L., Sun, C., Understanding the relationship between crystal structure, plasticity, and compaction behavior of theophylline, methyl gallate, and their 1:1 co-crystal. *CrystEngComm*, 12, 2466–2472, 2010.
20. Preethi, S. and Saral M., A., Screening of natural polysaccharides extracted from the fruits of Pithecellobium dulce as a pharmaceutical adjuvant. *Int. J. Biol. Macromol.*, 92, 347–356, 2016.

21. Wang, X., Wang, L., Yao, C., Xie, G., Song, S., Li, H., Qu, Y., Tao, X., Novel formulations of the antiviral drug favipiravir: Improving permeability and tabletability. *Cryst. Growth Des.*, 21, 7, 3807–3817, 2021.
22. Deka, P., Gogoi, D., Althubeiti, K., Rao, D.R., Thakuria, R., Mechanosynthesis, characterization, and physicochemical property investigation of a favipiravir cocrystal with theophylline and GRAS coformers. *Cryst. Growth Des.*, 21, 8, 4417–4425, 2021.
23. Wang, J., Dai, X.L., Lu, T.B., Chen, J.M., Temozolomide–hesperetin drug–drug cocrystal with optimized performance in stability, dissolution, and tabletability. *Cryst. Growth Des.*, 21, 2, 838–846, 2021.
24. Mannava, M.C., Gunnam, A., Lodagekar, A., Shastri, N.R., Nangia, A.K., Solomon, K.A., Enhanced solubility, permeability, and tabletability of nicorandil by salt and cocrystal formation. *CrystEngComm*, 23, 1, 227–237, 2021.
25. Karki, S., Friščić, T., Fábián, L., Laity, P.R., Day, G.M., Jones, W., Improving mechanical properties of crystalline solids by cocrystal formation: new compressible forms of paracetamol. *Adv. Mater.*, 21, 38–39, 3905–3909, 2009.
26. Singaraju, A., Bahl, D., Wang, C., Swenson, D., Sun, C., Stevens, L., Molecular interpretation of the compaction performance and mechanical properties of caffeine cocrystals: a polymorphic study. *Mol. Pharmaceutics*, 17, 21–31, 2020.
27. Mishra, M.K., Mishra, K., Narayan, A., Reddy, C.M., Vangala, V.R., Structural basis for mechanical anisotropy in polymorphs of a caffeine–glutaric acid cocrystal. *Cryst. Growth Des.*, 20, 10, 6306–6315, 2020.
28. Krishna, G.R., Shi, L., Bag, P.P., Sun, C.C., Reddy, C.M., Correlation among crystal structure, mechanical behavior, and tabletability in the co-crystals of vanillin isomers. *Cryst. Growth Des.*, 15, 4, 1827–1832, 2015.
29. Kale, D.P., Puri, V., Kumar, A., Kumar, N., Bansal, A.K., The role of cocrystallization-mediated altered crystallographic properties on the tabletability of rivaroxaban and malonic acid. *Pharmaceutics*, 12, 6, 546, 2020.
30. Sun, C.C., Decoding powder tabletability: Roles of particle adhesion and plasticity. *J. Adhes. Sci. Technol.*, 25, 483–499, 2011.
31. Bhatt, J.A., Bahl, D., Morris, K., Stevens, L.L., Haware, R.V., Structure-mechanics and improved tableting performance of the drug-drug cocrystal metformin: salicylic acid. *Eur. J. Pharm. Biopharm.*, 153, 23–35, 2020.
32. Shan, N. and Zaworotko, M.J., The role of cocrystals in pharmaceutical science. *Drug Discov. Today*, 13, 9-10, 440–446, 2008.
33. Tao, Q., Chen, J.M., Ma, L., Lu, T.B., Phenazopyridine cocrystal and salts that exhibit enhanced solubility and stability. *Cryst. Growth Des.*, 12, 6, 3144–3152, 2012.
34. Zhang, Y.N., Li, Y.J., Chang, L., Liu, L.X., Feng, Y.R., Wu, L.L., Zhang, L., Zhang, Y.J., Improving hygroscopic stability of palmatine chloride by forming a pharmaceutical salt cocrystal of palmatine chloride-gallic acid with neutral molecule. *J. Drug Deliv. Sci. Technol.*, 66, 102839, 2021.

35. Newman, A.W., Reutzel-Edens, S.M., Zografi, G., Characterization of the "hygroscopic" properties of active pharmaceutical ingredients. *J. Pharm. Sci.*, 97, 3, 1047–1059, 2008.
36. Thakur, T.S. and Thakuria, R., Crystalline Multicomponent Solids: An Alternative for Addressing the Hygroscopicity Issue in Pharmaceutical Materials. *Cryst. Growth Des.*, 20, 9, 6245–6265, 2020.
37. https://sciencenotes.org/hygroscopic-definition-and.
38. Edgar, G. and Swan, W.O., The factors determining the hygroscopic properties of soluble substances. I. The vapor pressures of saturated solutions. *J. Am. Chem. Soc.*, 44, 570–577, 1922.
39. Liu, F., Hooks, D.E., Li, N., Mara, N.A., Swift, J.A., Mechanical Properties of Anhydrous and Hydrated Uric Acid Crystals. *Chem. Mater.*, 30, 3798–3805, 2018.
40. Rajabi-Siahboomi, A.R., Levina, M., Upadhye, S.B., Teckoe, J., Excipient Selection in Oral Solid Dosage Formulations Containing Moisture Sensitive Drugs, in: *Excipient Applications in Formulation Design and Drug Delivery*, pp. 385–421, Springer, Cham, Switzerland, 2015.
41. Cheow, W.S., Kiew, T.Y., Hadinoto, K., combining inkjet printing and amorphous nanonization to prepare personalized dosage forms of poorly-soluble drugs. *Eur. J. Pharm. Biopharm.*, 96, 314–321, 2015.
42. Emami, S., Ghafari, R., Manafzadeh, E., Solid-State interaction of pharmaceutical cocrystals with water vapor. *Cryst. Growth Des.*, 21, 8, 4805–4820, 2021.
43. Blagden, N., de Matas, M., Gavan, P.T., York, P., Crystal engineering of active pharmaceutical ingredients to improve solubility and dissolution rates. *Adv. Drug Deliv. Rev.*, 59, 617–630, 2007.
44. Yokoyama, S., Sunohara, M., Fujie, T., Hygroscopicities of 3- Aminopyridine (3AP) and Fatty Acid Complexes (FA-3AP), and the Release of 3AP from FA-3AP. *Chem. Pharm. Bull.*, 41, 1876–1878, 1993.
45. Kale, D.P., Ugale, B., Nagaraja, C.M., Dubey, G., Bharatam, P.V., Bansal, A.K., Molecular Basis of Water Sorption Behavior of Rivaroxaban-Malonic Acid Cocrystal. *Mol. Pharmaceutics*, 16, 7, 2980–2991, 2019.
46. Lu, Q., Dun, J., Chen, J., Liu, S., Sun, C.C., Improving solid-state properties of berberine chloride through forming a salt cocrystal with citric acid. *Int. J. Pharm.*, 554, 14–20, 2019.
47. Qi, M.H., Li, H., Zhu, B., Hong, M., Ren, G.B., Cocrystals of Oxymatrine: reducing hygroscopicity and affecting the dissolution rate. *Cryst. Growth Des.*, 21, 7, 3874–3888, 2021.
48. Aitipamula, S., Vangala, V.R., Chow, P.S., Tan, R.B.H., Cocrystal Hydrate of an Antifungal Drug, Griseofulvin, with Promising Physicochemical Properties. *Cryst. Growth Des.*, 12, 5858–5863, 2012.
49. Duggirala, N.K., Smith, A.J., Wojtas, L., Shytle, R.D., Zaworotko, M.J., Physical stability enhancement and pharmacokinetics of a lithium ionic cocrystal with glucose. *Cryst. Growth Des.*, 14, 6135–42, 2014.

50. Cherukuvada, S., Babu, N.J., Nangia, A., Nitrofurantoin–paminobenzoic acid cocrystal: Hydration stability and dissolution rate studies. *J. Pharm. Sci.*, 100, 3233–3244, 2011.
51. Vangala, V.R., Chow, P.S., Tan, R.B.H., Characterization, physicochemical and photo-stability of a co-crystal involving an antibiotic drug, nitrofurantoin, and 4-hydroxybenzoic acid. *CrystEngComm*, 13, 759–762, 2011.
52. Nugraha, Y.P. and Uekusa, H., Suppressed hydration in metoclopramide hydrochloride by salt cocrystallisation. *CrystEngComm*, 20, 2653–62, 2018.
53. Chen, X., Li, D., Luo, C., Wang, J., Deng, Z., Zhang, H., Cocrystals of zileuton with enhanced physical stability. *CrystEngComm*, 20, 990–1000, 2018.
54. Pang, Z., Weng, X., Wei, Y., Gao, Y., Zhang, J., Qian, S., Modification of hygroscopicity and tabletability of l-carnitine by a cocrystallization technique. *CrystEngComm*, 23, 10, 2138–2149, 2021.
55. Nikam, V.J. and Patil, S.B., Pharmaceutical cocrystals of nebivolol hydrochloride with enhanced solubility. *J. Cryst. Growth*, 534, 125488, 2020.
56. Du, J.J., Lai, F., Váradi, L., Williams, P.A., Groundwater, P.W., Platts, J.A., Hibbs, D.E., Overgaard, J., Monoclinic paracetamol vs. Paracetamol-4, 4′-bipyridine co-crystal; what is the difference? a charge density study. *Crystals*, 8, 1, 46, 2018.
57. Chettri, A., Subba, A., Singh, G.P., Bag, P.P., Pharmaceutical co-crystals: A green way to enhance drug stability and solubility for improved therapeutic efficacy. *J. Pharm. Pharmacol.*, 76, 1, 1–12, 2024.
58. Babu, N.J., Sanphui, P., Nangia, A., Crystal engineering of stable temozolomide cocrystals. *Chem.- Asian J. Sci. Res.*, 7, 10, 2274–2285, 2012.
59. Xie, Y., Gong, L., Tao, Y., Zhang, B., Zhang, L., Yang, S., Yang, D., Lu, Y., Du, G., New Cocrystals of Ligustrazine: Enhancing Hygroscopicity and Stability. *Molecules*, 29, 10, 2208, 2024.
60. Wang, L., Liu, S., Gao, Z., Crystal structure, dissolution and hygroscopicity of a novel cocrystal hydrate of berberine hydrochloride with L (+)-lactic acid. *Die Pharmazie-Int. J. Pharm. Sci.*, 75, 10, 483–487, 2020.
61. Basavoju, S., Boström, D., Velaga, S.P., Indomethacin–saccharin cocrystal: design, synthesis and preliminary pharmaceutical characterization. *Pharm. Res.*, 25, 530–541, 2008.
62. Aitipamula, S., Wong, A.B., Chow, P.S., Tan, R.B., Cocrystallization with flufenamic acid: comparison of physicochemical properties of two pharmaceutical cocrystals. *CrystEngComm*, 16, 26, 5793–5801, 2014.
63. Nechipadappu, S.K., Ramachandran, J., Shivalingegowda, N., Lokanath, N.K., Trivedi, D.R., Synthesis of cocrystals/salts of flucytosine: Structure and stability. *New J. Chem.*, 42, 7, 5433–5446, 2018.
64. Singh, M., Barua, H., Jyothi, V.G.S., Dhondale, M.R., Nambiar, A.G., Agrawal, A.K., Kumar, P., Shastri, N.R., Kumar, D., Cocrystals by design: a rational coformer selection approach for tackling the API problems. *Pharmaceutics*, 15, 4, 1161, 2023.

65. Zhou, Z., Li, W., Sun, W.J., Lu, T., Tong, H.H., Sun, C.C., Zheng, Y., Resveratrol cocrystals with enhanced solubility and tabletability. *Int. J. Pharm.*, 509, 1–2, 391–399, 2016.
66. Lu, Z., Chen, H., Mo, J., Yuan, X., Wang, D., Zheng, X., Zhu, W., Cocrystal of phloretin with isoniazid: preparation, characterization, and evaluation. *RSC Adv.*, 13, 16, 10914–10922, 2023.
67. Joshi, T.V., Singaraju, A.B., Shah, H.S., Morris, K.R., Stevens, L.L., Haware, R.V., Structure-mechanics and compressibility profile study of flufenamic acid: nicotinamide cocrystal. *Cryst. Growth Des.*, 18, 10, 853–5865, 2018.
68. Vreeman, G., Guan, D., Cai, Y., Zhou, Q., Sun, C.C., Cocrystallization improves the tabletability of ligustrazine despite a reduction in plasticity. *Int. J. Pharm.*, 654, 123939, 2024.
69. Zhou, H., Duan, C., Qin, H., Huang, C., Hou, J., Chen, Y., Zhu, J., Xu, C., Jin, J., Zhuang, T., Synthesis and structural characterization of a novel palbociclib-kaempferol cocrystal with improved tabletability and synergistic antitumor activity. *J. Mol. Struct.*, 1281, 135101, 2023.
70. Ratih, H., Pamudji, J.S., Alatas, F., Soewandhi, S.N., Improving telmisartan mechanical properties through the formation of telmisartan and oxalic acid co-crystal by slow evaporation and ultrasound assisted co-crystallization from solution methods. *SJST*, 42, 1, 188–195, 2020.
71. Dai, X.L., Pang, B.W., Lv, W.T., Zhen, J.F., Gao, L., Li, C.W., Xiong, J., Lu, T.B., Chen, J.M., Improving the physicochemical and pharmacokinetic properties of olaparib through cocrystallization strategy. *Int. J. Pharm.*, 647, 123497, 2023.
72. Sun, C.C. and Hou, H., Improving mechanical properties of caffeine and methyl gallate crystals by cocrystallization. *Cryst. Growth Des.*, 8, 5, 1575–1579, 2008.
73. Chattoraj, S., Shi, L., Sun, C.C., Understanding the relationship between crystal structure, plasticity and compaction behaviour of theophylline, methyl gallate, and their 1: 1 co-crystal. *CrystEngComm*, 12, 8, 2466–2472, 2010.
74. Maeno, Y., Fukami, T., Kawahata, M., Yamaguchi, K., Tagami, T., Ozeki, T., Suzuki, T., Tomono, K., Novel pharmaceutical cocrystal consisting of paracetamol and trimethylglycine, a new promising cocrystal former. *Int. J. Pharm.*, 473, 1–2, 179–186, 2014.
75. Serrano, D.R., O'Connell, P., Paluch, K.J., Walsh, D., Healy, A.M., Cocrystal habit engineering to improve drug dissolution and alter derived powder properties. *J. Pharm. Pharmacol.*, 68, 5, 665–677, 2016.
76. Seo, J.W., Hwang, K.M., Lee, S.H., Kim, D.W., Park, E.S., Preparation and characterization of adefovir dipivoxil–stearic acid cocrystal with enhanced physicochemical properties. *Pharm. Dev. Technol.*, 23, 9, 890–899, 2018.
77. Hiendrawan, S., Veriansyah, B., Widjojokusumo, E., Soewandhi, S.N., Wikarsa, S., Tjandrawinata, R.R., Physicochemical and mechanical properties

of paracetamol cocrystal with 5-nitroisophthalic acid. *Int. J. Pharm.*, 497, 1–2, 106–113, 2016.

78. Sanphui, P., Mishra, M.K., Ramamurty, U., Desiraju, G.R., Tuning mechanical properties of pharmaceutical crystals with multicomponent crystals: voriconazole as a case study. *Mol. Pharmaceutics*, 12, 3, 889–897, 2015.

12

Patents and Regulatory Considerations of Pharmaceutical Cocrystals

Ritu Rathi[1], Ritamay Sau[1], Rakesh K. Sindhu[2], Harpreet Singh[3] and Inderbir Singh[1*]

[1]Chitkara College of Pharmacy, Chitkara University, Punjab, India
[2]School of Pharmacy, Sharda University, Greater Noida, Uttar Pradesh, India
[3]School of Pharmaceutical Sciences (Faculty of Pharmacy), IFTM University, Moradabad, Uttar Pradesh, India

Abstract

Pharmaceutical cocrystals are a crystalline substance that is a new addition to pharmaceutical drug design. Cocrystals offer improved physicochemical properties of drugs such as solubility, dissolution, mechanical properties, and more. However, the development and commercialization of pharmaceutical cocrystals have several legal and regulatory challenges. Hence, to preserve the innovative approach for medicinal applications, the European Medicines Agency (EMA) and the Food and Drug Administration (FDA) issued regulatory guidelines for pharmaceutical cocrystals. The EU and FDA publish guidelines for the creation, approval, and post-marketing surveillance of medicinal products based on cocrystals. The regulatory guidelines covered matters relating to manufacturing procedures, bioavailability, bioequivalence, and safety and quality aspects of both the coformer as well as the finished product. The current book chapter provides a brief overview of pharmaceutical cocrystals and their preparation techniques. The descriptive overview of the USFDA and EMA starting from its evolution to the current guidelines (NDA and ANDA applications) for pharmaceutical cocrystals is also been discussed. Later, various patent licenses for pharmaceutical cocrystals provided by the USFDA, EMA, and other countries are also mentioned. Cocrystals. Pharmaceutical developers can effectively promote cocrystal-based medicines while maintaining compliance with safety and legal norms by maneuvering through various regulatory and patent landscapes.

**Corresponding author*: inderbir.singh@chitakra.edu.in

Inderbir Singh, Subrata Mallick and Ritu Rathi (eds.) Cocrystals in Pharmaceutical Sciences: Design to Applications, (357–378) © 2025 Scrivener Publishing LLC

Keywords: Cocrystals, patents, regulatory, USFDA, EMA, CTD

12.1 Introduction

In this era of pharmaceutical development, the selection of a suitable solid form of an active pharmaceutical ingredient (API) is the most crucial phase during the design and development of a drug product. The developed product should be capable enough to possess desirable physicochemical properties such as improved solubility, dissolution, stability, and bioavailability [1]. In this context, pharmaceutical cocrystals are a promising strategy that is capable of altering the physicochemical properties of drugs without affecting their pharmacological properties. Cocrystals are multicomponent systems consisting of an active pharmaceutical ingredient (API) mixed with one or more cocrystal formers [2]. These are solid products, particularly at room temperature, and usually form bonds through noncovalent interactions such as hydrogen bonding, Pi stacking, and van der Waals forces. As a result, cocrystals possess different physicochemical properties than pure drugs involving altered solubility, dissolution, mechanical properties, etc. [3].

Researchers have extensively explored different methods for the preparation of pharmaceutical cocrystals. The selection of a suitable preparation technique is one of the most important steps during cocrystallization. The selection of technique is mainly based upon the solubility and thermal characteristics of API and coformer [4]. The different methods for the preparation of cocrystals are majorly classified as solvent-based methods and solvent-free methods. For the preparation of cocrystals, different methods allow better control over the size, shape, and purity of the cocrystals. Some methods are more suitable for large-scale production. The chemical nature of the components can influence the choice of method. Certain methods are faster than others, Grinding, for example, can produce cocrystals in a short time, whereas solution-based methods might take longer due to the need for solvent evaporation [5]. Various advanced techniques that are available for the preparation of cocrystals are solid-based cocrystal processing methods, mechanochemical grinding, hot-melt extrusion, contact formation, liquid-assisted cocrystal, liquid-assisted grinding, high shear granulation, spray drying, antisolvent cocrystallization, supercritical carbon dioxide processing, freeze drying, microfluidic and jet dispensing approaches, and ultrasound crystallization [6, 7]. Some of the widely used cocrystallization techniques are listed below in Table 12.1.

Table 12.1 Various preparation techniques of pharmaceutical cocrystals.

Techniques	Description	Advantage	Disadvantage
Solvent evaporation	Dissolving API and coformer in a common solvent, followed by evaporation to induce crystallization	Simple, versatile, suitable for a wide range of compounds	Solvent choice can affect crystal growth, may require optimization
Grinding (neat grinding)	Manually grinding API and coformer together in a mortar and pestle	Simple, cost-effective, does not require solvents	Lack of control over particle size, potential for polymorphic forms, scalability challenges
Solvent drop grinding	Mixing API and coformer with a small amount of solvent, followed by grinding to induce crystallization	Combines advantages of grinding and solution methods	Similar limitations to grinding methods, such as lack of particle size control
Antisolvent crystallization	Injecting a non-solvent into a solution containing API and coformer to induce precipitation and crystallization	Rapid crystallization, better control over crystal size	Requires careful selection of solvents and antisolvents, optimization of process parameters
Slurry conversion	Mixing suspension of API and coformer in a solvent, followed by controlled crystallization	Good control over crystal size and morphology	Time-consuming, may require optimization

(Continued)

Table 12.1 Various preparation techniques of pharmaceutical cocrystals. (*Continued*)

Techniques	Description	Advantage	Disadvantage
Coprecipitation	Simultaneous precipitation of API and coformer from a solution under controlled conditions	Good control over crystal size and morphology	Requires careful selection of solvents, optimization of process parameters
Hydrothermal/ solvothermal	Crystallization of API and coformer under high-temperature and high-pressure conditions in a closed vessel	Allows for synthesis of crystalline materials not easily obtained by other methods	Requires specialized equipment, careful control of reaction conditions

Cocrystals have been known for a long time but have received relatively little attention in research. In 1844, Wohler created the first cocrystal, a 1:1 ratio of benzoquinone and hydroquinone, known as quinhydrone. The definition of crystal engineering, as provided by Desiraju in 1989, entails understanding intermolecular interactions in crystal packing and utilizing this knowledge to design new solids with desired physical and chemical properties [7]. Over the past ten years, pharmaceutical cocrystals have attracted considerable interest from researchers. The growing attention to cocrystallization primarily arises from the advantages it offers, wherein supramolecular synthons and noncovalent interactions are utilized for accurately controlling the molecular arrangement within the crystal lattice. As a result, cocrystals offer better thermal stability, processing, and purity characteristics as compared to amorphous solids [8]. In 2004, a widely accepted definition of cocrystals was proposed by Almarsson and Zaworotko, describing them as solid formations involving an API and a coformer under ambient conditions. Multicomponent cocrystals are included in this definition, which highlights the importance of interactions such as hydrogen bonding, along with other noncovalent and non-ionic interactions like H-bonding or π-π interactions [9].

In 2011, a broadly accepted definition of cocrystals was established during a consensual meeting sponsored by the Indo-U.S. Science and Technology Forum (IUSSTF). Cocrystals were defined as single-phase crystalline materials composed of two or more different molecular and/or ionic compounds in a stoichiometric ratio, explicitly excluding solvates and simple salts. This definition was provided by the participants [10]. Cocrystals, as defined by the USFDA, are crystalline materials comprising two or more molecules within the same crystal lattice, linked via non-ionic and noncovalent interactions. In pharmaceutical cocrystals, one component acts as the API, while the other, known as the coformer, generally comprises a substance acknowledged as safe for use in drugs or food. These components interact within the crystal lattice through noncovalent interactions, such as hydrogen bonding, in specific stoichiometric ratios [11, 12]. Crystal engineering is rooted in supramolecular chemistry, emphasizing molecular recognition between complementary fragments that drive self-organization and enable the development of supramolecular functions. Cocrystallization, a pivotal aspect of crystal engineering, markedly alters the physicochemical properties of drug substances. This modification influences factors including solubility, dissolution, compaction, tableting, permeability, bioavailability, physical form, biochemical and hydration stability, and melting point [13].

The utilization of crystal engineering to fabricate cocrystals has emerged as a viable approach for formulating chemicals encompassing low water solubility, garnering significant attention [14]. The terminology and precise definition of cocrystal have been the subjects of ongoing disclosure and scrutiny for a significant period. In a collaborative effort involving the FDA, United States Department of Health and Human Services (HHS), and CDER (Center for Drug Evaluation and Research), a comprehensive description of cocrystals was established. The FDA issued a guidance document in April 2013, titled "Guidance for Industry; FDA 2013" which provided an official categorization and definition of pharmaceutical cocrystals [15]. According to this document, a cocrystal is a type of crystalline substance that consists of several molecules organized in the same crystal lattice structure. A pharmaceutical cocrystal is precisely described as a molecular complex comprised of a dissociable mixture consisting of an active pharmaceutical ingredient and an excipient, with the excipient acting as a neutral guest element [16].

Upon successful development of a pharmaceutical cocrystal yielded positive outcomes, the subsequent step involved securing regulatory approval, and facilitating the pathway to commercialization. For development to qualify for a patent, it must satisfy three conditions: novelty, non-obviousness, and utility/convenience [17]. To use pharmaceutical cocrystals in drugs marketed for the United States and the European Union, should strictly comply with stated regulatory laws. From a regulatory point of view, the different salt form of the drug is considered as a separate active pharmaceutical ingredient but a new cocrystal is not. Thus, to get approval for a drug product that includes either salt or an authorized active pharmaceutical ingredient, it is essential to file an Abbreviated New Drug Application [ANDA 505(b)(2)]. ANDA necessitates the regulatory body in assessing clinical data, to ascertain the safety and efficacy of the associated drug [18]. Nevertheless, if an individual intends to acquire authorization for a medicinal product that includes a cocrystal of a previously proved API, they just need to file an Abbreviated New Drug Application (ANDA) under section 505(j) [19]. The application does not need the FDA to assess clinical data.

In 2013, the USFDA became the inaugural regulatory agency to issue explicit guidelines about pharmaceutical cocrystals. As per the guidelines, pharmaceutical cocrystals are classified as drug product intermediates and are handled similarly to molecular complexes of APIs and excipients [20]. The standards demand that both API and the coformer must be in a state of neutrality, and their interaction must be noncovalent and non-ionic. If the pKa values of both API and its coformer are greater than 1 (pKa (acid) - pKa (the conjugate acid of base), there will be a significant transfer of protons, leading to ionization and possible formation of salt rather than cocrystal [21]. On the other hand, if both have a ΔpKa difference of less than 1, the transfer of protons will have little or no effect. On fulfilling this condition, the API coformer product must be characterized as a cocrystal. Before reaching the site of pharmacological activity, it is essential for both the API and the coformer to completely separate.

The FDA reclassified pharmaceutical cocrystals as a separate class of solvates and hydrates in their 2016 updated guidance. Pharmaceutical cocrystals were classified in the same regulatory category as a polymorph of the active pharmaceutical ingredient [22]. In addition, the FDA requires an *in-vitro* evaluation that specifically examines the ability of a substance to dissolve or to be dissolved. This evaluation is typically considered sufficient in demonstrating that the active medicine fully separates from the coformer. According to the latest definition by the FDA, cocrystals are crystalline substances consisting of two or more distinct molecules, usually a drug and coformers, arranged in the same crystal lattice [23]. Moreover, a cocrystal containing two or more APIs must be categorized as a fixed-dose combination product by the FDA rather than a new API [24].

The European Medicines Agency (EMA) holds a notably distinct viewpoint on pharmaceutical cocrystals in comparison to the Food and Drug Administration. According to the European Medicines Agency, the documentation of cocrystals entails identifying commonalities between the cocrystals and salts. Therefore, the guidelines outlined for documenting salts should be followed by the cocrystals also [25]. According to the EMA, the cocrystal dossier should include documents related to quality, manufacturing, reference standards, control, characterization, control, mechanism of container-closure, and stability of the drug ingredient. Ensuring the pharmacological suitability for coformers is essential, with commonly utilized components regarded as reagents in ICH Q11 [26]. Additionally, any complex unique coformers should be elucidated, accompanied by safety data. The approval process for the European Medicines Agency (EMA) includes a decentralized procedure, facilitating a mutual recognition procedure [27].

A "full application" is submitted for newer active substances (NAS). Organizations must provide a comprehensive dossier for approval, which comprises information about the clinical trials, pharmaceutical tests, and preclinical tests. The EMA classified solid-state compounds according to their intrinsic structures. Cocrystals are defined as highly ordered crystalline structures composed of two or more components in a specific stoichiometric ratio [28]. In contrast to salts, the arrangement of these structures in the crystal lattice is not determined by ionic bonding. It is also crucial to follow good manufacturing procedures when producing active compounds or finished pharmaceutical products that include the in-situ creation of cocrystals throughout the manufacturing process [29]. The submission of an active substance master file encompassing both cocrystals and API [30]. In such instances, it may be essential to create cocrystals consisting of many therapeutic entities. However, to confirm this potential, it is essential to present evidence based on a thorough examination of the dosage ratio, constituents of the cocrystals, and their influence on biopharmaceutical characteristics and therapeutic effectiveness.

This chapter will look into the complex system of patents and regulatory difficulties surrounding pharmaceutical cocrystals. It examines the fundamental principles, synthesis methods, and characterization techniques of cocrystals, providing a comprehensive understanding of their potential applications in the pharmaceutical industry. Furthermore, it will investigate the complex regulatory frameworks and intellectual property concerns that influence the development, manufacturing, and marketing of cocrystal-based medicinal products.

12.1.1 Comparison Between USFDA and EMA on the Aspects of Cocrystal

There are only two regulatory bodies, namely, the USFDA and EMA, to recognize the significance of cocrystals. The USFDA and EMA both stated cocrystals are composed of one or two components that are held together by non-ionic bonds in the same crystal lattice. Although on comparing the USDA and EMA regulatory guidelines we can spot several differences based on the API and co former's properties.

According to EMA's point of view, neither neutral nor charged molecules can develop crystals. On the other side, the FDA had not issued any remarks on that point. Additionally, some assertions are missing from the old proposed guideline. Some gaps throughout their guideline can create an issue for developing cocrystals. EMA also states that API and coformer can be neutral or charged (either positive or negative). However, as per FDA guidelines, all the components present in a crystal lattice should be in

Table 12.2 Comparison between USFDA and EMA on pharmaceutical cocrystals based on regulatory aspects.

	EMA	USFDA
Definition	Monophasic crystal structures are fabricated by two or more elements in a stoichiometric ratio	Crystal is made up of two or more molecules. One is API, and the other is coformer
Coformer role	Act as reagent	Act as excipient
Regulatory category	API	Polymorphism of API
Coformer	Non-active compound reagents	Neutral compound
Interaction	Non-ionic interaction	Non-ionic interaction
***Resemblance with active pharmaceutical ingredient**	Cocrystals depends on safety and efficacy	Yes
GMP regulation and guidelines	According to GMP regulation, part two of EU GMP guideline and ICH Q7 and in rare cases, part one of EU GMP guideline	cGMP for drug product
USDMF/ASMF	Can be filed	Non-practicable being DPI

a neutral state. Even the latest FDA guidelines do not mention any change in their document [31].

Similar to disagreement on the absence or presence of charges within the crystal, EMA and USFDA fail to agree on API and coformer aggregate state. In EMA's point of view, the aggregate state of API and coformer has less scientific value. The Food and Drug Administration (FDA) emphasizes that cocrystals are specific examples of hydrates and solvates in its latest guidance statement. Consequently, one could draw the conclusion that since the US agency distinguishes between hydrates, solvates, and cocrystals, each cocrystal component needs to be solid at room temperature.

In addition to these, the USFDA and EMA share disagreements over the scientific point of view of cocrystals. As per EMA, solvates, hydrates, and cocrystals allocate the same general principles for crystal curation. Hence, they did not classify them as all three belonging to the same solid-state class family. However, in accordance with the USFDA cocrystals are referred to as a special case of solvates and hydrates [32, 33]. For a better understanding, the comparison between the USFDA and EMA is represented in Table 12.2.

12.2 Generic Concept of Pharmaceutical Cocrystals

For European Union

After the establishment of EU legislation, EMA achieved a big accomplishment and focused on medicinal and pharmaceutical products. After that, a guideline is framed for the regulation of pharmaceutical products and post-marketing surveillance. All the EU member states took a resolution for market authorization which can be procured in a centralized, decentralized way, at that moment EMA easily executed the scientific evaluation and the commission gave access to the market. EMA issued a directive 2001/83/EC For generic products to get access to the market through a principal legislative document Directive 2001/83/EC.

Requirement for generic requisition using cocrystal API within EU:-

1. Regulatory Compliance: After the formation of cocrystal guidelines, the generic application for cocrystal should be under regulatory compliance as written in Directive 2001/83/EC.
2. Market Access: The generic application for cocrystals can enter the market under the directive 2001/83/EC. However, Article 10(1) gives relaxation for the application Under

10(1) which states that no requirements for CTD and non-CTD applications but the applicant should submit the market authorization application for full access to the market.

3. Reflection Paper: Another important parameter is the reflection paper which EMA issued under directive 2001/83/EC. Under the reflection paper, EMA mentioned salt and API should have the same regulatory status.

For the United States

USFDA published a directive named 21 U.S.C 505(j). In this directive, those medicines and APIs that have the same quality and properties as in RLD (Reference Listed Drug) are eligible for the application of ANDA. For a better understanding, a comparison between the USFDA and EMA requirements for generic cocrystal is represented in Table 12.3 below [34].

Table 12.3 EMA and USFDA requirements for generic cocrystals.

	EMA	**USFDA**
Reflection paper	EMA published	FDA also published
Incorporation of API	EMA gave permission for the direct incorporation of API in the European Union	USFDA's statement was unclear regarding the incorporation of API in the market
Definition	EMA definition was a clear point of view	But USFDA definition was an unclear point to view for that reason they had to tolerate criticism from scientific community
Qna	European medical agencies statement was clear, and they answered all the raised questions	FDA cannot keep answer which arose from various regulatory guidelines
Flexibility	The European medical agency gave flexibility to their guideline for generic drugs containing cocrystal	But the FDA did not give any flexibility

12.3 Data Consideration Using Pharmaceutical Cocrystals for Abridged Application

A clinical trial is the most important parameter that both agencies carry out to prove the product is safe and therapeutically effective. So, the agencies issued regulatory guidelines for clinical trials. To get market authorization, the clinical data reports are submitted in a fixed format, which is a common technical document (CTD). The USA, EU, and Japan were the first countries to adopt the CTD structure. The CTD structure was developed by the exertions of the International Conference on Harmonization (ICH) of Technical Requirements for Registration of Pharmaceuticals for Human Use for harmonizing application dossiers across different countries. CTD is comprised of modules 2–5. Module 1 contains administration and country/regional information, hence is not a part of CTD. Modules 2–5 contain all the scientific data gathered during the development and manufacturing of the drug product. Module 2 summarizes the data within modules 3, 4, and 5. Module 3 is composed of quality aspects during the development of pharmaceutical drug products. Module 4 is composed of non-clinical studies of pharmaceutical cocrystals and Module 5 contains data relating to clinical studies of pharmaceutical products [35].

12.3.1 Module 3: Quality Aspects for Pharmaceutical Cocrystals

Both agencies outlined quality aspects for pharmaceutical cocrystals and coformers. Keeping in mind the information given by competent authority some other quality aspects were crafted during the development of medicinal products using pharmaceutical cocrystals as an API source.

As per EMA, cocrystals are formed from the acid-base reaction of API and coformer and are linked via non-ionic interaction. The crystal lattice is formed through ion transfer. For such type of reaction, the pKa value is analyzed. However, this approach was not approved by EMA, but EMA suggested certain spectroscopic analytical techniques that can help in the determination of the formation of crystals. However, EMA did not suggest any spectroscopic technique for the determination of interactions in cocrystals, but generally, the X-ray diffraction technique may be the best technique to analyze the crystals. Other spectroscopic techniques like Raman Spectroscopy, IR spectroscopy, and solid-state NMR. The data obtained from such examinations should be mentioned in the drug substance section (3.2.S) of CTD. In addition, EMA states that the definition of cocrystals is for theoretical implication but practically the characterization of active solid substrate should be to find out whether the active ingredient meets proper quality, safety, and efficacy to be used for medicinal applications. And does not focus on the classification of solids as salts or cocrystals [36, 37].

As per the FDA regulatory perspective, a generic drug product developed by cocrystals should be registered via ANDA. For approval, the applicant should give supportive evidence for claiming that cocrystals are present in the product. In comparison to EMA, the USFDA gave a more comprehensive and detailed data list that goes along with the ANDA application of pharmaceutical cocrystals. The first aspect is through investigation of the presence of API and coformer in the crystal lattice. The applicant must carefully examine whether both the API and coformer should be present or not in the lattice.

The second aspect is that the cocrystal present in pharmaceutical products should be made up of non-ionic bonds but functional groups may attach with API and coformers. For that, the FDA issued a guideline for cocrystal determination. FDA outlined a statement for the analysis of the pKa value. If $\Delta pKa \geq 1$, then the ingredient of the crystal lattice can form an ionic bond. If $\Delta pKa \leq 1$, then it is considered the ingredient of a crystal lattice that can form a non-ionic bond. In the third aspect, supportive evidence should be given for the substantial dissolution of the active ingredient before it reaches its target site. FDA also mentioned that the interactions within the cocrystal lattice should behave like solvate and hydrate, as cocrystals are regarded as special cases of solvates and hydrates. Based on the dissolution results, the FDA determines that *in vitro* dissolution data demonstrating crystal lattice dissociation before reaching the target site is sufficient for submitting an ANDA. The data collected from the classification and characterization of cocrystals should be provided in drug substance format (3.2.S) of CTD.

In addition to quality aspects of pharmaceutical cocrystals, the USFDA and EMA provide quality considerations for coformers as well. A huge number of chemical substances, with different chemical characteristics, can act as a coformer. EMA states within their reflection paper that coformers should be pharmaceutical-acceptable to be used in medicinal products. EMA developed a citation to the ICH guideline Q11 (Development and Manufacture of Drug Substances-Chemical and Biotechnological/Biological Entities) and took a stand on behalf of the coformer which should be regarded as a reagent for the production of the drug substance. The European medical agency included all Q11 sections for coformers, as cocrystals have the same regulatory status as salts. Pharmaceutical cocrystals are discussed in guidelines materials from the FDA and the European Medicines Agency (EMA), but it is unclear how they are regulated. Meanwhile, the FDA categorized coformers as excipients, the EMA dictates that coformers must be pharmaceutically approved for use in a medicinal product. Coformers can be removed from categories like starting material and raw materials, as they are not required for the synthesis of active substances. A coformer may be deemed to have sufficient rationale if it has been used in a medicinal product before. If not, the applicant must present different data to justify its usage as a coformer. Excipients that have been previously approved by the FDA for use in medications or that are classified as generally regarded as safe (GRAS) are accepted. It is important to review all viable approaches with the FDA before ANDA application for any drug containing cocrystals [38–40].

12.3.1.1 Quality Aspects During Development of Generic Cocrystal API

As previously mentioned, the USFDA and EMA gave regulatory guidelines for the usage of pharmaceutical cocrystals. However, despite that guideline, some points should be mentioned in the guidelines. Both the regulatory authorities mentioned that the physiochemical properties of cocrystals should be thoroughly evaluated, whether the cocrystals are suitable for development of generic medicinal products. Since melting point is an important property for determining the purity and identification of can create a huge difference between API and pharmaceutical cocrystals. Hence, the melting point should be checked as there might be a large difference between the pure API and cocrystals. Other properties like stability and dissolution study should also be checked properly. Stability studies are necessary for determining the Shelf life of the pure drug and finished product. Dissolution studies are one of the important parameters that should be checked properly during cocrystal development and the dissolution of API in the reference medicinal product should match with the dissolution observed in cocrystals [41–43].

12.3.2 Module 4: Non-Clinical Safety Toxicological Studies

Non-clinical studies like toxicological and pharmacological safety studies are not necessary to be conducted for generic medicinal products. The applicant for Market Access Authorization (EU) or an ANDA (US) makes references to non-clinical studies conducted by the innovator. The same approach applies to pharmaceutical cocrystals as well. However, the literature available circumvents the need for non-clinical studies but the need for toxicological and safety pharmacological studies applies to novel coformers.

12.3.3 Module 5: Clinical Studies

During the production of the generic medicinal product, the manufacturers need not perform the whole clinical trials to define the safety and effectiveness of their product. The important parameters are the bioavailability study and bioequivalence study. By doing this study, the generic product manufacturer accumulates the data and proves that their product is equivalent to the innovator's product. Bioavailability studies show that the rate and extent of the drug in the unchanged form reaches systemic circulation for showing pharmacological activity. The bioequivalence study determines the pharmacokinetic parameter which helps to determine whether the product is bioequivalent or not. The pharmacokinetic parameters are C_{max}, AUC, and T_{max}. Apart from that factor, there is another factor that helps to demonstrate whether the product is clinically significant or not. The factor is the BCS classification system, both the regulatory authorities should outline the guidance on the data type that investigates the BCS class of API [44–46].

12.4 Patent on Cocrystals

After the successful evolution of cocrystal formulation, the regulatory authority took over for required approval to access the market. Regulatory procedures need to be improved to speed up patent filing, granting, and regulatory approval. The Patent should be novel, non-obviousness, and utility. There are two processes to file a patent, one is the national application and another is the international application. Currently, there are two ways available for approval of cocrystals, one is NDA(505(b)(2)) and another is ANDA(505(j)). For the NDA application, it is not required for the cocrystal to contain an API that is listed in the references listed drug, and for the ANDA application, the cocrystals should have contained the previously approved drug. Many research papers on pharmaceutical cocrystals and licenses that have been registered as pharmaceutical cocrystals are evolving daily. A few approved drug cocrystal licenses in the USA, EU, and other countries are listed in Table 12.4 below.

Table 12.4 Various cocrystals approved under different regulatory authorities.

Date	Publication no.	Title	Attorney	References
2014-01-02	US20140004051A1	Cocrystals and salts of contrast agents and imaging	University Of Iowa Research Foundation	[47]
2022-11-17	US 2022/0363661 A1	Cocrystals, pharmaceutical compositions thereof, and methods of treatment involving the same	Servier Pharmaceuticals LLC	[48]
2019-05-21	US10292951B2	Treatment methods and formulations for illnesses involving an inappropriate inflammatory response	First Wave Bio, Inc.	[49]
2021-02-04	WO/2021/022103	Cocrystals of posaconazole, methods of making and using the same	The Regents of the University of Michigan	[50]
2007-05-03	US20070100143A1	Crystalline alfuzosin base	Hetero Drugs Limited	[51]
2019-08-29	Wo 2019/162429	Cocrystals of ubiquinone and compositions comprising them	Center for Intelligent Research in Crystal Engineering	[52]

(Continued)

Table 12.4 Various cocrystals approved under different regulatory authorities. (*Continued*)

Date	Publication no.	Title	Attorney	References
2018-06-27	EP3339292A1	Cocrystals of apremilast	Zaklady Farmaceutyczne Polpharma SA 83-200 Starogard Gdanski (PL)	[53]
2016-03-23	EP2688575B1	Lithium compositions	University of South Florida	[54]
2011-07-06	EP2340818A1	Cocrystals of venlafaxine and celecoxib	Laboratories Del. Dr. Esteve, S.A.	[55]
2009-05-13	EP2056798A2	Sorbic acid analog cocrystals	Amgen, Inc	[56]
2015-08-13	AU2013201664B2	Intravenous formulation including water-soluble acetylsalicylic acid and theanine cocrystals.	Theaprin Pharmaceuticals Inc.	[57]
2007-06-21	JP2007516259A	Method for preparing mixed phase cocrystal with activator	Med Crystal Forms, LLC	[58]
2017-02-24	NZ710780A	New forms of rifaximin and uses thereof	Salix Pharmaceuticals Ltd	[59]

(*Continued*)

Table 12.4 Various cocrystals approved under different regulatory authorities. (*Continued*)

Date	Publication no.	Title	Attorney	References
2016-07-06	CN102952138B	The salt of a kind of pyrazole-pyrimidinone compound, polymorph and pharmaceutical composition, preparation method, and application	Shanghai Tehua Pharmaceutical Technology Co., Ltd.	[60]

12.5 Conclusion

According to various studies, cocrystals are crystalline substances that are a new addition to pharmaceutical drug design. According to some researchers, cocrystals can be a game changer for upcoming pharmaceutical drug design challenges. The major drug encounters issues with physiochemical properties so as with its absorption, distribution, metabolism, and excretion. Keeping this in mind, this problem can be resolved by cocrystal formation which can easily bypass the physicochemical issue of drugs. For this, authorized regulation of cocrystals is very important. In this scenario, the momentousness of cocrystals was accepted by the USFDA and EMA. However, the dissimilarities between both agencies regarding the classification of cocrystals are difficult for professionals working in the field of regulatory affairs.

Pharmaceutical cocrystals are of interest as they can help with pharmaceutical manufacturing processes by offering desirable manufacturing characteristics and avoiding parent API. Due to their non-infringement, generic industries are able to develop and market products ahead of the expiration of polymorph patents. Since the debut of these solids, regulatory affairs have facilitated plenty of discussion that has resulted in new regulatory strategies for novel pharmaceutical products as well as their possible inclusion into the USA and EU's existing legal frameworks. Given the regulatory challenges they will face, the future of novel drugs and generic goods containing cocrystal APIs will be fascinating. Cocrystals' debut shows how the pharmaceutical sciences and the regulatory landscape are always changing and how new APIs are attempting to find a home.

Acknowledgment

The authors are thankful to Chitkara College of Pharmacy, Chitkara University, Punjab, India for providing infrastructure and library facilities.

References

1. Rathi, R., Kaur, S., Singh, I., A Review on Cocrystals of Herbal Bioactives for Solubility Enhancement: Preparation Methods and Characterization Techniques. *Cryst. Growth Des.*, 22, 3, 2023–42, 2022.
2. Kumari, N. and Ghosh, A., Cocrystallization: cutting edge tool for physicochemical modulation of active pharmaceutical ingredients. *Curr. Pharm. Des.*, 26, 38, 4858–82, 2020.
3. Rathi, R. and Singh, I., Multicomponent crystal compromising dasatinib and selected cocrystals formers: a patent evaluation of EP2861589B1. *Pharm. Pat. Anal.*, 11, 1, 15–21, 2021.
4. Wong, S.N., Chen, Y.C., Xuan, B., Sun, C.C., Chow, S.F., Cocrystal engineering of pharmaceutical solids: Therapeutic potential and challenges. *CrystEngComm*, 23, 40, 7005–38, 2021.
5. Ringle, S., Regulatory Implications Using API Cocrystals for Generic Medicinal Products within the EU and US, Masters thesis, Wilhelms-University, Bonn, 2018.
6. Rathi, R., Sanshita, Singh, I., Hot Melt Extrusion Technique for Developing Pharmaceutical Cocrystals: A Review. *Drug Deliv. Lett.*, 14, 2, 97–108, 2024.
7. Jain, S., Rathi, R., Saini, S., Nagaich, U., Singh, I., Co-processed excipients developed by spray drying: A review. *AIP Conf. Proc.*, AIP Publishing, 2986, 1, 030119, 2024.
8. Kumar, A., Kumar, S., Nanda, A., A Review about Regulatory Status and Recent Patents of Pharmaceutical Cocrystals. *Adv. Pharm. Bull.*, 8, 3, 355–363, 2018.
9. Vishweshwar, P., McMahon, J.A., Bis, J.A., Zaworotko, M.J., Pharmaceutical cocrystals. *J. Pharm. Sci.*, 95, 3, 499–516, 2006.
10. Almarsson, Ö. and Zaworotko, M.J., Crystal engineering of the composition of pharmaceutical phases. Do pharmaceutical cocrystals represent a new path to improved medicines? *Chem. Commun. (Camb.)*, 0, 17, 1889–96, 2004.

11. Aitipamula, S., Banerjee, R., Bansal, A.K., Biradha, K., Cheney, M.L., Choudhury, A.R., Polymorphs, salts, and cocrystals: What's in a name? *Cryst. Growth Des.*, 12, 5, 2147–52, 2012.
12. Guidance for Industry: Regulatory Classification of Pharmaceutical Cocrystals, Center for Drug Evaluation and Research, United States Food and Drug Administration, https://www.fda.gov/downloads/Drugs/Guidances/UCM281764.pdf. Accessed on: 2 June 2024.
13. Guidance for Industry: Regulatory Classification of Pharmaceutical Cocrystals. Revision-1, Center for Drug Evaluation and Research, United States Food and Drug Administration, http://www.fda.gov/Drugs/GuidanceCompliance/Regulatoryinformation/Guidances/UCM516813.pdf. Accessed on: 2 June 2024.
14. Rathi, R., Kushwaha, R., Goyal, A., Singh, I., Oxaliplatin-flavone pharmaceutical cocrystal-CN111205332A: patent spotlight. *Pharm. Pat. Anal.*, 11, 5, 147–154, 2022.
15. Thayyil, A.R., Juturu, T., Nayak, S., Kamath, S., Pharmaceutical cocrystallization: Regulatory aspects, design, characterization, and applications. *Adv. Pharm. Bull.*, 10, 2, 203, 2020.
16. Douroumis, D., Ross, S.A., Nokhodchi, A., Advanced methodologies for cocrystal synthesis. *Adv. Drug Delivery Rev.*, 117, 178–195, 2017.
17. Cadden, J., Studies of cocrystal-excipient interaction by a combination of experimental and computational approaches, PhD thesis, University of Southampton, UK, 2022.
18. Trask, A.V., An overview of pharmaceutical cocrystals as intellectual property. *Mol. Pharmaceutics*, 4, 3, 301–309, 2007.
19. Ward, E.H., Congressional Research Service, Defining Active Ingredient: The US Food and Drug Administration's Legal Interpretation of Regulatory Exclusivities, 2023, https://crsreports.congress.gov/product/pdf/R/R46110.
20. Saal, C., Regulatory Frameworks Affecting Solid-State Development, in: *Solid State Development and Processing of Pharmaceutical Molecules: Salts, Cocrystals, and Polymorphism*, vol. 79, pp. 449–480, 2021.
21. Bandaru, R.K., Rout, S.R., Kenguva, G., Gorain, B., Alhakamy, N.A., Kesharwani, P., Dandela, R., Recent advances in pharmaceutical cocrystals: From bench to market. *Front. Pharmacol.*, 12, 780582, 2021.
22. Rathi, R., Guggal, Y., Twinkle, Sandhu, V., Singh, I., Exploring the Patent Landscape and Regulatory Prospective on Pharmaceutical Cocrystals. *J. Pharm. Technol. Res. Manag.*, 11, 49–60, 2023.
23. Gruss, M., Regulatory aspects for formulation design–with focus on the solid state, in: *Innovative Dosage Forms: Design and Development at Early Stage*, pp. 155–207, 2019.
24. Kumar, S. and Nanda, A., Approaches to design of pharmaceutical cocrystals: A review. *Mol. Cryst. Liq. Cryst.*, 667, 1, 54–77, 2018.

25. Almansa, C., Frampton, C.S., Vela, J.M., Whitelock, S., Plata-Salaman, C.R., Cocrystals as a new approach to multimodal analgesia and the treatment of pain. *J. Pain Res.*, 12, 2679–2689, 2019.
26. Rigilin, S., Kunjal, K.K., Thayyil, A.R., Shabaraya, R.A., FDA regulatory implications for cocrystals and recent cocrystal patents. *Int. J. Drug Regul. Aff.*, 10, 3, 10–18, 2022.
27. Xia, M.Y., Zhu, B.Q., Wang, J.R., Yang, Z.E., Mei, X.F., Superior dissolution behavior and bioavailability of pharmaceutical cocrystals and recent regulatory issues. *ACS Med. Chem. Lett.*, 13, 1, 29–37, 2021.
28. Gadade, D.D. and Pekamwar, S.S., Pharmaceutical cocrystals: regulatory and strategic aspects, design and development. *Adv. Pharm. Bull.*, 6, 4, 479, 2016.
29. Stirk, A.J., Souza, F.E., Gerster, J., Mir, F.M., Karadeolian, A., Rey, A.W., Formation of pharmaceutical salts and cocrystals via vapour-assisted tumbling (VAT)–a solvent efficient process. *Green Chem.*, 24, 4, 1505–1514, 2022.
30. Saraf, G.J., Burade, K., Gonjari, I., Hosmani, A., Pawar, A.A., A review on advances in Pharmaceutical cocrystal preparation routes, intellectual property perspective and regulatory aspects. *Int. J. Curr. Pharm. Res.*, 14, 5, 4–12, 2022.
31. Jafari, M.K., Padrela, L., Walker, G.M., Croker, D.M., Creating cocrystals: A review of pharmaceutical cocrystal preparation routes and applications. *Cryst. Growth Des.*, 18, 10, 6370–6387, 2018.
32. European of Medicines cocrystals, of Agency. active Reflection substances in paper medicinal on the products, Reflection paper on the use of cocrystals of active substances in medicinal products (europa.eu), 2014.
33. FDA, Guidance for Industry Regulatory Classification of Pharmaceutical Cocrystals, 2013, https://www.fda.gov/.
34. FDA, Guidance for Industry ANDAs : Pharmaceutical Solid Polymorphism, Guidance for Industry (fda.gov), 10th July 2007.
35. ICH, Organisation of the common technical document for the registration of pharmaceuticals for human use M4, vol. 5, ICH Official web site : ICH, pp. 1–44, 2004.
36. Kittel, C., *Introduction to Solid State Physics*, 8th Edition, Wiley, United States, 2004.
37. Gad, S.C., *Preclinical Development Handbook: ADME and Biopharmaceutical Properties, Preclinical Development Handbook*, Wiley Online Books, United States, 2008.
38. ICH, ICH Q11: Development and manufacture of drug substances-chemical and biotechnological/biological entities, vol. 1(3), ICH Official web site: ICH, p. 30, 2012.
39. Committee for Medicinal Products for Human Use, Guideline on Excipients in the Dossier for Application for Marketing Authorisation of a Medicinal Product, 2007, guideline-excipients-dossier-application-marketing-authorisation-medicinal-product-revision-2_en.pdf (europa.eu).

40. Food and Drug Administration, Guidance for Industry Nonclinical Studies for the Safety Evaluation of Pharmaceutical Excipients, FDA Guidance, Nonclinical Studies for the Safety Evaluation of Pharmaceutical Excipients | FDA, 2005.
41. Schultheiss, N. and Newman, A., Pharmaceutical Cocrystals and Their Physicochemical Properties. *Cryst. Growth Des.*, 9, 6, 2950–2967, 2009.
42. Lee, E.H., A practical guide to pharmaceutical polymorph screening & selection. *Asian J. Pharm. Sci.*, 9, 4, 163–75, 2014.
43. Aston, J.G., Fink, H.L., Tooke, J.W., Ones, M.R., Melting Point Calorimeter for Purity Determinations. *Anal. Chem.*, 19, 218–221, 1947.
44. Guideline on the Investigation of Bioequivalence (europa.eu), 2010.
45. U. S. Food and Drug Administration, Waiver of *In Vivo* Bioavailability and Bioequivalence Studies for Immediate Release Solid Oral Dosage Forms Based on a Biopharmaceutics Classification System, 2017, https://www.fda.gov/.
46. European Medicines Agency, Agomelatine oral tablet 25 mg product-specific bioequivalence guidance (europa.eu), 2017.
47. L.R. and Gillivray, M., Cocrystals and Salts of Contrast Agents and Imaging, US 2014/0004051A1, assigned to University of Iowa Research Foundation, United States, 2014.
48. Lane, B.S. and Gu, C.H., Cocrystals, pharmaceutical compositions thereof, and methods of treatment involving same, US 2022/0363661 A1, assigned to Servier Pharmaceuticals LLC, United States, 2022.
49. Glick, G.D., Franchi, L., Santus, G., Methods and compositions for treating conditions associated with an abnormal inflammatory response, US10292951 B2, assigned to First wave Bio, Inc., United States, 2019.
50. Hornedo, N.R., Kuminek, G., Cavanagh, K.L., Cocrystals of Posaconazole, Method of Making and Using Same, WO2021/022103 A1, assigned to The Reagents of The University Of Michigan, Switzerland, 2021.
51. Reddy, R., Reddy, R.R., Dasari, Ramakrishna, M., Crystalline alfuzosin base, US 2007/0100143 A1, assigned to Hetero Drugs Limited, United States, 2007.
52. P.L. Rafel, B.C. Rafael, L.B. Herrera, D.F. Lopez, Cocrystals Of Ubiquinol And Composition Comprise Them, WO2019/162429 A1, 2019.
53. Oracz, M. and Skoczen, P., Cocrystals of apremilast, EP3339292A1, assigned to Zaklady Farmaceutyczne Polpharma SA, Europe, 2018.
54. Zaworotko, M.J., Shytle, R.D., Ong, T.T., Kavuru, P., Cantwell, N.R., Nguyen, T., Smith, J.A., Lithium compositions, EP2688575B1, assigned to University of South Florida, Europe, 2016.
55. Salaman, C.R.P., Cés, S.V., Tesson, N., Castano, M.T., Cocrystals of venlafaxine and celecoxib, EP 2 340 818 A1, assigned to Esteve Pharmaceuticals SA, Europe, 2011.
56. Bak, A. and Ostovic, D., Sorbic Acid Analogue Cocrystals Patent, EP 2056798 A2, assigned to Amgen, Inc, Europe, 2009.

57. Brittain, H.G. and Felice, P.V., Intravenous Formulation with water soluble cocrystals of Acetylsalicylic Acid and theanine, AU2013201664 B2, assigned to Theaprin Pharmaceuticals Inc, Australia, 2015.
58. David, G., Methods for preparing mixed phase cocrystals with activator, JP2007516259A, assigned to Med Crystals Forms, LLC, Japan, 2007.
59. Parent, S.D., Schultheiss, N.C., Wu, Y., Bevill, M.J., Houston, T.L., Vlahova, P., New form of rifaximin and uses thereof, NZ710780A, assigned to Slix Pharmaceuticals Ltd, New Zealand, 2017.
60. Jianfeng, L., Guanghui, T., Zhen, W., Jin, S., Xiangrui, J., Zheng, L., Xiaojun, Y., Zhu, X., Xianguo, Z., Weiliang, Z., Hualiang, J., Jingshan, S., The salt of a kind of pyrazolo pyrimidinone compound, polymorph and pharmaceutical composition, preparation method and application, CN102952138B, Assigned to Topharman Shandong Co Ltd, China, 2016.

Index

Printed in the USA/Agawam, MA
September 16, 2026

910661.130